Polymer Mixing and Extrusion Technology

PLASTICS ENGINEERING

Series Editor

Donald E. Hudgin

Princeton Polymer Laboratories

Plainsboro, New Jersey

1. Plastics Waste: Recovery of Economic Value, *Jacob Leidner*
2. Polyester Molding Compounds, *Robert Burns*
3. Carbon Black-Polymer Composites: The Physics of Electrically Conducting Composites, *Edited by Enid Keil Sichel*
4. The Strength and Stiffness of Polymers, *Edited by Anagnostis E. Zachariades and Roger S. Porter*
5. Selecting Thermoplastics for Engineering Applications, *Charles P. MacDermott*
6. Engineering with Rigid PVC: Processability and Applications, *Edited by I. Luis Gomez*
7. Computer-Aided Design of Polymers and Composites, *D. H. Kaelble*
8. Engineering Thermoplastics: Properties and Applications, *Edited by James M. Margolis*
9. Structural Foam: A Purchasing and Design Guide, *Bruce C. Wendle*
10. Plastics in Architecture: A Guide to Acrylic and Polycarbonate, *Ralph Montella*
11. Metal-Filled Polymers: Properties and Applications, *Edited by Swapan K. Bhattacharya*
12. Plastics Technology Handbook, *Manas Chanda and Salil K. Roy*
13. Reaction Injection Molding Machinery and Processes, *F. Melvin Sweeney*
14. Practical Thermoforming: Principles and Applications, *John Florian*
15. Injection and Compression Molding Fundamentals, *Edited by Avraam I. Isayev*
16. Polymer Mixing and Extrusion Technology, *Nicholas P. Cheremisinoff*

Other Volumes in Preparation

Polymer Mixing and Extrusion Technology

NICHOLAS P. CHEREMISINOFF

MARCEL DEKKER, INC. New York and Basel

Library of Congress Cataloging-in-Publication Data

Cheremisinoff, Nicholas P.
Polymer mixing and extrusion technology.

(Plastics engineering ; 16)
Includes bibliographies and index.
1. Plastics—Mixing. 2. Plastics—Extrusion.
I. Title. II. Series: Plastics engineering (Marcel Dekker, Inc.) ; 16.
TP1175.M5C48 1987 668.4 87-3469
ISBN 0-8247-7793-X

MARCEL DEKKER, INC.
270 Madison Avenue, New York, New York 10016

Current printing (last digit):
10 9 8 7 6 5 4 3 2 1

PRINTED IN THE UNITED STATES OF AMERICA

In Memory

This volume is dedicated in loving memory of my mother, Louise, whose meticulous assistance and unending commitment continue to be a major source of inspiration.

Preface

Plastics and elastomers are among the most extensively used modern engineering materials. They are utilized in a staggering variety of applications, a few examples of which are consumer household articles, automotive parts, electrical cable insulation, mechanical goods products, construction materials, food packaging, pharmaceuticals packaging and containers, beverage containers, aerospace and aviation parts, and ornamental objects of art. Over the past two decades polymer science has grown exponentially, bringing forth new materials, processes, and applications. In many consumer-oriented markets a variety of new polymeric materials have been introduced which in fact have begun to test the limits of today's technology in manufacturing high-quality end-use articles.

This book is devoted to two major processing operations used in the handling and manufacturing of conventional plastics and rubber products. The major unit operations treated in this volume, mixing and extrusion, are fundamental to the processing of many engineering polymers. Despite the wide practice of these operations, considerable empiricism exists in both designing equipment and applying it to polymer processing. The book is intended as a practical guide for the mixing and extrusion of conventional polymeric materials. It provides a fundamental approach by matching equipment capabilities to polymer structural characteristics to ensure optimum processing and manufacturing performance and by scaling up product performance parameters. Plastics engineers, rubber compounders, and product development engineers will find the volume of use, as will students and researchers.

Gratitude is extended to Marcel Dekker, Inc., for the production of this book.

Nicholas P. Cheremisinoff

Contents

Polymer Mixing and Extrusion Technology

1
Fundamental Concepts

DEVELOPMENT OF CONSTITUTIVE EQUATIONS

Rheology is a subject which addresses the deformation of fluids. By this it is meant that whenever surface forces, called stresses, are applied to a body of fluid, that body deforms or flows. We can describe this behavior mathematically by means of a set of *constitutive equations*. A constitutive equation defines the cause and effect relationships in terms of the properties or characteristics of the fluid material. To develop such relationships *balance equations* must be derived. A balance equation is merely a mathematical statement of the universal laws of conservation of mass, energy, and momentum that are specific to the system of interest. By way of a simple example found in many fluid dynamic textbooks (Cheremisinoff, 1982; Azbel and Cheremisinoff, 1983), consider a shearing force applied to a rectangular body of incompressible fluid as illustrated in Figure 1. For steady-state, isothermal flow conditions we may write the following mass balance (continuity equation) and momentum balance (equation of motion):

Mass balance:

$$\frac{d}{dx}\,\rho V_x = 0 = \rho\,\frac{dV_x}{dx} \tag{1a}$$

and if the body undergoes deformation,

$$V_x \neq f(x) \tag{1b}$$

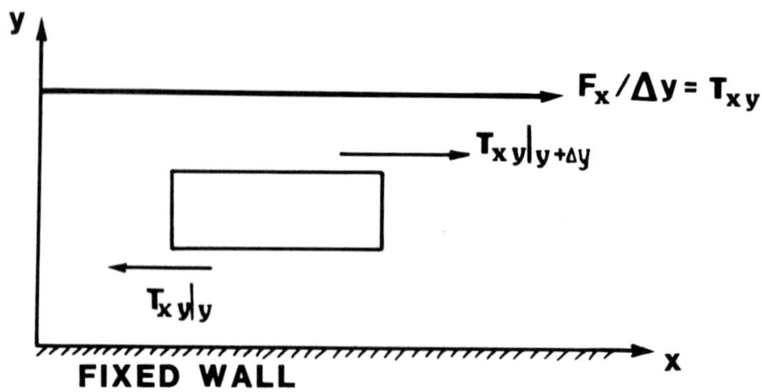

Figure 1 Fluid body subjected to constant force.

Momentum balance:

$$-\frac{d\tau_{xy}}{dy} = 0 \tag{2}$$

and $\tau_{xy} = C$ for any body, where C is the applied stress.

Now let us consider an extreme case where the body is a *Hookean elastic solid.* The constitutive equation for this system is

$$\tau_{xy} = G\dot{\gamma}_{xy} \tag{3}$$

where $\dot{\gamma} = C/G$ and G is the shear modulus. The coordinate system for this model is defined in Figure 2.

The preceding examples are simplified representations of a fluid's or body's behavior when it is subjected to a force. From a practical standpoint a material's ability to undergo changes in shape or movement can be characterized by measuring its resistance to undergoing deformation. The quantity that provides a measure of deformation is the shear strain $\dot{\gamma}_{xy}$. When $G = G(\dot{\gamma}_{xy})$, a solid body is non-Hookean but rather is *elastic.* The shear modulus G is a rheological constant that characterizes the body or material. Similarly, $G(\dot{\gamma}_{xy})$ is a material rheological function.

Now let us consider the body to be a Newtonian fluid; the constitutive equation for this system is Newton's law of viscosity:

$$\tau_{xy} = -\mu\frac{dV_x}{dy} = -\mu\frac{d\gamma_{xy}}{dt} = -\mu\dot{\gamma}_{xy} \tag{4}$$

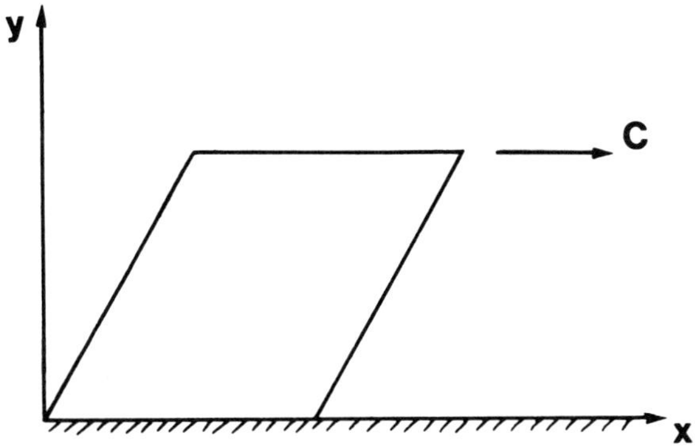

Figure 2 Body undergoing deformation.

The system is defined as shown in Figure 3. It is obvious that

$$\frac{dV_x}{dy} = -\frac{C}{\mu} \tag{5}$$

Integrating this expression gives

$$V_x(y) = \frac{-C_y}{\mu} + C_1 \tag{6a}$$

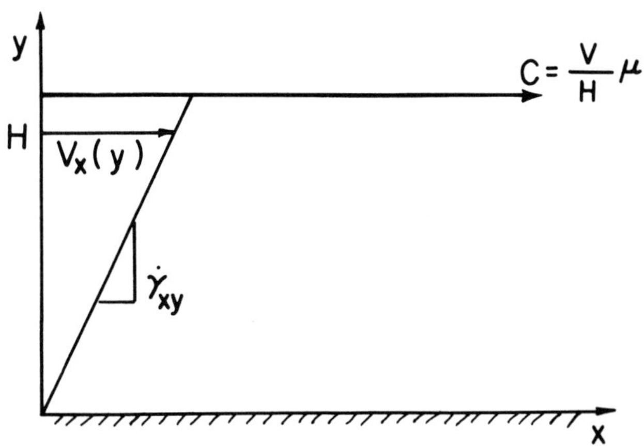

Figure 3 Coordinate system for developing constitutive equation for generalized Newtonian fluid.

where $C_1 = 0$ and hence,

$$V_x(y) = \frac{-C_y}{\mu} \tag{6b}$$

or

$$V = \left(\frac{-C}{\mu}\right)H \tag{6c}$$

and

$$C = -\left(\frac{V}{H}\right)\mu = \tau_{xy} = -\mu\dot{\gamma} \tag{7}$$

The variables in the preceding relationships and Figure 3 are defined as follows:

μ = Newtonian viscosity (i.e., resistance to flow)
$\dot{\gamma}_{xy}$ = shear rate (i.e., measure of flow)
G = material rheological constant
$G(\dot{\gamma})$ = material rheological function

Constitutive equations can be generalized into two categories: linear and nonlinear. For a Newtonian material, Eq. (4) is the governing expression and it is noted that $\mu \neq \mu(\dot{\gamma})$. In other words, viscosity—the resistance to deformation—depends only on those property characteristics of the material that affect intermolecular forces of the fluid. These property characteristics include the chemical structure of the fluid, temperature, and pressure. From thermodynamic principles, the viscosity of a Newtonian fluid can be approximately defined by the following relation:

$$\mu(T, P) = \mu_0 \exp\left\{\frac{\Delta E}{R}\left(\frac{T - T_0}{T_0 T}\right)\right\}\exp\{\beta(P - P_0)\} \tag{8}$$

where

T = absolute temperature
P = absolute pressure
ΔE = flow activation energy
β = viscosity pressure coefficient
R = universal gas law constant

In contrast, a nonlinear form of the constitutive equation is

$$\tau = f(mf_i \cdot \cdot \cdot mf_{nj}\dot{\gamma}) \tag{9}$$

Non-Newtonian materials are described by nonlinear constitutive equations where the most general statement is

$$\tau = -\eta(\dot{\gamma}) \cdot \dot{\gamma} \tag{10}$$

Equation (10) is often referred to by rheologists as the generalized Newtonian fluid equation, where η is a scalar quantity. For simplicity, we shall attempt to relate this behavior by analogy to linear systems. For example, Eq. (4) can be restated for an incompressible fluid as

$$\tau_{xy} = -\frac{\mu}{\rho}\frac{d(V_x\rho)}{dy} \tag{11}$$

The quantity $V_x\rho$ can be thought of as a momentum concentration since it has units of momentum per unit volume. Since $d(V_x\rho)/dy$ is a gradient, we may regard this quantity as a flux of the x-directed momentum in the y-direction. For the engineer, then, this constitutive equation is analogous to the well-known Fourier's law of heat conduction and Fick's law of diffusion.

Fourier's law: $$\frac{q}{A} = -k\frac{d\theta}{dy} \tag{12a}$$

$$\frac{q}{A} = \frac{-k}{C_p\rho}\frac{d(\theta C_p\rho)}{dy} = -\alpha\frac{d(\theta C_p\rho)}{dy} \tag{12b}$$

Fick's law: $$\frac{N_A}{A} = -D_{AB}\frac{dC_A}{dy} \tag{13}$$

where

A = area
q = heat flow
θ = temperature
k = thermal conductivity
α = thermal diffusivity
C_p = specific heat
C_A = concentration of component A
D_{AB} = binary molecular diffusivity for the system $A + B$

Equation (12b) is Fourier's law stated for an incompressible fluid and the quantities $\theta C_p \rho$ and $d(\theta C_p \rho)/dy$ are the volumetric thermal concentration and thermal concentration gradient, respectively. Fick's law [Eq. (13)] is stated for steady-state equimolal counterdiffusion of components A and B. The analogies between Eqs. (4), (12), and (13) are striking: μ/ρ, α, and D_{AB} all have the same primary units of length^2/temperature. As pointed out by Skelland (1967), the quantity μ/ρ can be thought of as momentum diffusivity. Standard texts on transport phenomenon (e.g., Bird et al., 1967) conventionally term μ/ρ the kinematic viscosity.

CLASSIFICATION OF NON-NEWTONIAN BEHAVIOR THROUGH FLOW CURVES

Equation (4) is a linear expression and hence a plot of τ_{xy} versus $\dot{\gamma}$ results in a straight line passing through the origin and having slope μ. The resultant plot is known as a *flow curve*. Fluid materials which display nonlinear behavior through the origin at a given temperature and pressure are non-Newtonian (refer to Figure 4A). Non-Newtonian fluids are broadly classified time-independent fluids, time-dependent fluids, and viscoelastic fluids. The first category is comprised of fluids for which the shear rate at any point is a function of only the instantaneous shear stress. In contrast, time-dependent fluids are those for which shear rate depends on both the magnitude and duration of shear. Some fluids in this second class also show a relationship between shear rate and the time lapse between

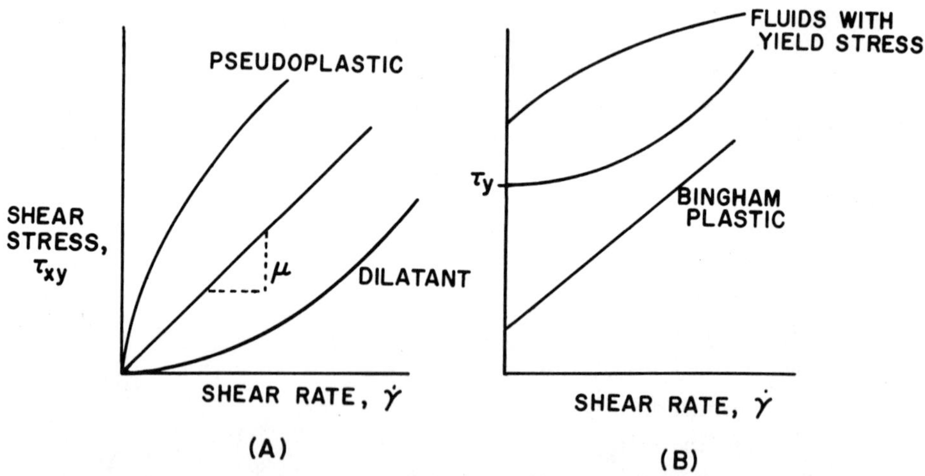

Figure 4 Flow curves for non-Newtonian, time-independent fluids.

consecutive applications of shear stress. Less common viscoelastic fluids display the behavior of partial elastic recovery upon the removal of a deforming shear stress.

Time-dependent non-Newtonian fluids can exhibit the property of a yield stress. Yield stress τ_y can be thought of as a minimum stress value that must be exceeded for deformation to occur. In other words, when $\tau_{xy} < \tau_y$ the fluid's internal structure remains intact, and when $\tau_{xy} > \tau_y$, shearing movement takes place. Flow curves for this type of materials are illustrated in Figure 4B. An idealistic fluid is the Bingham plastic, in which an attempt to describe non-Newtonian behavior with a linear constitutive equation is made. The rheological flow curve in Figure 4B for this fluid can be stated as:

$$\mu_a = \eta + \frac{\tau_y}{\dot{\gamma}} \tag{14}$$

where μ_a is the fluid's *apparent viscosity* and is analogous with the Newtonian apparent viscosity as restated in Eq. (4):

$$\mu_a = \frac{\tau_{xy}}{\dot{\gamma}} \tag{15}$$

The Bingham plastic constitutive equation contains a yield stress τ_y and a term η referred to as the *plastic viscosity*. Equation (14) states that the apparent viscosity decreases with increasing shear rate. In practical terms this means that a value of the apparent viscosity can only be related as a flow property with a corresponding shear rate. Although the Bingham plastic fluid itself is idealistic, it can be applied to modeling a portion of a real non-Newtonian's flow curve. Also, as shown in subsequent chapters, many materials show only small departure from exact Bingham plasticity and therefore can be approximately described by Eq. (14).

A large variety of industrial fluidlike materials may be described as being *pseudoplastic*. The flow curve is illustrated in Figure 4A. Examining this curve we note that it is characterized by linearity at very low and very high shear rates. The slope of the linear region of the curve at the high shear rate range is referred to as the *viscosity at infinite shear* (μ_∞), whereas the slope in the linear portion near the origin is the *viscosity at zero shear rate* (μ_0). A logarithmic plot of τ_{xy} versus $\dot{\gamma}$ is found to be linear over a relatively wide shear rate range and hence may be described by a power law expression (known as the Ostwald–de Waele model):

$$\tau_{xy} = K\dot{\gamma}^n \tag{16}$$

The slope n and intercept K are given the names *flow behavior index* (or pseudoplasticity index) and *consistency index,* respectively. The power law exponent ranges from unity to zero with increasing plasticity (i.e., at $n = 1$, the equation reduces to the constitutive equation of a Newtonian fluid). The value of the consistency index is obtained from the intercept on the τ_{xy} axis and hence represents the viscosity at unit shear rate. As shown later for a variety of polymers, K is very sensitive to temperature, whereas n is much less sensitive. By analogy to Newton's law, the apparent viscosity of a power law fluid is

$$\mu_a = K\dot{\gamma}^{n-1} \tag{17}$$

Since $n < 1$, the apparent viscosity of a pseudoplastic fluid decreases with increasing shear rate, and hence these materials are often referred to as *shear thinning*.

The next class of time-independent fluids are *dilatant* materials. Volumetric dilatancy refers to the phenomenon whereby an increase in the total fluid volume under application of shear occurs. Rheological dilatancy refers to an increase in apparent viscosity with increasing shear rate. The flow curve is illustrated in Figure 4A. As in the case of a pseudoplastic fluid, a dilatant material is usually characterized by zero yield stress. Hence, the power law model may also be used to describe this fluid behavior but with n-values greater than unity.

Time-dependent non-Newtonian fluids are classified as thixotropic and rheopectic. Thixotropic fluids display a reversible decrease in shear stress with time at a constant shear rate and temperature. The shear stress of such a material usually approaches some limiting value. Both thixotropic and rheopectic fluids show a characteristic hysteresis as illustrated by the flow curves in Figure 5. The flow curves are constructed from data generated by a single experiment in which the shear rate is steadily increased from zero to some maximum value and then immediately decreased toward zero. The arrows on the curves denote the chronological progress of the experiment. An interesting complexity of these materials is that the hysteresis is time history dependent. In other words, changing the rate at which $\dot{\gamma}$ is increased or decreased in the experiment alters the hysteresis loop. For this reason, generalized approaches to defining an index of thixotropy have not met with much success.

Rheopectic fluids are sometimes referred to as antithixotropic fluids because they exhibit a reversible increase in shear stress over time at a constant rate of shear under isothermal conditions. The location of the hysteresis loop shown in Figure 5B is also dependent on the material's time history, including the rate at which $\dot{\gamma}$ is changed.

The final category of non-Newtonians is viscoelastic fluids, which are materials exhibiting both viscous and elastic properties. We shall describe viscoelasticity in some detail later. Basically, it should be kept in mind that for purely

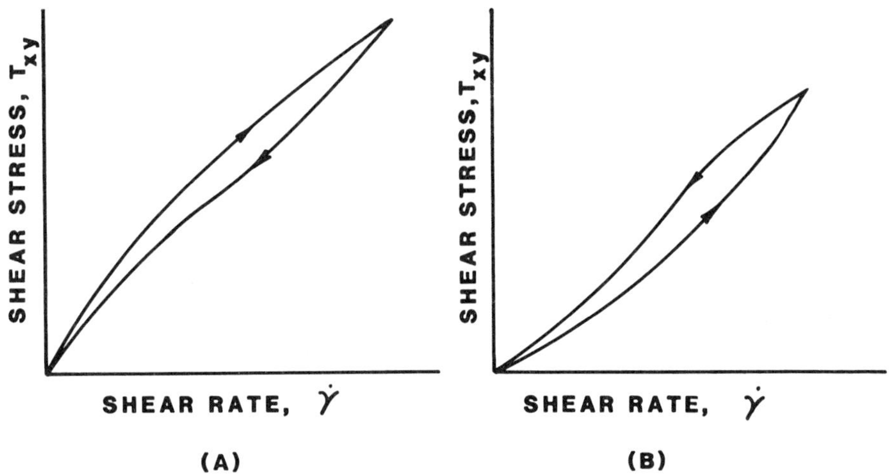

Figure 5 Hysteresis loops for time-dependent fluids. (A) Thixotropic hysteresis. (B) Rheopectic hysteresis.

Hookean elastic solids the stress corresponding to a given strain is time independent, but with viscoelastic materials the stress dissipates over time. Viscoelastic fluids undergo deformation when subjected to stress; however, part of their deformation is gradually recovered when the stress is removed. We may therefore fancifully refer to viscoelastic substances as fluids that have memory. Viscoelasticity is frequently observed in the processing of various polymers and plastics. For example, in the production of synthetic fibers such as Nylon for ultrafine cable, material is extruded through a die consisting of fine perforations. In these examples the cross section of the fiber or cable may be considerably larger than that of the perforation through which it was extruded. This behavior is a result of the partial elastic recovery of the material. Table 1 provides examples of the various types of non-Newtonian fluids.

In this book we shall attempt to relate rheological characteristics observed in laboratory studies to commercial processing problems. The materials discussed here are largely polymer melts and solutions, and blends of various industrial non-Newtonian materials widely used in industry. These materials show the properties of pseudoplasticity, exhibit normal forces in excess of hydrostatic pressure in simple shear flows, and often display viscoelastic behavior (i.e., stress relaxation, creep recovery, and stress overshoot). Each of these rheological features will be discussed separately but the reader should bear in mind that many materials show all three characteristics. Engineering practices have not yet advanced to the point where we may develop a single generalized constitutive

Table 1 Examples of Non-Newtonian Materials

Time-independent fluids	
Fluids with a yield stress	Various plastic melts, oil well drilling mids, oils, sand suspensions in water, coal suspensions, cement and rock slurries, peat slurries, margarine, shortenings, greases, aqueous thorium oxide slurries, grain water suspensions, toothpaste, soap, detergent slurries, paper pulp
Pseudoplastic fluids (without yield stress)	Rubber solution, adhesives, polymer solutions and melts, greases, starch suspensions, cellulose acetate, mayonnaise, soap, detergent slurries, paper pulp, paints, certain pharmaceutical suspensions, various biological fluids
Dilatant fluids (without yield stress)	Aqueous suspensions of titanium dioxide, cornflour/sugar solutions, gum arabic in water, quicksand, wet beach sand, deflocculated pigment dispersions with suspended solids such as mica and powdered quartz, iron powder dispersed in low-viscosity liquids
Time-dependent fluids	
Thixotropic fluids	Melts of high-molecular-weight polymers, some oil well drilling mids, various greases, margarine and shortening, printing inks, various food substances, paints
Rheopectic fluids	Betonite clay suspensions, vanadium pentoxide suspensions, gypsum suspensions, various soils, dilute suspensions of ammonium oleate
Viscoelastic fluids	Bitumens, flour dough, napalm, various jellies, polymers and polymer melts (e.g., Nylon), various polymer solutions

equation to describe the entire range of rheological features. Clearly, the more features that are incorporated into a constitutive equation, the more complex it becomes, and hence the more difficult it is to apply its use in a balance equation needed to solve a problem in fluid mechanics.

In this monograph we rely heavily on the use of rheological flow curves to understand and project the processing characteristics of non-Newtonian materials. Some general features to be kept in mind about these flow curves for later discussions are as follows. First, flow curves, especially for many types of polymers, display a Newtonian plateau or region, then a transition to non-Newtonian, and finally a power law region. The upper limit of shear rate of the Newtonian plateau depends on the polymer's molecular weight and temperature.

This upper limit for many polymers is typically 10^{-2} s^{-1} (Nylon and PET are exceptions). This limit decreases with molecular weight and molecular weight distribution, as well as decreasing temperature. The onset of the transition region between the Newtonian plateau and power law can be related through the dimensionless Deborah number:

$$D_b = \frac{\lambda}{\text{temp}} = \lambda\dot{\gamma} \tag{18}$$

where λ is the material's relaxation time.

The transition region is usually quite distinct for monodispersed polymeric melts, and broad for polydisperse melts. The magnitude of viscosity for each of the regions depends on both material and state variables, as will be illustrated later. A final characteristic of pseudoplastic flow curves to keep in mind is that the power law index n in fact is not an absolute constant for a specific material. For many materials, however, it often reaches some constant value for a reasonably wide range of conditions.

CONSTITUTIVE EQUATIONS FOR PSEUDOPLASTICITY

The constitutive equation for a power law fluid was stated in Eq. (17), where apparent viscosity depends on two parameters, K and n. The consistency index K has units of (N-s^n/m^2). K is temperature sensitive and is found to follow the Arrhenius relationship:

$$K = K_0 \exp\left[\frac{\Delta E}{R_G}\left(\frac{1}{T} - \frac{1}{T_0}\right)\right] \tag{19}$$

where

ΔE = activation energy of flow per mole
R_G = gas constant per mole
T, T_0 = absolute and reference temperatures
K_0 = constant, characteristic of the material

For relatively small temperature ranges, Eq. (19) can be approximated by

$$K = K_0 \exp[-a(T - T_0)] \tag{20}$$

It should be noted that because of thermodynamic considerations K also depends exponentially on pressure.

The apparent viscosity of a power law fluid is a function of all the velocity gradients in nonsimple shearing flows. However, from the standpoint of practicality most laboratory viscometric techniques measure viscosity in only one velocity gradient. This is a general fluid dynamic problem treated in a later chapter. An important point to realize is that the power law equation is an empirical model and should therefore not be used for extrapolation of viscosity data. It does not level off to a Newtonian plateau, but instead keeps on increasing. Table 2 provides values of the power law coefficients *K* and *n* for different materials.

The so-called Carreau model provides a constitutive equation which in fact does level off to a limiting viscosity η_0, which is either measured or estimated. It is, however, a more complex equation having additional model parameters:

$$\frac{\eta(\dot{\gamma}) - \eta_\infty}{\eta_0 - \eta_\infty} = [1 + (\lambda\dot{\gamma})^2]^{(n-1)/2} \tag{21}$$

Table 2 Values of Power Law Model Coefficients for Various Materials

Material	Temperature (K)	Shear rate (s^{-1})	K (10^3) (n-s^2/m^2)	*n*
High-density polyethylene (HDPE)	453	100–1000	6.19	0.56
HDPE	473	100–1000	4.68	0.59
Alathon (TM) 7040 (E. I. Dupont de Nemours & Co.)	473	100–6000	4.31	0.47
Low-density polyethylene (LDPE)	433	100–4000	9.36	0.41
LDPE	453	100–6500	5.21	0.46
Alathon (TM) 1540 (E. I. Dupont de Nemours & Co.)	473	100–6000	4.31	0.47
Polypropylene (PP)	483	100–3000	32.1	0.25
E 612 (TM) (Exxon Chemical Co.)	513	50–3000	22.4	0.28
Nylon	498	100–2500	2.62	0.63
Capron 8200 (TM)	503	100–2000	1.95	0.66
(Allied Chemical Corporation)	508	100–2300	1.81	0.66
Polymethyl methacrylate (PMMA)	493	100–6000	88.3	0.19
PMMA	513	100–6000	42.7	0.25
Lucite 147 (TM) (E. I. Dupont de Nemours & Co.)	533	100–7000	26.2	0.27

η_∞ is the solvent viscosity for solutions. For polymer melts it is common practice for $\eta_\infty = 0$. Note that we have changed our notation for apparent viscosity to the more conventional symbol η.

Another model which also incorporates the Newtonian plateau viscosity (η_0) is the Ellis model:

$$\frac{\eta}{\eta(\tau)} = \left(\frac{\tau}{\tau_{1/2}}\right)^{\alpha-1} \tag{22}$$

where

$$\eta\Big|_{\tau=\tau_{1/2}} = \frac{\eta_0}{2} \tag{23}$$

The exponent $(\alpha - 1)$ is the slope obtained from a plot of $(\eta_0/\eta - 1)$ versus $\log(\tau/\tau_{1/2})$.

The Ellis model is sometimes more conveniently expressed as

$$\tau_{xy} = \frac{1}{A + B\tau_{xy}^{\alpha-1}}\dot{\gamma} \tag{24}$$

where coefficients A and B have units of $L^2\tau^{-1}F^{-1}$ and $L^{2\alpha}\tau^{-1}F^{-\alpha}$, respectively. Term α is dimensionless. Figure 6 provides a comparison of the power law and Ellis models for ABS at two temperatures. Note that the power law

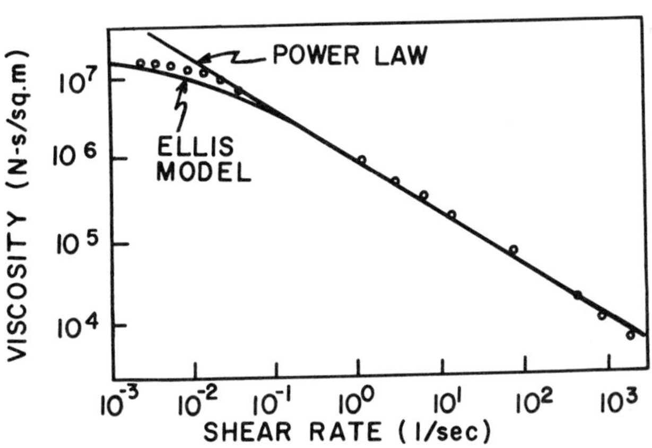

Figure 6 Comparison of power law and Ellis model for ABS polymer melt.

model provides an adequate fit of the data for a large portion of the shear rate range, but departs from the measured values at the low shear rates.

BASIS FOR DETERMINING RHEOLOGICAL FLOW PROPERTIES

In Chapter 2 we discuss the principles used in obtaining viscosity measurements using standard viscometric techniques. Viscometers are devices used for determining relationships between τ_{xy} and $\dot{\gamma}$. Viscoelastic materials require measurement of both normal and shearing stresses as functions of shear rate, and consequently require more sophisticated measurement techniques. We must first consider a general problem of fluid mechanics which will help us to understand the operational principles of standard viscometers.

We consider the classic problem of fluid rod climbing in Couette flow as illustrated in Figure 7. The mass balance for this system is

$$\frac{\partial V_\theta}{\partial \theta} = 0 \tag{25}$$

And for the θ-momentum:

$$-\frac{1}{r^2}\frac{\partial}{\partial r}(r^2\tau_{r\theta}) = 0 \tag{26}$$

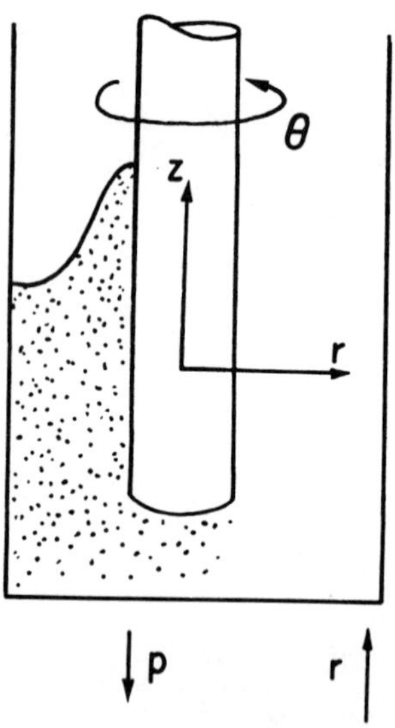

Figure 7 Couette flow system.

For a generalized Newtonian fluid we may write:

$$\tau_{r\theta} = -\eta(\dot{\gamma}_{r\theta})\dot{\gamma}_{r\theta}$$
$$\dot{\gamma}_{r\theta} = r\frac{d}{dr}\left(\frac{V_\theta}{r}\right) \tag{27}$$

Combining the above expressions gives

$$\frac{1}{r^2}\frac{d}{dr}\left[r^3\eta\frac{d}{dr}\left(\frac{V_\theta}{r}\right)\right] = 0 \tag{28}$$

We can solve this expression by specifying ($\dot{\gamma}_{r\theta}$). Additionally, we may write a torque balance on the rotating rod:

$$\mathrm{T} = R_i H 2\pi R_i \tau_{r\theta}|_{r=R_i} \tag{29}$$

In the preceding set of constitutive equations derived for a generalized Newtonian fluid it is clear that only a torque T is needed to sustain Couette-type flow. In contrast, a non-Newtonian such as a polymer melt flowing in Couette flow shows $p = p(r)$. In other words, we have a nonzero dp/dr term. Consequently, the preceding analysis does not adequately relate to non-Newtonian behavior.

To try to understand what happens with a polymer melt or similar material in this type of flow, we must examine the r-component momentum equation and keep track of all the stress terms that can be nonzero. This results in

$$-\rho\frac{V_\theta^2}{r} = -\frac{dP}{dr} - \frac{1}{r}\frac{d}{dr}(r\tau_{rr}) + \frac{\tau_{\theta\theta}}{r} \tag{30}$$

In creeping Couette flows it is observed that

$$\rho\frac{V_\theta^2}{r} \simeq (10^{-1} \sim 10^{-2})\frac{dP}{dr}$$

This condition would result in a depression of the liquid level nearest the rod and not a climbing effect. We can therefore neglect this term (which is the centrifugal force). Hence, our model becomes

$$\frac{dP}{dr} = -\frac{1}{r}\left(r\frac{d}{dr}\tau_{rr} + \tau_{rr}\right) + \frac{\tau_{\theta\theta}}{r}$$

$$\frac{d}{dr}(P + \tau_{rr}) = \frac{\tau_{\theta\theta} - \tau_{rr}}{r}$$

Solving:

$$(P + \tau_{rr})|_{R_o} - (P + \tau_{rr})|_{R_i} = \int_{R_i}^{R_o} \left(\frac{\tau_{\theta\theta} - \tau_{rr}}{r} \right) dr \tag{31}$$

Experimentally it is observed that the integral term is less than zero, and hence

$$\tau_{\theta\theta} - \tau_{rr} < 0 \tag{32}$$

The analysis reduces to attempting to define which stress component $\tau_{r\theta}$ and $(\tau_{\theta\theta} - \tau_{rr})$ contributes to shearing Couette flow. That is, we must now introduce an appropriate rheological constitutive equation in order to make this assessment. One approach is to use the Criminale–Ericksen–Filby (1956) equation, which is general for steady shearing flows with one gradient.

$$\boldsymbol{\tau} = -\eta(\dot{\gamma}) \cdot \dot{\boldsymbol{\gamma}} - [\tfrac{1}{2}\Psi_1(\dot{\gamma}) + \Psi_2(\dot{\gamma})]\dot{\boldsymbol{\gamma}} \cdot \dot{\boldsymbol{\gamma}} + \tfrac{1}{2}\Psi_1 \frac{\mathscr{D}\dot{\boldsymbol{\gamma}}}{\mathscr{D}t} \tag{33}$$

where $(\mathscr{D}\dot{\boldsymbol{\gamma}}/\mathscr{D}t)$ is the corotational time derivative defined as:

$$\frac{\mathscr{D}\dot{\boldsymbol{\gamma}}}{\mathscr{D}t} = \frac{\mathscr{D}\dot{\boldsymbol{\gamma}}}{\mathscr{D}t} + \tfrac{1}{2}[\boldsymbol{\omega} \cdot \dot{\boldsymbol{\gamma}} - \dot{\boldsymbol{\gamma}} \cdot \boldsymbol{\omega}] \tag{34}$$

and

$\boldsymbol{\omega}$ = vorticity tensor
$\eta(\dot{\gamma})$ = viscosity function
$\Psi_1(\dot{\gamma})$ = first normal stress difference coefficient function
$\Psi_2(\dot{\gamma})$ = second normal stress difference coefficient function

Applying the condition of Couette flow:

$$\dot{\boldsymbol{\gamma}} = \begin{bmatrix} 0 & \dot{\gamma}_{\theta r} & 0 \\ \dot{\gamma}_{r\theta} & 0 & 0 \\ 0 & 0 & 0 \end{bmatrix}$$

Equation (33) can now be solved to give:

$$\begin{aligned} \tau_{r\theta} &= -\eta(\dot{\gamma}) \cdot \dot{\gamma}_{\theta r} \\ \tau_{\theta\theta} - \tau_{rr} &= -\Psi_1(\dot{\gamma}) - \dot{\gamma}_{\theta r}^2 \\ \tau_{rr} - \tau_{zz} &= -\Psi_2(\dot{\gamma}) \cdot \dot{\gamma}_{\theta r}^2 \end{aligned} \tag{35}$$

The fluid modeled in this approach exhibits shear as well as normal stresses over and above P_{hyd} in Couette flow. Furthermore,

$$\dot{\gamma} = \begin{bmatrix} 0 & \dot{\gamma}_{12} & 0 \\ \dot{\gamma}_{12} & 0 & 0 \\ 0 & 0 & 0 \end{bmatrix}$$

and hence

$$\begin{aligned} \tau_{12} &= -\eta\dot{\gamma}_{12} \\ \tau_{11} - \tau_{12} &= -\Psi_1\dot{\gamma}_{12}^2 \\ \tau_{22} - \tau_{33} &= -\Psi_2\dot{\gamma}_{12}^2 \end{aligned} \tag{36}$$

We may conclude that the proper application of rheometry must be directed at evaluating all three functions: η, Ψ_1, and Ψ_2. From the standpoint of expediency, as is often the case in industrial environments, the tools of rheology (i.e., viscometers) do not always measure all the appropriate functions. Consequently, many viscometers are applied in a very limited fashion as crude quality control devices and generally do not provide suitable scientific or engineering data applicable to projecting end-use material processability. In later chapters we shall attempt to point out some of these shortcomings in detail. For now, Figure 8 provides a summary of the types of shear flow geometries non-Newtonians can be subjected to in order to evaluate the appropriate rheological functions. Some of the more common geometries are discussed in the following chapter.

NOTATION

A	area
A,B	coefficients in Ellis rheological model
C	applied stress
C_A	concentration of component A
C_p	specific heat
D_{AB}	binary molecular diffusivity for the system $A + B$
D_b	dimensionless Deborah number
E	activation energy of flow
fv	material rheological constant
G	shear modulus
K	consistency index
K_0	material constant in Arrhenius relationship
k	thermal conductivity
n	power law exponent
P	pressure
q	heat flow
R	universal gas law constant
R_G	gas constant per mole
r	radius
T	absolute temperature
V_x	velocity
x,y	distances

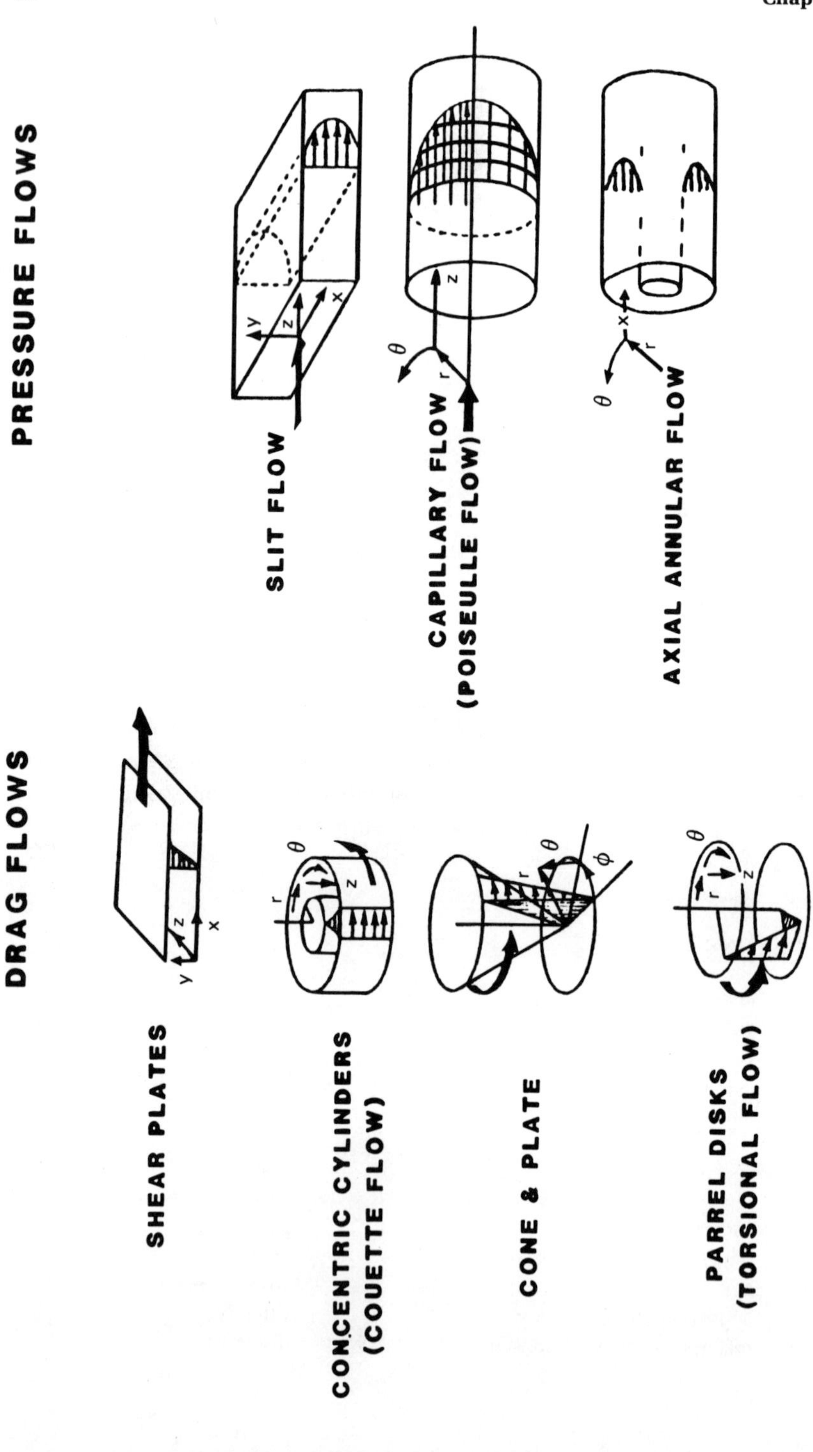

Figure 8 Common shear flow geometries.

Greek Symbols

α	thermal diffusivity	μ_a	apparent viscosity
β	viscosity pressure coefficient	ρ	density
		T	torque
γ_{xy}	shear strain	$\boldsymbol{\tau}$	stress tensor
$\dot{\gamma}$	shear rate	τ_{xy}	wall shear stress
$\dot{\boldsymbol{\gamma}}$	shear tensor	$\Psi_1(\dot{\gamma})$, $\Psi_2(\dot{\gamma})$	first and second normal stress difference coefficient functions, respectively
η	viscosity		
$\eta(\dot{\gamma})$	viscosity function		
η_∞	solvent viscosity		
θ	temperature	ω	vorticity tensor
λ	relaxation time		

REFERENCES

Azbel, D. S., and N. P. Cheremisinoff, *Fluid Mechanics and Unit Operations,* Ann Arbor Science Publishers, Ann Arbor, MI, 1983.

Bird, R. B., W. E. Stewart, and E. N. Lightfoot, *Transport Phenomenon,* Wiley, New York, 1967.

Cheremisinoff, N. P., *Fluid Flow: Pumps, Pipes and Channels,* Ann Arbor Science Publishers, Ann Arbor, MI, 1982.

Criminale, Erickson, and Filby, *Arch. Rat. Mech. Anal., 1,* 410–417 (1956).

Skelland, A. H. P., *Non-Newtonian Flow and Heat Transfer,* Wiley, New York, 1967.

2

Viscometric Techniques

A variety of viscometers are used in the polymers and plastics industries. These devices fall into two general categories, namely, scientific instruments designed to make basic measurements of rheological properties and instruments that are primarily employed for industrial quality control. This chapter outlines the operating principles and basis for both general classes of instruments. It should be noted that a wide variety of rheological test methods and instruments are used in other industries. A summary of the major rheological measurement techniques for different industries is given in the last section of this chapter.

ANALYSIS OF CAPILLARY FLOW

The capillary tube viscometer is one of the simplest and most widely used instruments for rheological characterization. Before describing its operation we first consider the problem of capillary flow. Consider the system in Figure 1, where a fluid is forced at a steady rate from a large reservoir into a small diameter capillary tube of length L. We shall assume steady, isothermal flow of an incompressible fluid and $L/R \gg 1$. The coordinate system is defined along the r, θ, z axes. If we ignore entrance and exit losses in the capillary tube, then

$$\begin{aligned} &V_\theta = 0; \qquad V_r = 0 \\ &V_z \neq 0; \qquad V_z \neq f(\theta) \qquad \text{(axisymmetry)} \end{aligned} \tag{1}$$

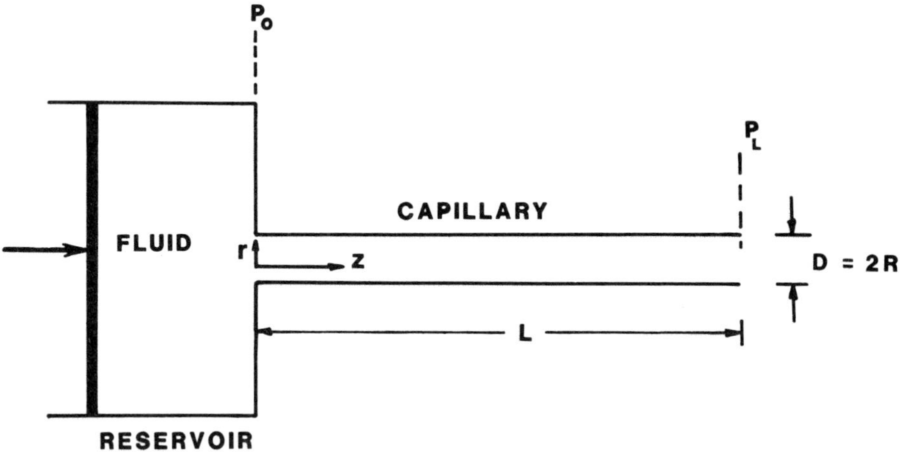

Figure 1 Capillary tube flow.

Writing the continuity equation

$$\frac{\partial P}{\partial t} + \frac{1}{r}\frac{\partial}{\partial r}(\rho r V_r) + \frac{1}{r}\frac{\partial}{\partial \theta}(\rho V_\theta) + \frac{\partial}{\partial z}(\rho V_z) = 0 \tag{2}$$

where density ρ is assumed to be constant.

Since the system is at steady state $(\partial P/\partial t) = 0$, and since $V_r = 0$, $V_\theta = 0$, the second and third terms on the left-hand side of Eq. (2) drop out as well, leaving:

$$\frac{dV_z}{dZ} = 0 \tag{3}$$

The momentum equation for the z-component is

$$P\left[\frac{\partial V_z}{\partial t} + V_r\frac{\partial V_z}{\partial r} + \frac{V_\theta}{r}\frac{\partial V_z}{\partial \theta} + V_z\frac{\partial V_z}{\partial_z}\right] = -\frac{\partial P}{\partial z} - \left(\frac{1}{r}\frac{\partial}{\partial r}(r\tau_{rz}) + \frac{1}{r}\frac{\partial \tau_{\theta r}}{\partial \theta} + \frac{\partial \tau_{zz}}{\partial z}\right) + \rho g_z \tag{4}$$

From the constraints of the system imposed, the z-component momentum equation reduces to

$$\frac{dP}{dz} = -\frac{1}{r}\frac{d}{dr}(r\tau_{rz}) \tag{5}$$

Both sides of this expression are a function of r [i.e., $f(r)$] and both sides are constant. We can therefore assume that $dP/dz \simeq \Delta P/L$ (note that $dP/dz < 0$), and can therefore integrate Eq. (5):

$$r\tau_{rz} = -\frac{\Delta P}{L}r^2 + C_1$$

$$\tau_{rz} = \frac{-\Delta P}{L}r + \frac{C_1}{r} \tag{6}$$

At $r = 0$, $\tau_{zr} \neq \infty$, therefore $C_1 = 0$, and hence:

$$\tau_{zr} = -\left(\frac{\Delta P}{2L}\right)r \tag{7}$$

Equation (7) describes the shear stress profile for any incompressible fluid, and the profile across the capillary tube will have the form shown in Figure 2.

Now, if the fluid's rheological properties follow that of a power law constitutive equation, that is, $\tau = -K|\dot{\gamma}|^{n-1}\dot{\gamma}$ where we note that $\tau = \tau_{zr}$, $\dot{\gamma} = dV_z/dr$, and K and n are constants, then

$$-K\left|\frac{dV_z}{dr}\right|^{n-1}\frac{dV_z}{dr} = -\left(\frac{\Delta P}{2L}\right)r \tag{8}$$

Note that

$$\left|\frac{dV_z}{dr}\right| = -\frac{dV_z}{dr}$$

since $dV_z/dr < 0$.

We may now integrate this expression and impose the "no-slip" at the wall condition [i.e., $V_z(r = R) = 0$]:

$$V_z(r) = \left(\frac{nR}{n+1}\right)\left[\frac{\Delta PR}{2KL}\right]^{1/n}\left[1 - \left(\frac{r}{R}\right)^{(n+1)/n}\right] \tag{9}$$

Equation (9) is the velocity profile of a power law fluid in tube flow. Recall from Chapter 1, for $n = 1$, $K = \mu$; that is, the limit of a Newtonian fluid is reached

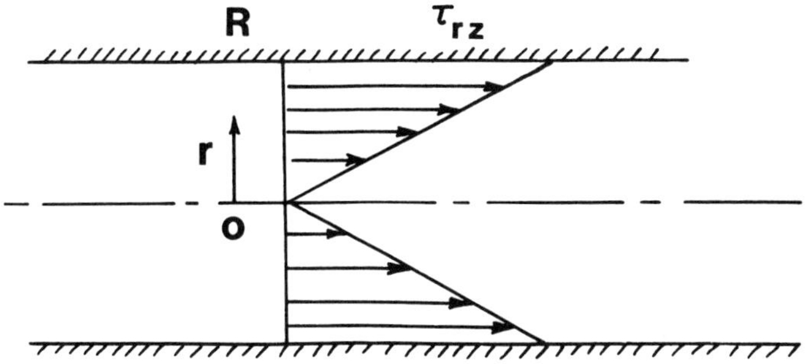

Figure 2 Shear stress profile in tube flow.

and the velocity profile of Eq. (9) reduces to the familiar parabolic velocity profile expression

$$V_z(r) = \frac{\Delta P R^2}{4\mu L}\left[1 - \left(\frac{r}{R}\right)^2\right] \tag{10}$$

The theoretical velocity profiles computed from Eq. (9) have the shapes shown in Fig. 3: for $n = 1$ the parabolic shape [Newtonian, i.e., Eq. (10)]; for $n > 1$ a dilatant fluid; for $n < 1$ a pseudoplastic fluid. For polymer melts and solutions, $n < 1$, and hence velocity profiles are typically "pluglike." This means that there are very high velocity gradients in the vicinity of the tube wall, as shown in Figure 3.

In the above analysis, the constitutive equation for the velocity profile of a

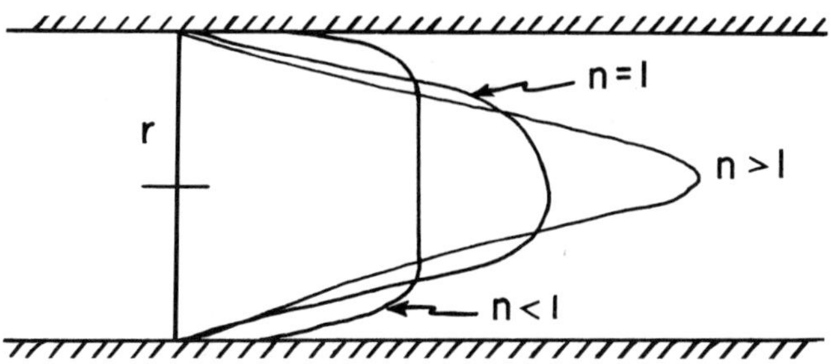

Figure 3 Velocity profiles generated by Eq. (9): n = Newtonian fluid; $n > 1$ dilatant fluid; $n < 1$ pseudoplastic fluid.

power law fluid was derived from first principles. This implies that we can calculate the volumetric flow rate of the fluid through the tube knowing $V_z(r)$.

$$Q = 2\pi \int_0^R V_z(r) r \, dr \tag{11}$$

Substituting in the velocity profile expression [Eq. (9)] and integrating, gives

$$Q = \frac{\pi n R^3}{3n + 1} \left[\frac{R \, \Delta P}{2KL} \right]^{1/2} \tag{12}$$

Equation (12) can be written in a linear form as:

$$\log Q = \log\left(\frac{n\pi R^3}{3n + 1}\right) + \frac{1}{n} \log\left(\frac{R}{2KL}\right) + \frac{1}{n} \log \Delta P \tag{13}$$

The striking observation of Eq. (13) is that the first two sets of expressions on the right-hand side are constant. Hence, $Q = f(\Delta P)$, and a logarithmic plot of Q versus P will give a straight line. Nature and mathematics do not always agree. Actual flow data on pseudoplastic materials tend to be nonlinear as shown in Figure 4, the reason for this being that $n = f(dV_z/dr)$, that is, n tends to depend on Q. Although from a qualitative standpoint, the power law constitutive equation shows proper trends, quantitatively it is a poor predictive tool. From the standpoint of rheological characterization, there is no substitute for experiments and we must therefore design a capillary tube flow experiment to be general enough to evaluate all types of time-independent fluids.

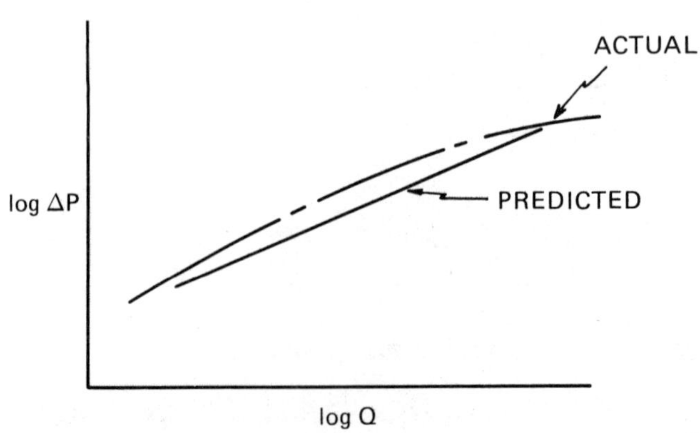

Figure 4 Comparison of actual to theoretical flow curve prediction.

CAPILLARY VISCOMETRY

Orienting the capillary tube so that the flow is vertically downward, we may write the force balance

$$\frac{\pi D^2}{4}\Delta P = \pi D L \tau_w$$

or

$$(\tau_{rx})_{r=R} = \tau_w = \frac{D\,\Delta P}{4L} \tag{14}$$

Following the development of the Rabinowitsch-Mooney equation, for steady, laminar flow of a time-independent fluid through a vertical tube, we consider the flow in a slice of fluid constituting an annulus between r and $r + dr$:

$$dQ = V 2\pi r\, dr$$

where V is the linear velocity at r. Hence,

$$Q = \pi \int_0^R V 2r\, dr = \pi \int_0^{R^2} V\, d(r^2)$$

Applying the no-slip condition ($V = 0$, $r = R$), the expression becomes

$$\frac{Q}{\pi R^3} = \frac{8Q}{\pi D^3} = \frac{1}{\tau_w^3}\int_0^{\tau_w} \tau_{rx}^2 f(\tau_{rx})\, d\tau_{rx} \tag{15}$$

where $f(\tau_{rx}) = -dV/dr$. By appropriate manipulation and replacing τ_w with $D\,\Delta P/4L$, we obtain the Rabinowitsch-Mooney equation of shear rate at the tube wall for steady, laminar flow of a time-independent fluid:

$$3\left(\frac{8Q}{\pi D^3}\right) + \frac{D\,\Delta P}{4L}\,\frac{d(8Q/\pi D^3)}{d(D\,\Delta P/4L)} = \left(-\frac{dV}{dr}\right)_w \tag{16}$$

Since $Q = VA = \frac{1}{4}\pi D^2 V$, Eq. (16) can be rearranged:

$$3\left(\frac{8Q}{\pi D^3}\right) = \left(-\frac{dV}{dr}\right)_w = \frac{3}{4}\left(\frac{8V}{D}\right) + \left(\frac{8V}{D}\right)\frac{d[\frac{1}{4}(8V/D)]/(8V/D)}{d(D\,\Delta P/4L)/(D\,\Delta P/4L)}$$

or

$$-\dot{\gamma}_w = \left(-\frac{dV}{dr}\right)_w = \frac{8V}{D}\left[\frac{3}{4} + \frac{1}{4n}\right] = \frac{3n+1}{4n}\left(\frac{8V}{D}\right) \tag{17a}$$

where

$$n = \frac{d \ln(D \, \Delta P/L)}{d \ln(8V/D)}$$

We may also state this expression as

$$-\dot{\gamma}_w = \tfrac{3}{4}\Gamma + \frac{\tau_w}{4} \frac{d\Gamma}{d\tau_w} \tag{17b}$$

where $\Gamma = 4Q/\pi R^3$ is the shear rate for a Newtonian fluid.

The Poiseuille equation for a Newtonian fluid (Bird et al., 1967) is

$$\Delta P = \frac{8\mu LV}{R^2} = \frac{32\mu LV}{D^2}$$

which can also be written as

$$\ln\left(\frac{D \, \Delta P}{4L}\right) = \ln\left(\frac{8V}{D}\right) + \ln(\mu)$$

Since μ is constant,

$$\frac{d \ln(D \, \Delta P/4L)}{d \ln(8V/d)} = n = 1.0$$

Substituting this expression into Eq. (17) shows that $(-dV/dr)_w = 8V/D$ [see Skelland (1967) for details]. This statement is true only for a Newtonian fluid since viscosity is a constant. In the case of a non-Newtonian n varies along the log-log plot of $D \, \Delta P/4L$ versus $8V/D$. To apply a capillary tube viscometer then to develop a non-Newtonian fluid's flow curve, data must be converted into a logarithmic plot of $D \, \Delta P/4L$ versus $8V/D$, where n is determined as the slope of the curve at a particular value of the wall shear stress (i.e., $\tau_w = D \, \Delta P/4L$). The shear rate corresponding to this wall shear stress value is determined from Eq. (17). It is important to note that Eq. (17) is derived from laminar flow; hence one must ensure that capillary tube flow is in this regime by checking the Reynolds number. Metzner and Reed (1955) have derived the generalized Reynolds number for a power law fluid as:

$$\mathrm{Re}_{\mathrm{gen}} = \frac{D^n V^{2-n} \rho}{\dot{\gamma}} \tag{18}$$

or

$$\mathrm{Re}_{\mathrm{gen}} = \frac{D^n V^{2-n} \rho}{K 8^{n-1}} \tag{19}$$

This expression applies to all fluids which are not time dependent. In fact, Metzner and Reed (1955) have shown that such non-Newtonian fluids obey the conventional Newtonian friction factor versus Reynolds number. Therefore, laminar flow prevails for $\mathrm{Re}_{\mathrm{gen}} < 2100$.

Data generated in any capillary tube viscometer will require corrections for four contributions to frictional losses: (1) the head of fluid above the tube exit; (2) kinetic energy effects; (3) tube entrance effects; (4) effective slip near the tube wall. The last effect is not always present and can generally be detected as explained later.

The first three corrections can be made through application of the total mechanical energy balance. Skelland (1967) outlines a correction approach and provides an illustrative example. An alternative approach which generally gives good results is the Bagley (1954) shear stress correction method.

This method was originally developed using a variety of polyethylenes. The resulting flow curve is independent of the capillary's L/D ratio; however, the procedure requires the assumptions of zero slip at the tube walls and time independency of the fluid material. The method uses a fictitious length of tubing $N_f R$ that is added to the actual length of tube L, such that the measured total ΔP across L is that which would be obtained in fully developed flow over the length $(L + N_f R)$ at the flow rate used in the experiment. The procedure is analogous to the application of an effective length in estimating kinetic resistances in pipe flow problems (Cheremisinoff et al., 1983, 1984). The shear stress at the wall for fully developed flow over the length $(L + N_f R)$ is then

$$\tau_w = R\,(\Delta P/2)(L + N_f R) \tag{20}$$

If the fluid is time independent and there is no slip at the tube wall, then τ_w is a unique function of $4Q/\pi R^3$ ($\simeq \Gamma$) in laminar flow [see Skelland (1967) for derivation]. Hence

$$\tau_w = \frac{R}{2}\left(\frac{\Delta P}{L + N_f R}\right) = f\left(\frac{4Q}{\pi R^3}\right) \tag{21}$$

from which

$$\frac{L}{R} = -N_f + \frac{\Delta P}{2f(4Q/\pi R^3)} \tag{22}$$

The procedure involves obtaining a series of ΔP measurements made on several tubes of various L/R (or L/D) ratios, while maintaining $4Q/\pi R^3$ constant. Table 1 summarizes wall shear stress and pressure drop measurements obtained for a polymer sample at one temperature.

A plot of L/R versus ΔP is shown in Figure 5 for different shear rates. Equation (22) is a straight line and hence the intercept on the L/R ordinate at $\Delta P = 0$ gives $-N_f$. A nonlinear plot as shown by the data for 3.6 and 10.8 s^{-1} would suggest one of our model assumptions is incorrect (i.e., the presence of time dependence or effective slip).

From a linear regression of each of the plots in Figure 5 the correction term N_f (intercept) can be derived and values for this example are given in Table 2. A plot of N_f versus Γ ($\equiv 4Q/\pi R^3$) is constructed in Figure 6 and further regressed to provide an analytical solution to Eq. (21). That is, substitution in the expression $\tau_w = R\,\Delta P/2(L + N_f R)$ allows calculation of τ_w free from end effects and should lead to a flow curve which is independent of the L/D ratio used in the capillary viscometer. Illustrations of the results are shown in Figures 7 and 8 and Table 3. Figure 7 shows a plot of τ_w versus Γ using the raw data obtained from the capillary tube. Table 3 summarizes calculations for the wall shear stress values corrected for entrance losses and Figure 8A shows a plot of this corrected data. As shown in Figure 8, τ_w data for L/D ratios of 2, 5, and 10 collapse to a single correlation. The data for $L/D = 0.25$ are in closer agreement for the high shear rate range ($\Gamma > 36\ s^{-1}$) but data points fall away from the average curve at

Table 1 Raw Data Obtained from a Capillary Tube Viscometer (Data for 100°C)

		Shear rate (s^{-1})							
		3.6	10.8	36.1	36.1	180.7	542.2	1084.4	1807.4
D_0 (in.)	L/D	Shear stress (kPa)							
0.0592	0.25	793	1034	1724	1310	2310	3344	3999	5240
0.0592	2	147	181	302	315	565	781	931	1060
0.0592	5	110	121	224	210	372	495	589	686
0.0592	10	117	148	234	209	352	446	507	569
L/R		Pressure (kPa)							
0.5	0.25	793	1034	1724	1310	2310	3344	3999	5240
4	2	1176	1448	2416	2520	4520	6248	7448	8480
10	5	2200	2420	4480	4200	7440	9900	11780	13720
20	10	4680	5920	9360	8360	14080	17840	20280	22760

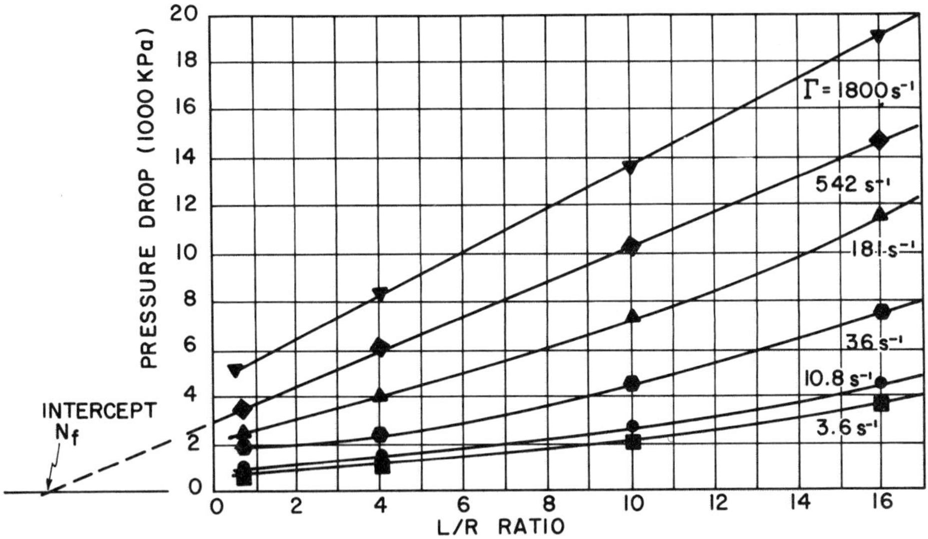

Figure 5 Plot of L/R versus ΔP for evaluation of Bagley correction term.

Table 2 Bagley Correction Term Computed from Eq. (22)

Shear rate	Correction factor ($-N_f$)	Coefficient of fit
3.6	2.07	0.99
10.8	1.54	0.98
36.1	2.40	1.00
36.1	2.70	0.99
180.7	3.19	1.00
542.2	4.06	1.00
1084.4	4.57	1.00
1807.4	5.38	1.00

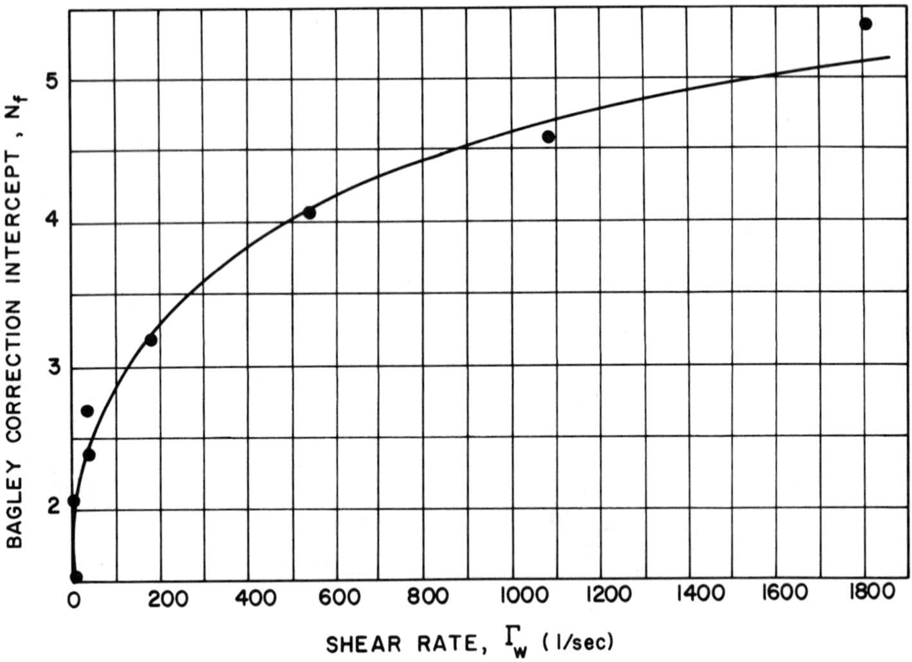

Figure 6 Plot of Bagley correction term versus Newtonian shear rate Γ.

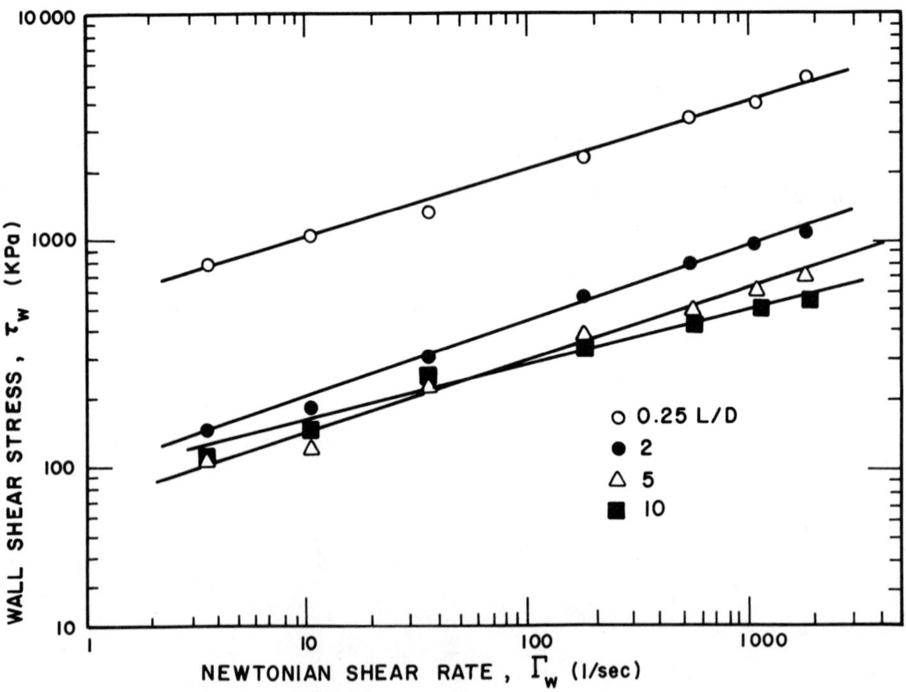

Figure 7 Plot of uncorrected flow curve data.

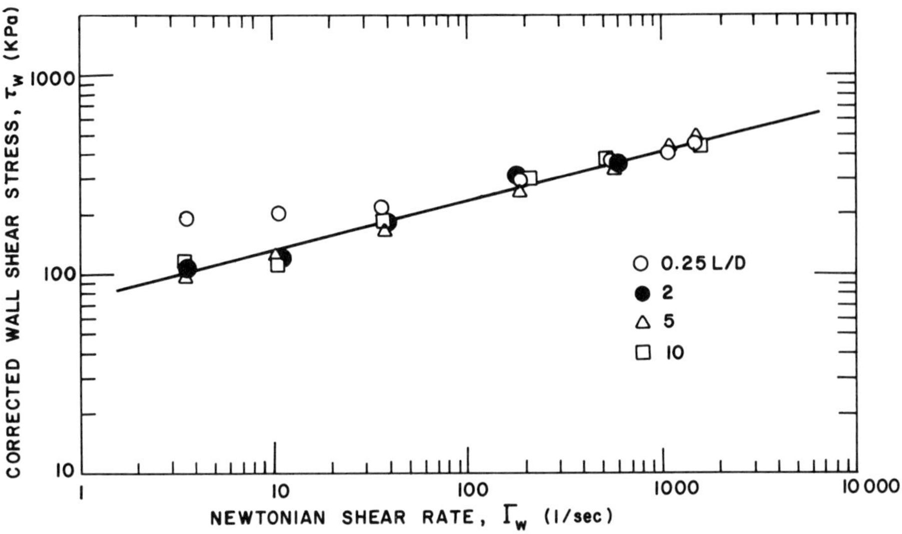

(A)

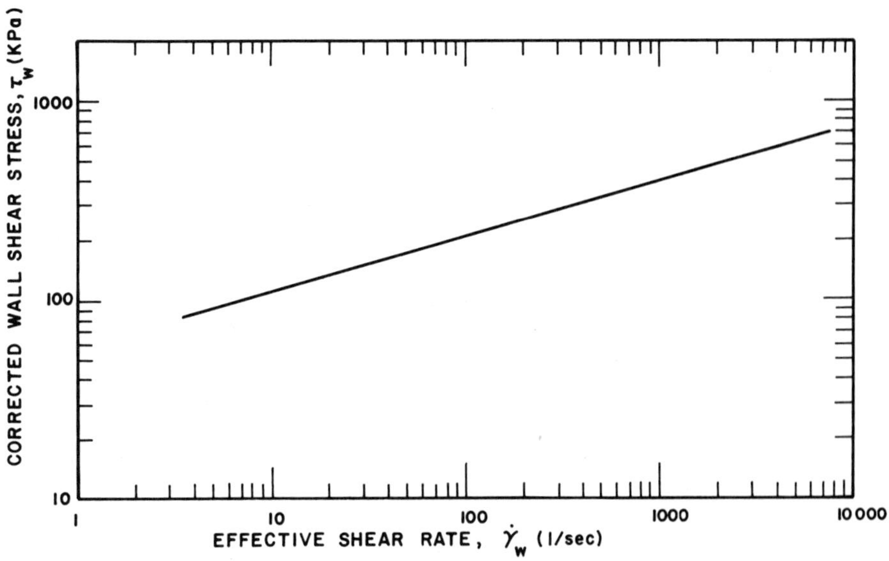

(B)

Figure 8 (A) Plot of corrected flow curve data using Newtonian shear rate. (B) Plot of corrected flow curve data using effective shear rate data.

Table 3 Corrected Wall Shear Stress Values (Data for 100°C): Corrected Wall Shear Stress (kPa)

True L/D	Shear rate (s^{-1}) 3.6	10.8	36.1	36.1	180.7	542.2	1084.4	1807.4
0.3	183.4	204.7	285.7	217.1	298.9	363.5	388.7	468.6
2.0	103.8	120.2	185.4	193.3	306.9	385.7	430.8	466.4
5.0	94.3	100.6	179.0	167.8	278.4	351.1	402.2	454.6
10.0	108.0	134.4	207.8	185.6	301.3	370.1	411.5	453.5
Avg 2–5 L/D	102.1	118.4	190.7	182.2	295.5	369.0	414.8	458.2

the low shear range. Since the resulting flow curve is linear, we may assume there is no time dependency for the material; hence the lack of agreement in τ_w data obtained at $L/D = 0.25$ is probably due to slippage. The power law coefficients K, n, can be readily regressed from the single line in Figure 8A ($\tau_w = K\Gamma^n$). The corrected shear rate at the wall can now be computed using Eq. (17). A plot of τ_w versus the effective shear rate $\dot{\gamma}$ computed from Eq. (17) is shown in Figure 8B. To complete this example, Table 4 provides a comparison of the viscosity data derived from the capillary ignoring entrance losses to the values obtained by the correction procedure outlined above. As shown, the effect of entrance losses is significant. It is interesting to note that data for $L/D = 10$ are in close agreement with the corrected values, indicating frictional losses to be small for this portion of the experiment. The polymer studied in this case is a crystalline EPDM rubber; however, no generalization can be formulated from the study of a single sample. To minimize entrance losses and therefore minimize lengthy correction calculations, the L/D ratio should be made as large as is practically possible. Thomas (1963), for example, had to resort to an L/D ratio of 1000 in work on suspensions. At excessive L/D ratios, frictional corrections become negligible in viscometric data. The remaining column in Table 4 reports the generalized Reynolds number [Eq. (18)], showing that the experiments were clearly performed in the laminar flow regime.

In the preceding example we were fortunate in that the data resulted in a linear flow curve. If the data showed a nonlinear response, it would have been necessary to employ standard tangent-drawing procedures. By constructing tangents to the curve on Figure 8A, we would obtain corresponding values of n from the tangent slopes, and K-values from the tangent intercept at $\Gamma = 1$. We would then recalculate $\dot{\gamma}_w$ from the Rabinowitsch-Mooney equation.

Table 4 Comparison of Corrected Viscosities to Values Obtained Ignoring Entrance Losses

Shear rate (1/s)	Corrected shear rate (1/s)	Viscosity (kPa-s), Measured viscosity (kPa) [0.25]	[2]	[5]	[10]		Corr. generalized visc. Reynolds no.
3.6	6.3	219.4	40.5	30.5	32.4	27.79	0.000057
10.8	18.9	95.5	16.7	11.1	13.7	12.19	0.000394
36.1	63.175	36.2	8.7	5.8	5.8	4.93	0.003259
36.1	63.175	47.7	8.3	6.2	6.5	4.93	0.003259
180.7	316.225	12.8	3.1	2.1	2	1.47	0.054591
542.2	948.85	6.2	1.4	0.9	0.8	0.64	0.373447
1084.4	1897.7	3.7	0.9	0.6	0.5	0.38	1.256123
1807.4	3162.95	2.9	0.6	0.4	0.3	0.26	3.071111

[a][ ] = L/D ratio.

It is important to note that some non-Newtonian fluids display a peculiar orientation of their molecules in the vicinity of the tube walls. In an aqueous suspension, the discrete phase may actually move away from the wall, leaving a thin layer of the continuum phase in the immediate proximity of the wall. In this situation there is a reduction in the apparent viscosity in the vicinity of the wall, known as the phenomenon of *effective slip* at the wall. For a capillary tube viscometer, an effective slip coefficient β can be evaluated as a function τ_w using the following relationship:

$$\frac{Q}{\pi R^3 \tau_w} = \frac{\beta}{R} + \frac{1}{\tau_w^4} \int_0^{\tau_w} \tau_{rx}^2 f(\tau_{rx})\, d\tau_{rx} \tag{23}$$

A series of capillary tube measurements is needed in order to evaluate β. For a range of tubes of various R but constant L, a plot of $Q/R^3\tau_w$ versus τ_w can be constructed. If no slip occurs with the material (i.e., $\beta = 0$), all the curves will coincide in accordance with Eq. (23). If, however, we obtain a family of curves, then β has some value. From the family of curves, select some fixed value of τ_w and obtain values of $Q/\pi R^3\tau_w$ for each corresponding R-value curve. Now prepare a plot of $Q/\pi R^3\tau_w$ versus $1/R$. The slope of this curve provides a β-value for the selected τ_w, according to Eq. (23). By repeating this procedure for different τ_w-values a relationship between τ_w and β can be established. Figure 9 illustrates the method. Once a relationship between the effective slip coefficient

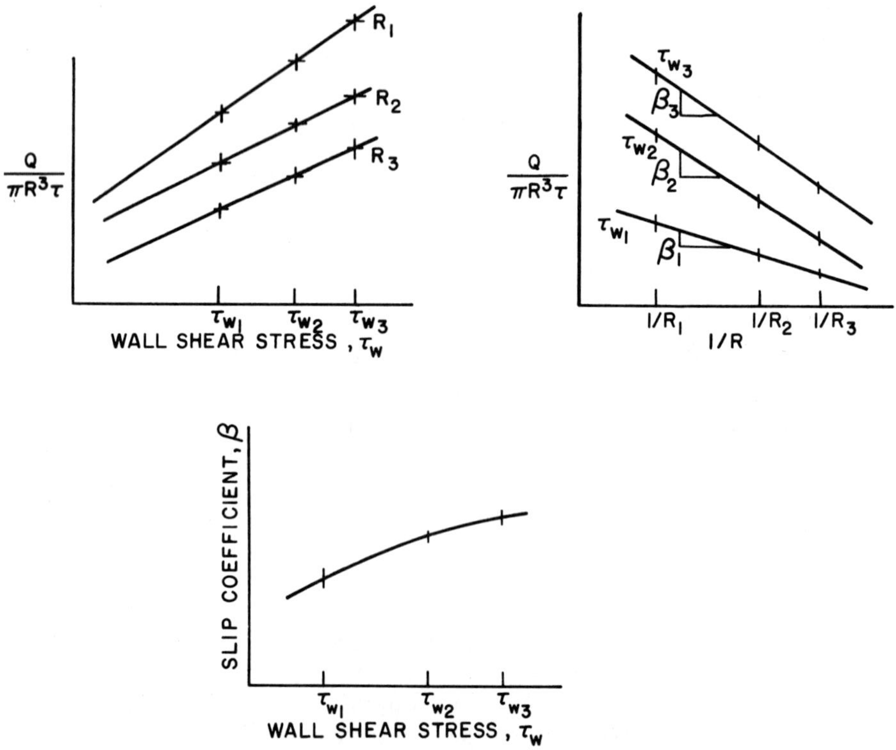

Figure 9 Procedure for correcting for effective wall slip.

and τ_w is established, the measured volumetric flow rate from the capillary tube can be corrected for slippage:

$$Q' = Q_{\text{measured}} - \beta\tau_w\pi R^2 \tag{24}$$

where Q' is the volumetric flow rate corrected for slippage and $\tau_w = D\,\Delta P/4L$ corresponding to Q_{measured}.

CONCENTRIC CYLINDER ROTARY AND ROTATING CYLINDER VISCOMETERS

Concentric cylinder rotary and rotating cylinder viscometers are most often applied to solution viscosity measurements and are usually limited to shear rates of less than 100 s^{-1}.

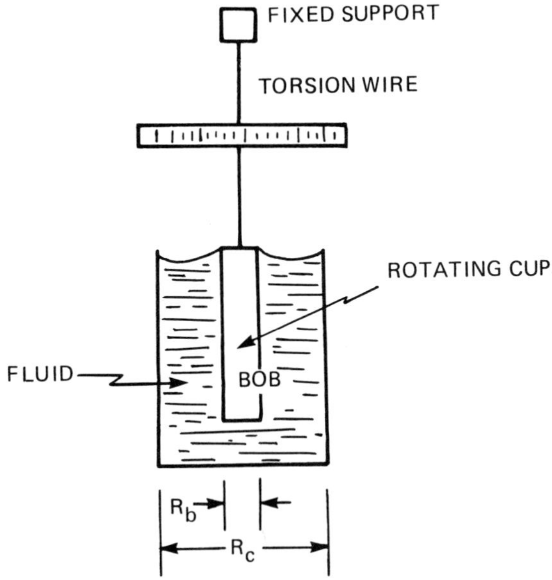

Figure 10 Basic elements of concentric cylinder rotary viscometer.

The basic elements of the concentric cylinder rotary viscometer are shown in Figure 10. The unit operates by applying shear to a fluid located in the annulus between the concentric cylinders. One cylinder (usually the cup) rotates while the other is fixed in space. From a series of measurements of the angular speed of the rotating cup and of the torque applied to the stationary cylinder, a flow curve for the fluid under shear can be derived.

The development of the constitutive equations for this type of viscometer is as follows. From a torque balance about the surface of the fixed bob while the cup revolves at a steady angular velocity, we write

$$\tau_b(2\pi R_b l)R_b = 2\pi R_b^2 l \tau_b = \mathrm{T} \tag{25}$$

where

T = torque
R_b = bob radius
τ_b = shearing stress at the bob surface

The balance equation neglects end effects at the base of the bob and assumes that shear occurs only at the cylindrical surface of the bob. In practice, however, end effects are important and can be accounted for by including an effective bob length l_{eff} in Eq. (25). Solving Eq. (25) for the bob surface shear stress, then

$$\tau_b = \frac{T}{2\pi R_b^2 l} \tag{26}$$

The shear rate at the bob surface is a complex relationship, which for brevity is not included here. An approximate relationship proposed by Calderbank and Moo-Young (1959) based on the analytical derivation of Kreiger and Maron (1954) is as follows:

$$\left(\frac{dV}{dr}\right)_b = \frac{4\pi N}{1 - (R_c/R_b)^2} C_R \tag{27}$$

Equation (27) is applicable for cup-to-bob radii ratios smaller than 1.75. The term C_R represents a series expansion of the shear rate–geometric relationship developed by Kreiger and Maron for time-independent power law fluids. Table 5 provides values of C_R that can be used in Eq. (27). The rheological flow curve can be developed by a logarithmic plot of τ_b versus $(dv/dr)_b$. By employing different bob or cup sizes, the different ranges of shear rates can be studied. In the extreme, for very small annular gaps (i.e., $R_c/R_b \rightarrow 1.0$), the shear rate approaches a constant value across the annulus and can be approximated simply by

$$\frac{dV}{dr} = \frac{2\pi RN}{R_c - R_b} \tag{28}$$

Close clearance viscometers such as this are not suitable for use with suspensions where particle sizes are comparable to the annular gap.

End effects can be corrected for by substituting an effective length l_{eff} for the true bob length l in Eq. (26). The effective length can be determined by calibrating the viscometer using a Newtonian fluid of known viscosity μ and applying the following formula:

$$l_{eff} = \frac{T}{2\pi R_b^2 \mu (dV/dr)_b} \tag{29}$$

To correct for effective slip in rotary viscometers, the correction method of Mooney (1931) can be used. The method involves obtaining measurements using three different R_c/R_b ratios. For R_1, R_2, R_3 radii, such that $R_1 < R_2 < R_3$, three radii ratios can be defined ($S_a = R_2/R_1$, $S_b = R_3/R_2$, $S_c = R_3/R_1$). The rota-

Table 5 Values of Calderbank–Moo-Young Shear Rate Term (C_R)[a]

	R_c/R_b ratio →					
n^H	1.070	1.150	1.166	1.250	1.400	1.746
0.050	2.722	4.999	5.435			
0.100	1.708	2.617	2.801	3.735	5.184	
0.200	1.287	1.622	1.689	2.031	2.593	3.615
0.300	1.162	1.342	1.377	1.554	1.843	2.382
0.400	1.102	1.213	1.234	1.340	1.511	1.826
0.500	1.068	1.139	1.153	1.220	1.327	1.522
0.600	1.045	1.091	1.100	1.144	1.212	1.335
0.700	1.029	1.058	1.064	1.091	1.134	1.209
0.800	1.017	1.034	1.037	1.053	1.077	1.119
0.900	1.007	1.015	1.016	1.023	1.034	1.052
1.000	1.000	1.000	1.000	1.000	1.000	1.000
1.250			0.971		0.941	0.910
1.500			0.952		0.903	0.853
1.750			0.939		0.877	0.814
2.000			0.929		0.857	0.785
2.250			0.921		0.842	0.763
2.500			0.915		0.830	0.746
2.750			0.910		0.820	0.732
3.000			0.906		0.812	0.720
3.250			0.902		0.806	0.711
3.500			0.899		0.800	0.702
3.750			0.897		0.795	0.695
4.000			0.894		0.791	0.689
Coefficient (A)	0.903	0.861	1.144	0.898	1.169	1.155
Exponent (B)	−0.304	−0.505	−0.320	−0.547	−0.418	−0.485
Coef. of fit (r^*)	0.946	0.963	0.891	0.974	0.928	0.957

[a]Each set of R_c/R_b data sets have been regressed using the least squares to a power law expression. For approximate C_R-values apply above model coefficient: $C_R = A\ (n^H)_B$.

tional speed of the cup required to obtain the same torque for each of these cases can be measured as N_{12}, N_{23}, and N_{13}. The effective slip coefficient corresponding to the stress at the surface $2\pi R_2 l$ is

$$\beta_2 = \frac{2\pi^2 R_2^3}{\mathrm{T}} (N_{12} + N_{23} - N_{13}) \tag{30}$$

Repeating this procedure for different torque values provides a correlation between β_2 and the surface shear stress τ_2. To correct for slippage, for a given torque value, τ_b can be computed from Eq. (26). The corresponding expression for wall shear stress at the cup surface is

$$\tau_c = \frac{\mathrm{T}}{2\pi R_c^2 l} \tag{31}$$

The effective slip velocities at the bob and cup surfaces are $V_{s,b} = \beta_b \tau_b$ and $V_{s,c} = \beta_c \tau_c$, respectively. The effective velocity of the fluid at the rotating cup surface will be less than that of the cup by the amount $V_{s,c}$. The effective velocity of the fluid at the stationary bob surface is $V_{s,b}$. The true shear rate corresponding to τ_b can be calculated by multiplying either Eq. (27) or (28) by the ratio of the actual velocity difference (i.e., from cup to bob) to the velocity difference without slip (i.e. $(V_{sc} - V_{sb})/V_{\mathrm{meas}}$).

A modification of the concentric cylinder viscometer is the Brookfield viscometer, in which the cup radius is effectively extended to infinity. Flow curves are generated from measurements of the torque needed to rotate a cylindrical rod at various speeds when immersed in the fluid of "infinite" volume. Basically, the fluid is contained in a cup whose radius is much larger than the rod. With this arrangement, the walls of the retaining cup do not influence the shearing movement of the fluid. Equation (26) defines the shearing stress at the rod surface (R_b = radius of the cylindrical rod, l = length of rod immersed in the fluid). As before, an effective rod immersion length, l_{eff}, can be developed to correct for end effects. The shear rate at the surface of the rotating rod for any time-independent fluid is given by

$$\left(-\frac{dV}{dr}\right)_b = \frac{4\pi N}{n''} \tag{32}$$

where n'' is the slope of a logarithmic plot of torque versus rotational speed N, evaluated at the particular N for which T was measured. A plot of τ_b versus $(-dV/dr)_b$ provides the flow curve. With either device we may also generate a plot of $D\ \Delta P/4L$ versus $8V/D$ (Γ – Newtonian shear rate) for laminar tube flow.

In general, both devices enable studies over a very limited shear rate range. Commercial units are usually equipped with an electrical heating jacket and appropriate heat exchange to remove heat generated by friction in the fluid.

CONE-AND-PLATE VISCOMETER

The cone-and-plate viscometer is a widely used instrument for shear flow rheological properties studies. The device can be used to measure both viscosity and normal stresses. Applications and analysis of this type of viscometer for studying polymer melts are covered by Lee and White (1975, 1974), Lobe and White (1979), Chapman and Lee (1970), Minagawa and White (1976), King (1966), Meissner (1972), and Weissenberg (1948). The principal features of this viscometer are illustrated in Figure 11, consisting essentially of a flat, horizontal plate and an inverted cone, the apex of which is in near contact with the plate. The angle between the plate and cone surface is small (typically <2°). The fluid sample is placed in this small gap between the cone and plate. A flow curve is developed from measurements of the torque needed to rotate the cone at different speeds. For a constant speed N, the linear velocity at r is $2\pi rN$. The gap height at r is $r \tan \phi$. Hence, the magnitude of the shear rate at r is

$$\frac{2\pi rN}{r \tan \phi} = \frac{2\pi N}{\tan \phi} \tag{33}$$

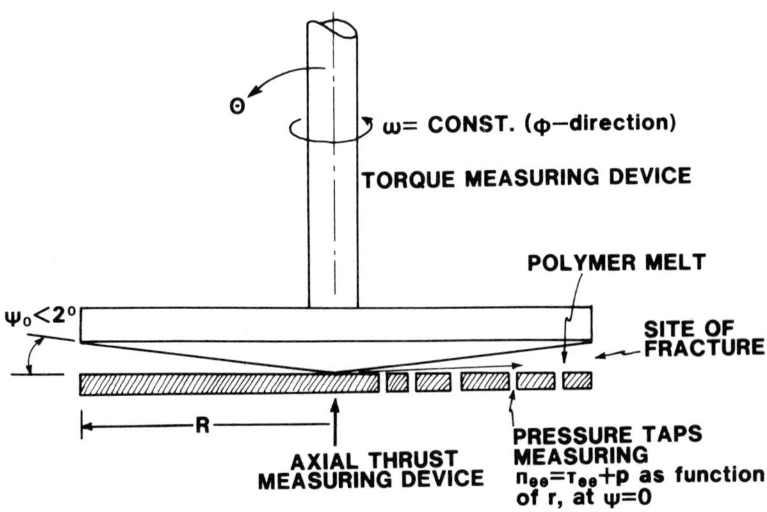

Figure 11 Schematic of cone-and-plate viscometer.

The shear rate is constant between $0 > r > R$, which means that τ_{yx} is also constant over this range. For small and ϕ, tan $\phi \simeq \phi$, and hence the magnitude of the shear rate equals approximately $2\pi N/\phi$. The following expression defines the relationship between measured torque and shear stress:

$$T = 2\pi\tau_{yx}\int_0^R r^2\,dr = \tfrac{2}{3}\pi R^3\tau_{yx}$$

or

$$\tau_{yx} = \frac{3T}{2\pi R^3} \tag{34}$$

Equation (34) enables τ_{yx} to be obtained at different measured torque values. Shear rate can be obtained from Eq. (33), and hence τ_{yx} can be plotted against shear rate for the flow curve.

Commercial instruments that are capable of measuring normal stresses as well as the shear stress are the Weissenberg rheogoniometer (manufactured by Sangamo Controls, Bognor Regis, U.K.) and the Mechanical Spectrometer (manufactured by Rheometrics Inc., Union, NJ).

SANDWICH VISCOMETER

Another type of shear flow instrument primarily used for polymer melts is the sandwich viscometer (also known as the parallel-plate viscometer). The measurement principal is illustrated in Figure 12. A viscous material such as an elastomer is sheared between two parallel steel plates or, as in Fig. 12, in a sandwich between three plates. The shear rate is

$$\dot{\gamma} = \frac{V}{H} \tag{35}$$

where H = interplate distance.

The applied force can be determined as a function of time, which after normalization with area, provides shear stress data as a function of time:

$$\tau = \frac{F}{2A} \tag{36}$$

It is possible to measure transient start-up stresses of the material with this device. After a sufficiently long time and if the material is stable, the steady-state stress can be measured, at which point the viscosity can be determined as a

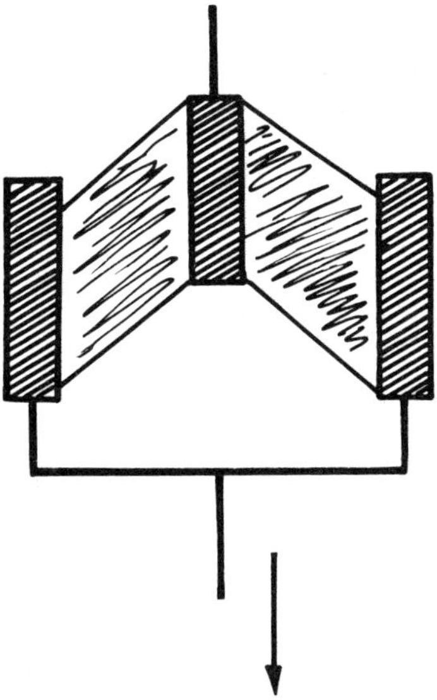

Figure 12 Schematic of sandwich viscometer.

function of shear rate. With many materials such as elastomers, both large times and equipment strains are needed. These conditions often result in slippage. To minimize slippage contact surfaces are usually knurled.

This type of viscometer is capable of making measurements at very low shear rates on materials which usually exhibit slippage in nonpressurized rotational devices (e.g., cone-and-plate type viscometers). The instrument is described in detail by Zakharenko et al. (1962), Middleman (1969), Goldstein (1974), and Furuta et al. (1976).

PARALLEL-DISK VISCOMETER

The parallel-disk viscometer (also known as a shearing-disk viscometer) measures viscosity in torsional flow between a stationary and a rotating disk. Design variations are illustrated in Figure 13. The operation involves the use of a serrated disk which is rotated in a sample fixed in a pressurized cavity. This system was developed for rubber applications where both pressurization and serrated surfaces are used to avoid slippage. Properties are measured based on the shear rate at the outer radius of the disk, where

$$Y_a = \frac{a\omega}{H} \tag{37}$$

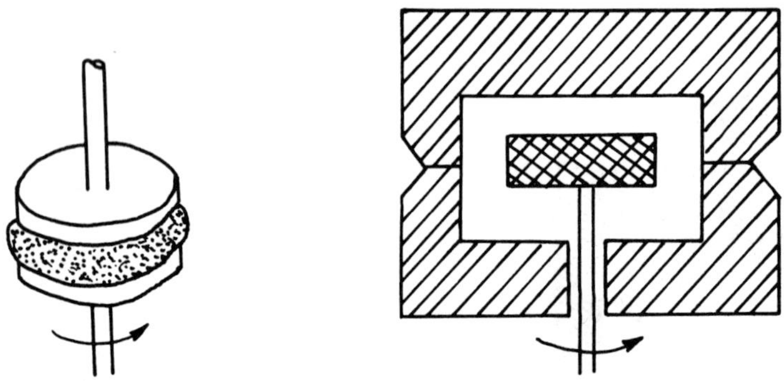

(A) (B)

Figure 13 (A) Parallel-disk viscometer. (B) Mooney shearing-disk viscometer.

Shear stress and normal stress differences are given by the following relationships:

$$\tau|_a = \frac{3T}{2\pi a^3}\left[1 + \frac{1}{3}\frac{d\ \ln T}{d\ \ln \gamma_a}\right] \tag{38}$$

$$P_{11} - 2P_{22} + P_{33}|_a = \frac{2F'}{\pi a^2}\left[1 + \frac{1}{2}\frac{d\ \ln F'}{d\ \ln \gamma_a}\right] \tag{39}$$

where F' = normal thrust. The parallel disk apparatus is limited to low shear rates.

The Mooney viscometer consists of a rotating disk in a cylindrical cavity. This instrument has become the standard quality control instrument of the rubber industry. The American Society for Testing Materials (ASTM) specifies the diameter of the disk to be 19.05 mm and thickness 5.54 mm. A single-point viscosity measurement is made at a standardized rotor speed of 2 rpm, after a period of 4 min and 100°C. In the elastomers industry some manufacturers still report measurements at 8 min and 127°C; however, considerable effort is underway to standardize to the ASTM guidelines. Measurements are reported in terms of a calibrated torque measurement on a standard dial (referred to as ML-4 reading). The designation ML-4, means: M, *Mooney* viscometer; L, *large* disk size (a smaller diameter disk (S) is used for vulcanization scorch tests); and 4, 4 minutes of test.

ML-4 readings can be converted to torque, shear stress, and viscosity values. Nakajima and Harrel (1979) provide the following correlations:

$$T = 8.30 \times 10^{-2} D_R \tag{40a}$$

$$\tau_a = 0.382\ (D_R)f' \times 10^4 \tag{40b}$$

where D_R is the Mooney unit dial reading. T and τ_a have units of N/m and Pa, respectively. f' is a function defined by the following relationship for shear stress at the disk's outer radius:

$$\tau|_a = \frac{T}{\pi a^3} f' \tag{41}$$

$$f' = \frac{n' + 3}{4}\left[1 + \frac{[(2/n')^{n'}(n' + 3)hH^{n'}]}{2a^{n'+1}\{1 - [a/(a + \delta)]^{2/n'}\}^{n'}}\right]^{-1} \tag{42}$$

where

δ = distance between the outer disk radius and cavity
h = disk thickness
n' = power law exponent, $\simeq (d \ln T/d \ln \dot{\gamma}_a)$

Function f' typically has a value of around 0.55 for the standard measurement conditions.

SLIT RHEOMETER

A slit rheometer is an instrument analogous to the capillary viscometer. The primary difference between the devices is the orifice cross section. Han (1976, 1974, 1971) and Wales et al. (1965) describe the use of this instrument in the study of polymer melts. The device makes use of a series of flush-mounted transducers located along the flow tube. These transducers measure the pressure gradients along the direction of flow, which can then be converted to wall shear stress values via

$$\tau_w = \delta_c \frac{dP}{dZ} \tag{43}$$

where δ_c = half-thickness of the channel.

Han (1976) gives the following expression for the shear rate at the channel wall:

$$\dot{\gamma}_w = \frac{3Q}{4ab^2}\left[\frac{2}{3} + \frac{1}{3}\frac{d(\ln(3Q/4ab^2))}{d(\ln\tau_w)}\right] \tag{44}$$

where a = half-width.

In general, the instrument is capable of operating over comparable shear rate ranges to the capillary viscometer.

MELT FLOW INDEXER

The melt indexer is widely accepted as a standard quality control instrument in the plastics industry. The device, developed by DuPont, is typically used to specify polyethylenes, polypropylene, and polystyrene. It is essentially an extrusion rheometer that is in fact similar to a capillary rheometer. The instrument employs a single capillary (diameter of 0.655 mm and L/D ratio of 12) in which extrusion of material is accomplished through pressure applied by a dead weight. The melt index is given as a measure of the weight in grams extruded for a 10-minute test.

The ASTM specifies a series of test conditions involving temperature and applied dead load settings. These conditions are designations A through K, and they cover the following ranges; for temperature: 125–275°C, for applied dead load: 325–21,600 g (or in terms of pressures: 0.5–29.5 atm). Specifications are selected so that melt indices lie between the values of 0.15 and 25 g per 10 min test. The most frequently used condition in the industry is specification E (190°C, 2.95 atm).

Although this is a widely used instrument, its value is that of a quality control tester which can only empirically relate to end-use processability of products. The melt index (MI) is related to the inverse of viscosity; however, no end loss corrections have been developed nor can the MI be easily related to the Weissenberg–Rabinowitsch shear rate expression.

ELONGATIONAL FLOW DEVICES

Up to this point instruments based on shear flow have been discussed; however, there are numerous polymer processing operations that involve uniaxial elongational flows. An example in processing where elongational flow occurs is in the entrance die region of an extruder. This phenomenon can be studied in a capillary die. Pressure losses through such dies can be used to determine elongational viscosity in polymer melts [see Cogswell (1969, 1978), for example]. For this reason a variety of laboratory test devices and techniques have been devised over the years in order to conduct rheological investigations and/or quality control testing in this type of flow. In general, measurements of elongational viscosity are complex to interpret and more difficult than measurements made in shear flow. To make elongational viscosity measurements requires maintaining a constant elongation rate or stress while reaching steady-state conditions. These type of tests are typically applied to polymer melts that include high-molecular-weight polyolefins, styrenics, elastomers, and their compounds. A few of the most frequently used tests are briefly described below.

The most common method of measuring elongational viscosity is to stretch a filament of the material. The fluid obviously must have a sufficiently high viscosity to be maintained under controllable deformation. Materials suitable for this type of testing are elastomers and polyolefins where melt viscosities are typically 10,000 Pa-s or greater. These types of test measurements are limited to deformation rates below 5 s^{-1}. Ballman (1965), Vinogradov et al. (1970), and Stevenson (1972) describe different types of filament stretching methods. A very simple arrangement involves clamping both ends of a vertical thermostated filament, which is then stretched at a rate dl/dt in such a manner as to maintain a constant rate of deformation. In this manner the rate of deformation ϕ' is

$$\phi' = \frac{1}{l}\frac{dl}{dt} \tag{45}$$

and

$$l = l(0)e^{Et} \tag{46}$$

In other words, the length of a material undergoing stretching follows an exponential law. The test is illustrated in Figure 14A.

Another approach, illustrated in Figure 14B, involves supporting a horizon-

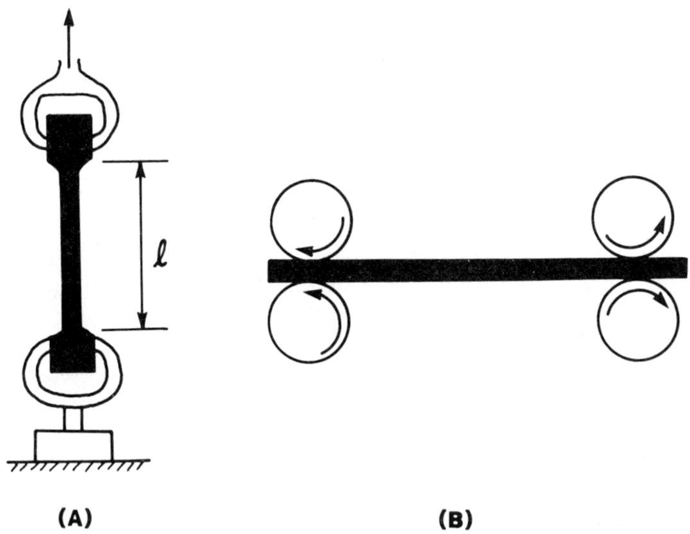

Figure 14 (A) Vertical method of elongational viscosity measurement at constant elongation rate. (B) Horizontal filament stretching test.

tal filament on the surface of a hot immiscible oil. The polymer filament is held at both ends between pairs of toothed or serrated wheels that turn at linear velocity $V/2$. The filament is drawn at both ends through the wheels. One of these wheels contains a tension-sensing instrument so that deformation rate is proportional to the velocity ($\phi' = (V/2)/(L/2) = V/L$).

Different variations of the filament stretching technique for the measurement of elongational viscosity are described by Mocosko and Lornsten (1973), Ide and White (1978), Cotten and Thiele (1979), and Dealy (1978).

STRESS RELAXATION AND SMALL STRAIN MEASUREMENTS

These classes of rheological measurements are carried out on bulk polymers and are usually aimed at obtaining information on the material's response to small strains. Strain can be applied as a step function and the stress relaxation that takes place after its application is measured.

Stress relaxation measurements are routinely made on elastomers. The test involves application of a strain to a polymer sample that is in extension, shear, or compression. The measurable parameter is the force response over time, $F(t)$, to the applied strain. Some investigators report the material property in terms of a relaxation modulus defined by

$$E(t) = \lim_{\Delta l \to 0} \left[\frac{F(t)/A_0}{\Delta l / l_0} \right] \tag{47}$$

where

A_0 = initial cross section of sample
l_0 = initial length of sample
Δl = amount sample is stretched in extension

For measurements made under shear, a surface of the sample is subjected to a displacement in a direction normal to its thickness. The shear modulus is given by:

$$G(t) = \lim_{\Delta l \to 0} \left[\frac{F(t)/A}{\Delta l / H} \right] \tag{48}$$

where H = sample thickness.

These types of tests are normally coupled with mechanical properties testing. A variety of polymers, particularly elastomers, are widely used in consumer- and industry-related products because of their good mechanical properties. Tensile

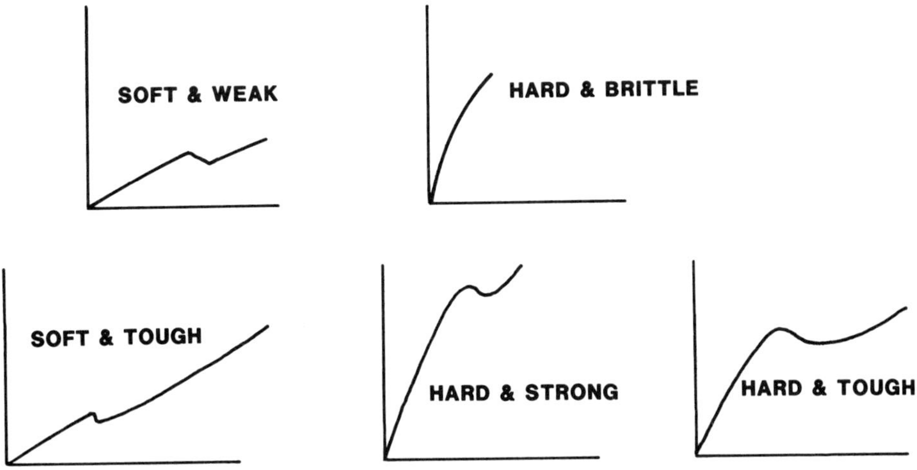

Figure 15 Typical tensile stress–strain curves for polymers.

measurements at several strain rates are often made as a quality control check of the polymers' integrity, even though their behavior in compound states may be quite different. The stress–strain curve provides information on both the rheological and mechanical behaviors of a polymer. Figure 15 illustrates typical features of tensile stress–strain curves for polymers. The classification of curves in Fig. 15 shows that polymers can be classified in general terms as rubbery, glassy, or variations in between. Compounding can greatly change the stress–strain response curve of a given polymer.

Tensile testing in rubber labs is normally done on a tensile testing machine such as an Instron and performed according to ASTM standard methods (ASTM D638, ASTM D882). As a product quality control test, many elastomer manufacturers are no longer using tensile testing of neat rubber, but rely more heavily on rheometer–cure–response data. In product development work, however, tensile testing is very much an integral part of polymer characterization work. A minimal testing scheme involves obtaining a stress–strain curve at room temperature conditions and at perhaps three or more strain rates. From each curve various quantities related to the mechanical and rheological properties of the sample can be derived. Figure 16 shows the generalized tensile stress–strain curve for plastics and the various parameters of interest. The principal information obtained from these curves includes:

1. The modulus of elasticity (the initial slope of the curve)
2. The yield stress (peak stress in the early portion of the curve)

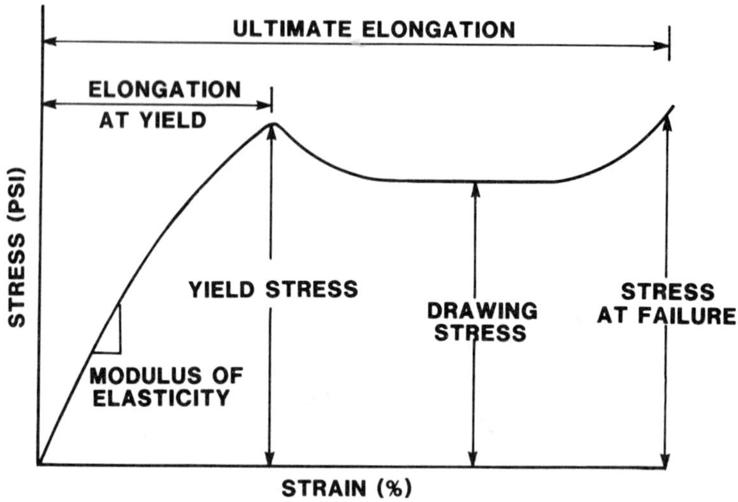

Figure 16 Tensile stress–strain curve for plastics.

3. The stress at failure
4. The elongation (or strain) at yield (also called the yield strength)
5. The ultimate elongation or strain at failure (also referred to as the *elongation at break*)
6. The energy to break; measured by the area under the stress–strain curve up to the point of failure

The last feature, the energy to break, is truly a measure of sample toughness.

An important class of experiments originally introduced by Philippoff (1934) involves the determination of stress response to an imposed sinusoidal oscillation. Results are analyzed in terms of the dynamic viscosity function η' (not to be confused with the dynamic viscosity of classical fluid mechanics) and the dynamic rigidity function G'. In oscillatory shear flow, we may represent the velocity field using rectangular Cartesian coordinates as follows:

$$V_1 = \epsilon\omega x_2 e^{i\omega t}, \qquad V_2 = V_3 = 0 \tag{49}$$

where ϵ is small enough for second- and higher-order terms to be neglected and ω is frequency. The corresponding stress field is

$$P_{12} = \eta^*(\omega)\epsilon\omega e^{i\omega t}, \qquad P_{11} - P_{22} = P_{22} - P_{33} = 0 \tag{50}$$

where η^* is the complex dynamic viscosity expressed by convention as:

$$\eta^* = \eta' - i\,\frac{G'}{\omega} \tag{51}$$

Other functions typically defined are

$$\eta'' = \frac{G'}{\omega} \tag{52a}$$

$$G'' = \omega\eta' \tag{52b}$$

The most widely used method for measuring $G'(\omega)$ and $G''(\omega)$ on polymer melts systems involves placing the melt in the gap between a cone and plate and oscillating one of the elements relative to the other with a shear strain. Figure 17 illustrates this technique.

Another arrangement, illustrated in Figure 18, involves rotating two parallel eccentric disks. The polymer sample undergoes periodic sinusoidal deformation. The forces exerted on the disk are interpreted as $G'(\omega)$ and $G''(\omega)$.

Many industrial research laboratories use the Weissenberg–Rheogoniometer and Rheometrics Mechanical Spectrometer for cone-and-plate sinusoidal oscillation measurements of $G'(\omega)$ and $G''(\omega)$. The reader is referred to the works of the following investigators for discussions of sinusoidal experiments: Vinogradov et al. (1972), Lobe and White (1979), Munstedt (1976), Aoki (1979), Andrews et al. (1948), Tobolsky (1960), Maxwell and Chartoff (1965), Chen and Bogue (1972) and Walters (1975).

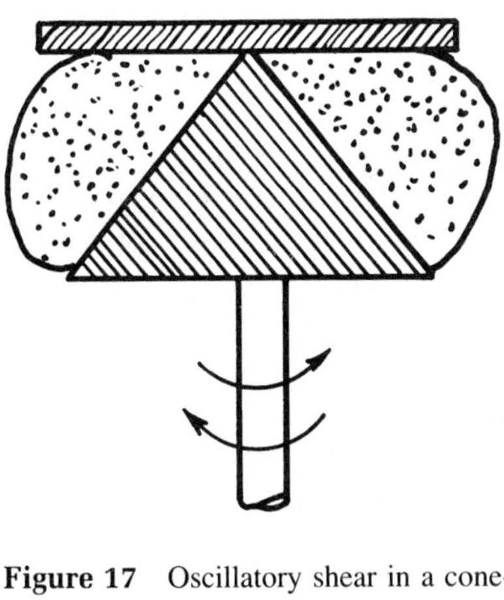

Figure 17 Oscillatory shear in a cone-and-plate device.

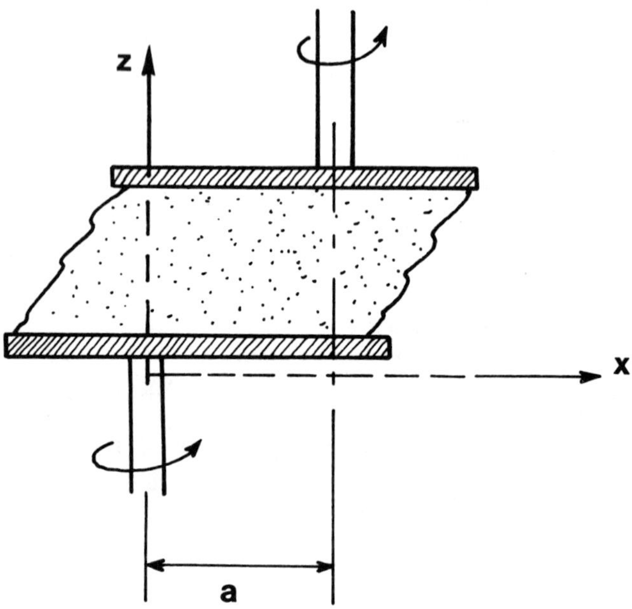

Figure 18 Principle of orthogonal rheometer.

FLOW BIREFRINGENCE MEASUREMENTS

Determination of birefringence during flow is a rheological property measurement technique applied almost exclusively to homogeneous systems of flexible polymer chains. This technique is based on a relationship between flow birefringent behavior and stresses and is determined by the rheo-optical law established by Philippoff (1956) and others (Dexter, Miller, and Philippoff, 1961). Basically, the law relates the difference in the principal refractive indices to the difference in the principal stresses as developed during flow. The principal stresses represent the magnitudes of the principal axes of the ellipsoid of the stress tensor. The technique has been applied in purely scientific studies on flexible chain polymer melts such as polyolefin, polystyrene, and elastomers. The major difficulty in its application is that the technique is limited to amorphous polymers. Even trace amounts of crystallinity have a dramatic effect on measurements. Measurements can be performed on polymer solutions using a coaxial cylinder viscometer, whereas polymer melt measurements are normally taken on parallel-plate slit extrusion devices. These instruments seek to use flow birefringence to measure shear flow properties.

Measurement of birefringence is accomplished by studying the transmission of a polarized monochromatic light beam that is passed through the fluid medium to some depth. Since the medium in the plane perpendicular to the transmitted wave has different refractive indices in different directions, the wave splits into two parts. Each refracted wave has a different velocity, which corresponds to the two refractive indices. The waves are out of phase with each other and the beam exiting the instrument passes through a polarizer crossed normal with the initial direction of polarization. Birefringence is determined either through the reduction in intensity of the exiting beam, through the use of compensators, or through characterization of the fringe patterns produced by the external polarizer. The fringes indicate conditions under which the contributions of the two waves to the direction allowed passage by the analyzer cancel. The method based on the use of compensators is described by Wales and Philippoff (1973). Theory and application of flow birefringence measurements of rheological properties are treated in the following references: Wales (1969), Han and Drexler (1973), Brizitsky et al. (1978), Andrews et al. (1948), Wales and Janeschitz-Kriegle (1967), Gortemaker et al. (1976), Meissner (1969), Dexter et al. (1961), Lodge (1956), Chen and Bogue (1972), Gortemaker and Janeschitz-Kriegle (1976), and Treloar (1940, 1941, 1947).

SWELL AND SHRINKAGE TESTS

Another rheological test more often used for product quality control is measurement of the elastic recovery in polymer melts and their compounds. This is usually done by measuring the swell (or shrinkage) of materials undergoing extrusion. The tendency of polymers, whether thermoplastic or rubber, to enlarge when emerging from an extruder die is called *die swell.* Die swell normally refers to the ratio of extruded size to die size. This behavior is a measure of the relative elasticity in the flowing polymer stream. Die swell is caused by the release of the residual stresses when the sample emerges from the die. Measurement of this behavior has become widely recognized in the rubber and plastics industries as an important indication of polymer processibility.

Traditional measurements reported by various investigators have been made on extrudates either with a micrometer or by weight per unit length. The Monsanto Industrial Chemicals Co. (Akron, OH) has introduced an automated capillary rheometer (called the Monsanto Processability Tester) which employs an extrudate swell detector based on interaction with a scanning laser beam positioned immediately below the capillary die exit. This enables measurement of the running die swell. Figure 19 shows running die swell data for a compound correlated against wall shear for different L/D dies. It is important to note that with some materials a distinction between running die swells and relaxed die swells should be made. Highly elastic materials will tend to swell more over time depending on their elastic recovery energy.

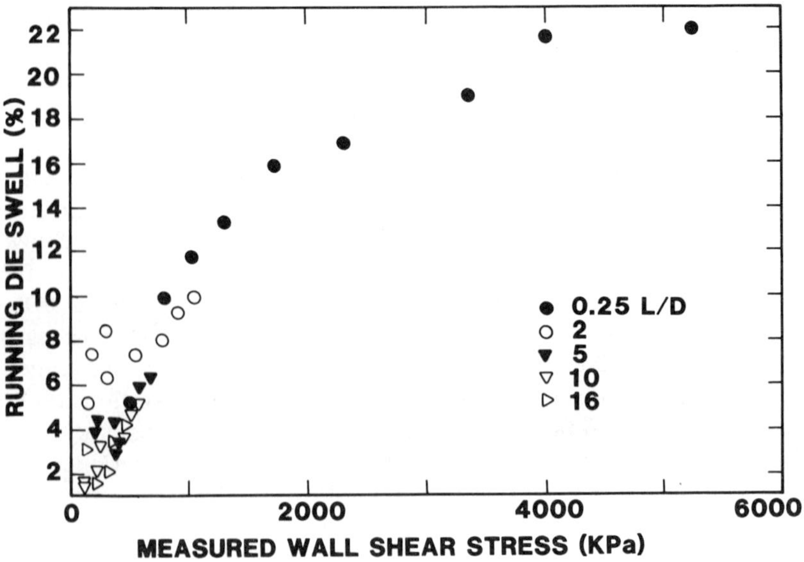

Figure 19 MPT data on running die swell.

C. W. Brabender Instruments, Inc. (Hackensack, NJ) markets an optical-electronic die swell measuring system that can be used for continuous control and monitoring of extruders and plasticorders. The basic design features are illustrated in Figure 20. An infrared, long-life diode (A) emits homogeneous infrared light through a filter (B), diffuser (C), and lens (D); then through a window (E). The light impinging on the object (F) is scattered. Only that light reaching the ambient light filter (G) is measured by the photodetector H1. Light reaching photodetector H2 is diverted through the prism (1) and is uninterrupted. This acts as a reference beam. The electronic instrumentation measures the differences

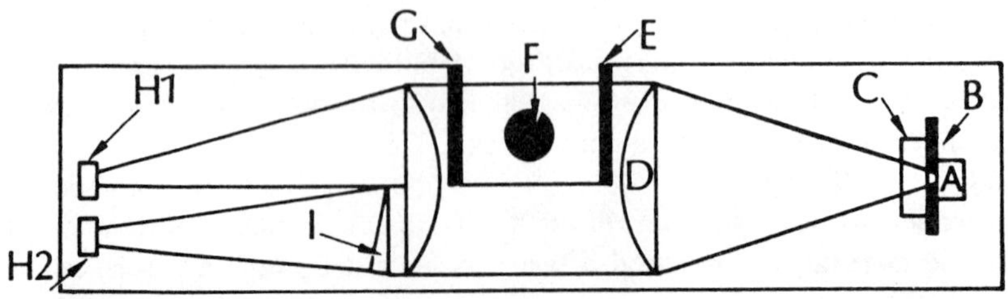

Figure 20 Schematic of optical-electrical die swell tester.

between photodetector H1 and H2 and calculates the object thickness to an accuracy of 0.2% of full scale. The object to be measured must be located vertically within the measuring area. The maximum permissible deviation from vertical is 2.5°. Operation at angles greater than 2.5° from vertical will yield greater errors than 0.1%. If the measuring angle is more than 2.5° from vertical, the measurement will be increased by the factor $1/\cos a$, where a is deviation from vertical.

While the upper collimated light serves as the measuring beam, the lower portion impacts on the second photodetector and acts as the reference beam. Thus the temperature and age affected by drift which occurs on all semiconductor elements is negated.

In some applications such as hose and tubing manufacturing, it is important to know both the extrudate diameter and cross section. The device described above can provide two diameter measurements, which will give an out-of-round indication (eccentricity).

Experimental studies reported in the literature using various methods of swell measurement are reported by Lobe and White (1979), Graessley et al. (1970), Pliskin (1973), and White and Roman (1976).

VISCOMETRIC TECHNIQUES FOR DIFFERENT INDUSTRIES

A variety of other instruments are employed by various industries for either product development work or product quality control. Although we shall review these methods in relationship to the industries and products they are most frequently applied to, these devices are by no means restricted to those applications. One type of instrument may have become a standard tool for a particular industry largely through an evolution of preference and in some cases convenience, rather than primarily for technical reasons. Most of these techniques can be applied to any fluid characterization; however, the techniques can greatly vary in terms of the specific information they provide.

One industry where rheological characterization has been applied for many years in a subjective manner is the manufacture of liquid detergents. Consumer applications for these products include dishwashing, hard surface cleaning liquid scourers for walls and floors, fabric softeners, thickened bleaches, liquid abrasive cleaners, combination liquid detergent–wax products for cars, and liquid detergents for washing fabrics. Other examples include personal washing products such as hair shampoos and conditioners, body shampoos, shower gels, toothpaste, skin cleansers. The particular rheological properties imparted to these various products is governed in a number of cases by consumer preference or attitudes as well as by technical benefits derived from certain flow properties. The consumer often makes his or her own rheological assessment through sensory perception of the product. From touch and sight, the consumer describes different products as watery, thin, thick, jellylike, pastelike, and so on. It is the rheologist's job in part to be able to quantify this subjective terminology so that product development can proceed.

Among the most frequently used viscometers in this industry are the U-tube viscometer, the rolling ball (or Hoeppler viscometer), the Ford cup, and the Brookfield viscometer. These basic devices are illustrated in Figure 21. The first three devices have often been used in a totally empirical manner for quality control purposes. For example, in each of these cases the time for the material to flow through the tube or from a vessel, or for a ball to roll between two points, would be measured over some conveniently specified time interval. With some devices a scale might be included to display a reading of variables, as in the case of the Brookfield, showing readings between say 10 and 90% of full-scale deflection. Many devices have been adopted, not just in this industry, on the basis of convenience and expediency rather than on a scientific or engineering basis. A disadvantage of many of the simpler instruments is their inability to generate more than one value to characterize a fluid. Often this reading is converted into a viscosity value based on a Newtonian calibration.

Barnes (1980) describes several modifications to the various techniques used in characterizing detergents. The first of these modifications concerns the Hoeppler viscometer, which normally operates using a fixed tube angle and several balls. He proposes to use a variable angle of rolling. By altering the angles, the average stress in the fluid can be made to vary because of the change in magnitude of the gravitational force on the ball in the tube direction. For this arrangement, viscosity data can be fitted to a relation:

$$\eta = k_1 (\sin \alpha')\theta(A + Bn) \tag{53}$$

where

α' = angle of the tube from the horizontal
θ = time for ball to roll between two fixed points
n = slope of curve generated by plotting $\ln(\sin \alpha')$ versus $\ln(1/\eta)$

Note that shear rate in this experiment is $\dot{\gamma} = k_2/\theta$, and A, B, k_1 and k_2 are constants. Also, $A + B = 1$. Usually two experiments (i.e., tests at two angles) provide sufficient information such that a power law relation can be derived from η-values.

The Ford cup provides a measure of the time for a fixed volume of fluid (e.g., 50 ml) to flow from the cup into a graduated cylinder. One problem in using this technique with such materials as detergents is the formation of foam which can make accurate flow time determinations difficult. This can be over-

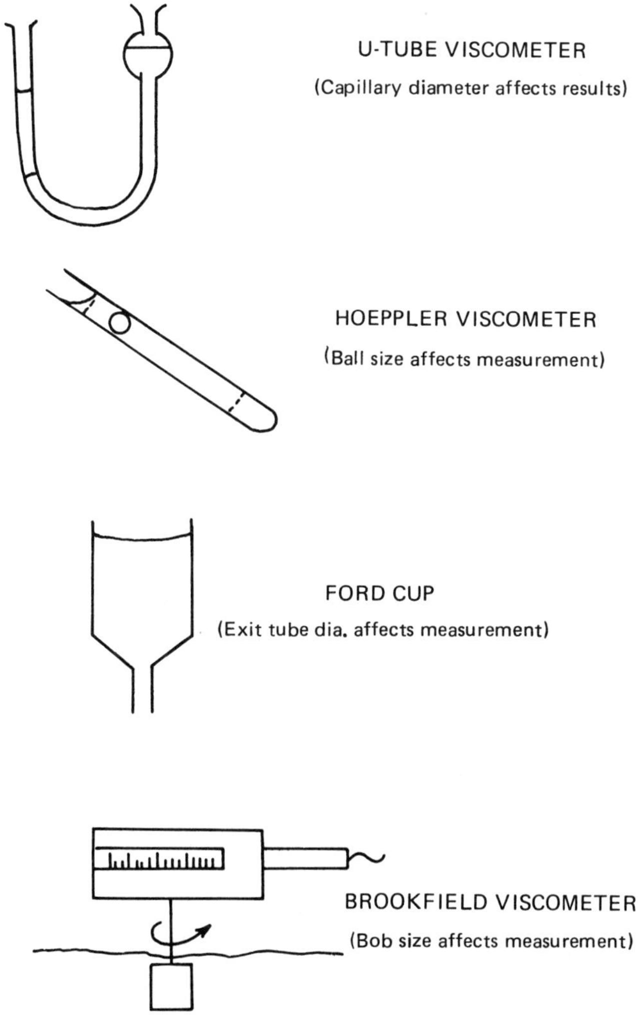

Figure 21 Various viscometers used in detergents industry.

come by employing electroresistivity probes along the inside walls of the cup (see Figure 22). This provides accurate measurement of the flow times involved. (See Cheremisinoff (1986) for a discussion of electroresistivity probes.) A sharp edge orifice fitted to the outlet tube provides steady calibrated flow conditions. Barnes (1980) gives the following expressions for viscosity and average shear rate using this technique:

$$\eta = \mu\left(\frac{1+n}{2}\right), \quad \text{N-s/m}^2 \tag{54}$$

$$\dot{\gamma}_{ij} = \frac{2(h_i - h_j)}{\theta_{ij}} \frac{R^2}{r^2}, \quad \text{s}^{-1} \tag{55}$$

where i, j refer to different heights h_1, h_2, h_3, . . . , etc.

Measurements using a modified Ford cup are quite comparable to those obtained with concentric cylinder viscometers. Materials that are highly thixotropic generate flow times which can be much longer than predicted if steady-state concentric cylinder data are used (Barnes, 1980) for comparison, because the total shear in the Ford cup is small compared with the steady-state rotational data. The Ford cup is not applicable to materials where elongational viscosity is much greater than shear flow at the same shear rates. Many detergents show anomalous elastic effects and hence the Ford cup is not suitable for viscosity measurement in all cases.

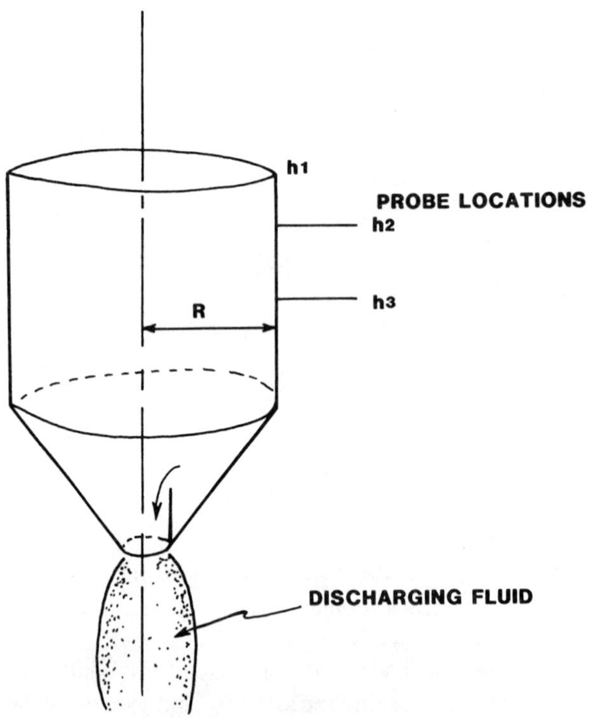

Figure 22 Operation of the Ford cup.

Viscosity measurements in shear flow of detergent products can be made in commercial rheometers based on cone and plate instruments or concentric cylinder viscometers. Typical commercial units employed are the Haake Rotovisko, the Ferranti–Shirley viscometers, and the Contraves Rotary viscometer. These instruments cover shear rates typically between 1 to over 1000 s^{1}. Other instruments employed to a lesser extent are the Deer rheometer and the Weissenberg Rheogoniometer.

The Deer rheometer is designed to subject the sample to a constant stress whereby the subsequent deformation is measured. This instrument uses a frictionless air bearing so that it is capable of applying a variety of stress patterns with variable time of application and rate of application of stress. Variations of this instrument are capable of applying oscillatory stress to the fluid sample. Deformation can be studied to measure several parameters: creep, steady-state shear rate, recoverable shear, and various oscillatory parameters. Through geometry changes the instrument can cover a range of applied stresses. Barnes (1980) describes a modified Deer rheometer employed in the study of soap rheology. The system is illustrated in Figure 23 where the rheometer has been modified to measure the extensional properties by simple bending performed at very small deformation. Torque is applied to two vertically mounted rods of soap which are fixed at both ends. Torque is applied to the samples through the sample end retaining sleeves by magnetic knife edges mounted on radial arms. The knife edges ensure that no twisting of the rods takes place during bending. The extensional compliance is calculated from bending beam theory analysis.

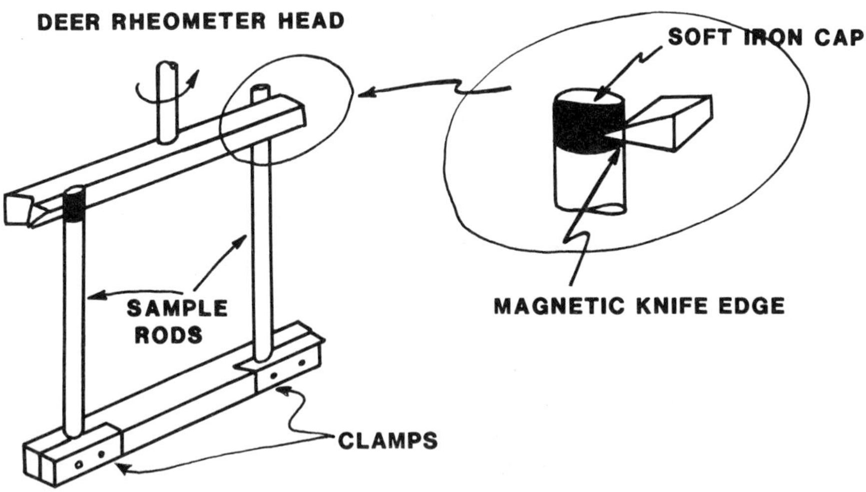

Figure 23 Modified Deer rheometer used by Barnes (1980).

The Weissenberg Rheogoniometer, as noted earlier, can be used to measure G' and G'' as a function of frequency and deformation. These parameters can also be measured for a range of superimposed values of steady shear rate. It is most often applied in industrial laboratories to measure for a range of shear rates, and as a function of shear/time at initiation or cessation of shear.

Other frequently used viscometers in the detergents manufacturing industry include capillary and orifice viscometers and penetrometers. As noted earlier in this chapter, a capillary viscometer operates by forcing the fluid through a straight circular tube. This instrument is capable of shear rates up to $10^5\ s^{-1}$. For fluids obeying power law behavior, the Mooney–Rabinowitz correction makes possible a definition of the average viscosity–average shear rate parameters. Orifice viscometers are often used in the plastics industry to measure parameters important in pumping fluids through circular orifices. The detergents industry has applied such devices to handle very viscous semisolid materials such as soap. The last noteworthy device is the penetrometer, which monitors the movement of a weighted penetrator into the surface of a very viscous sample. The penetrometer is most often used for testing such materials as butter or bitumen, in which the penetrator is a sharp truncated pin. Within the detergents industry, this device has proven useful for semisolid soaps.

There are a variety of other instruments and/or variations of devices described above that are applied in lubrication technology. Lubricating oils are manufactured from the distillation of crude oil. Primary distillation is performed at atmospheric pressure to remove the lighter gasoline, kerosene, and gas oil fractions, and the lubricating oils remain in the residue that boils above about 370°C. The residue must be distilled under high-vacuum conditions, resulting in several lubricating-oil fractions and an asphaltic bitumen. The second-stage residue from those crudes low in asphalt is also a source of lubricating oils. These fractions are referred to as residual oils to distinguish them from distillate oils, which are boiled off in a high-vacuum still. Lubricating-oil fractions can constitute up to 5% of the total crude, depending on the crude source. A further refinement step in manufacturing the lubricating oils is usually performed with acid or solvents to remove a variety of unwanted constituents such as sulfur compounds. Hydrotreating at moderate temperatures and pressures saturates the unsaturated hydrocarbons and removes sulfur-bound molecules. Paraffinic constituents separate out as wax at low temperatures. These are removed by chilling and filtering or by a solvent treatment process. Polar compounds and dark coloring constituents are removed from the oil in a final treatment stage using active clays. The refinery streams in these processes are referred to as base oils in the petroleum and allied industries. Base oils are characterized by the origin of the crude, the specific refining process used, viscosity, their viscosity index (VI), and various other properties. The VI is a measure of the variation of viscosity with temperature, and, in general, the higher the VI the smaller the variation.

An important class of products is mineral oils, examples of which are paraffins and aromatics. Oil additives are usually supplied as concentrates in mineral oil and are readily added during blending operations. Additives can have a variety of purposes for the end-use product, for example, oxidation inhibitors, additives to neutralize acids, surface-active agents to disperse and maintain other additives in suspension, ultra-high-pressure lubricating additives. Polymers are also added to oils to modify rheological properties at extreme temperature conditions (e.g., for motor oils, to increase pourability at very low temperatures). A general example of a pour-point depressant is an additive which inhibits the formation of a wax-crystal network at low temperatures. This type of product extends the liquid property range of the lubricant. In lubrication technology, the Navier–Stokes equations for incompressible, viscous inelastic fluids are often applied to engineering problems. Information on viscosity and proper application of these equations is usually sufficient to estimate oil film thickness and the friction. These equations are usually solved isothermally for computation simplicity and do not account for viscous heating effects. Hence it is necessary to fully define the viscosity–temperature dependence of these products in order to correct engineering formulations from the Navier–Stokes equations. This is done, as noted earlier, through the use of the VI (viscosity index); standardized procedures are described in ''Kinematic Viscosity,'' ISO Method 3104, IP Method 71, ASTM Method D445, and ''Industrial Liquid Lubricants''—ISO Viscosity Classification—ISO 3448.

The capillary viscometer is the standard instrument used for these measurements. Temperature conditions are standardized in these test methods so that oils can be compared on a relative basis (e.g., 40°C and 100°C). In the ISO Viscosity Classification system, a 10-grade oil has a viscosity between 9.0 and 11.0 cSt at 40°C. In total there are six grades for each decade in viscosity.

Many commercial oils contain polymeric additives. Adding small amounts of polymer to an oil reduces the rate of decrease in viscosity with increasing temperature. Examples of the types of polymers used are polyisobutene, polyalkylmethacrylates, and copolymers such as styrene–butadiene and styrene–isoprene. Molecular weights of these additives range from about 50,000 to several hundred thousand. In an automobile engine, motor oil is exposed to effective shear rates ranging from 10^4 to 10^6 s^{-1} over wide temperature ranges. Considerably lower shear rates are involved during engine starting. In starting an engine, the cranking speed must be high enough to enable a cylinder to fire. Furthermore, the power the engine delivers when it does fire and the assistance of the starter motor are disengaged during this process. Both these actions depend on the viscosity of the motor oil. One important industry test which is a low-temperature, high-shear test method is the Cold-Cranking Simulator (ASTM Method D2602). The U.S. Society of Automotive Engineers (SAE) provides a classification of engine oil viscosity based on this test method. Table 6 lists this

Table 6 SAE Viscosity Grades for Engine Oils

SAE viscosity grade	Viscosity range: cP at −18°C (ASTM D 2602) (max.)	cSt at 100°C (ASTM D 445) (min.)	cSt at 100°C (ASTM D 445) (max.)
5W	1,250	3.8	—
10W	2,500	4.7	—
20W	10,000	5.6	—
20	—	5.6	<9.3
30	—	9.3	<12.5
40	—	12.5	<16.3
50	—	16.3	<21.9

classification, which is based on a low-shear viscosity measurement at 100°C and a high-shear viscosity measurement at −18°C. It should be noted that the industry is in the process of revising test methods and the classification system. Current evolution is toward a high-shear Cold-Cranking Simulator (CCS) Viscosity and a yield stress value to be obtained in terms of specified temperatures.

The design of motor oil products requires detailed knowledge of the rheological demands of the oil made by the engine. For example, high temperatures normally achieved in an automobile engine (e.g., 100–150°C) causes shear thinning, which can result in increased wear. Also, however, at low temperatures shear thinning provides the advantage of easier starting conditions. It was noted earlier, however, that base oils are dewaxed in order to remove paraffinics which crystallize out of solution at low temperatures. However, engine oils are never completely dewaxed because of prohibitive refining costs. Consequently, commercial oils can encounter temperatures in which residual waxes will separate out. This important property is tested for by the so-called cloud point, based on ISO Method 3015, IP Method 219, ASTM Method D2500. The cloud point is the temperature at which wax first begins to separate. Another rheological test of motor oil products is the pour point (ISO Method 3016, IP Method 15, ASTM Method D97). At lower temperatures oils acquire a yield stress due to the network structure formed by the wax crystals. Considerable research effort has gone into the development of pour-point depressants, which are additives that interfere with the formation of wax crystal structure.

Illustrations of viscometric instruments used in this industry are given in Figures 24 through 27. Figure 24 shows a reverse-flow capillary viscometer.

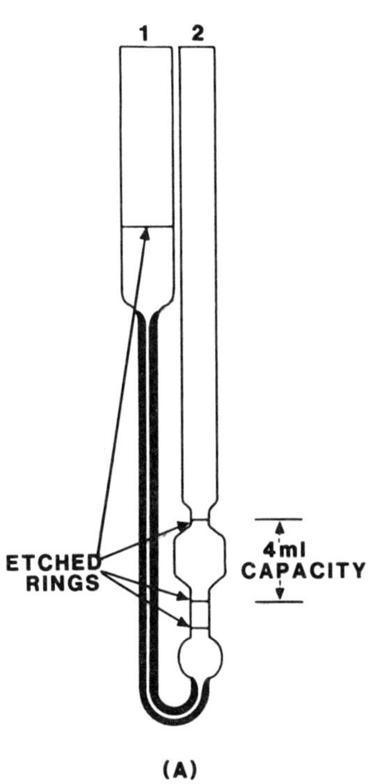

Figure 24 (A) Reverse-flow capillary viscometer. (B) Suspended level viscometer.

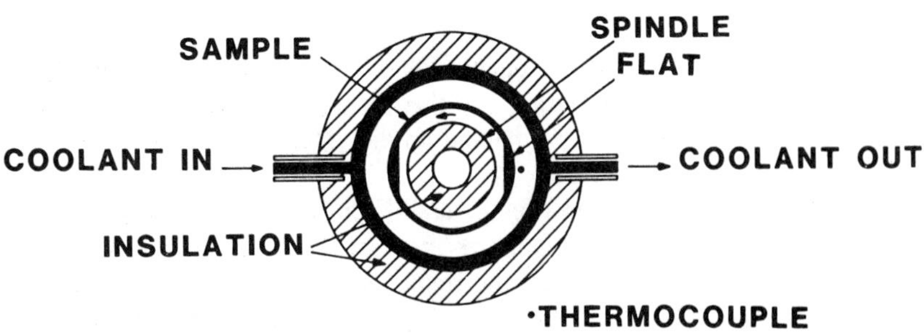

Figure 25 Schematic of cold cranking simulator viscometric cell.

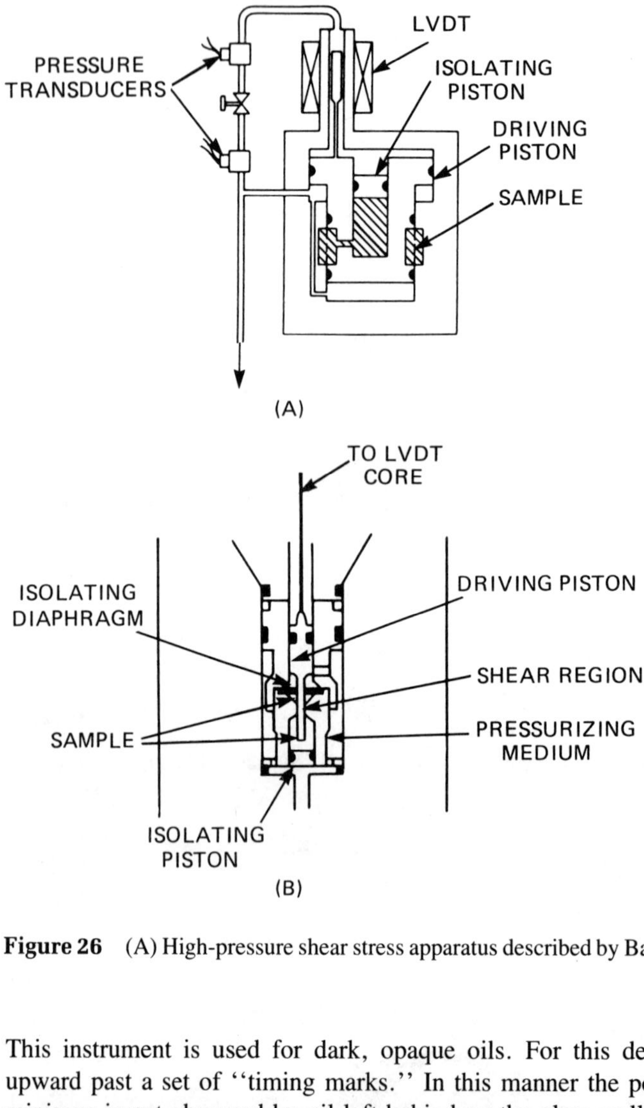

Figure 26 (A) High-pressure shear stress apparatus described by Bair and Winer (1979).

This instrument is used for dark, opaque oils. For this device the oil flows upward past a set of "timing marks." In this manner the position of the fluid miniscus is not obscured by oil left behind on the glass wall. Figure 24B illustrates a suspended-level viscometer.

Viscosity measurements at low temperatures for transmission and gear oils is usually made with a Brookfield viscometer (methods IP/267 and ASTM D.2983). Other devices equally applicable are the Haake Rotovisko RV2 (i.e., constant-speed concentric-cylinder instrument), the Hoeppler rolling ball viscometer, and the U-tube capillary viscometer.

The cold-cranking viscometer cell is illustrated in Figure 25. This device

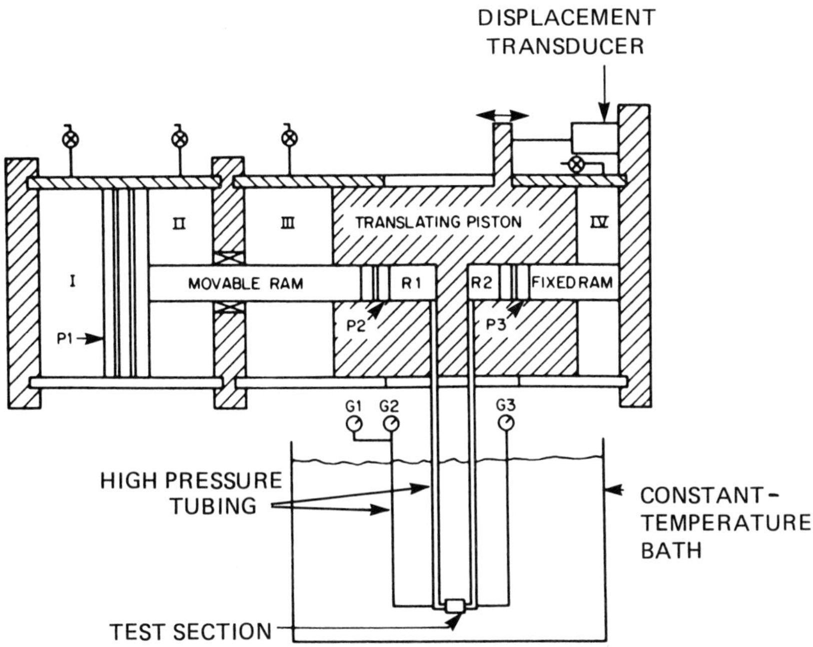

Figure 27 Schematic of high-pressure capillary viscometer described by Novak and Winer (1963).

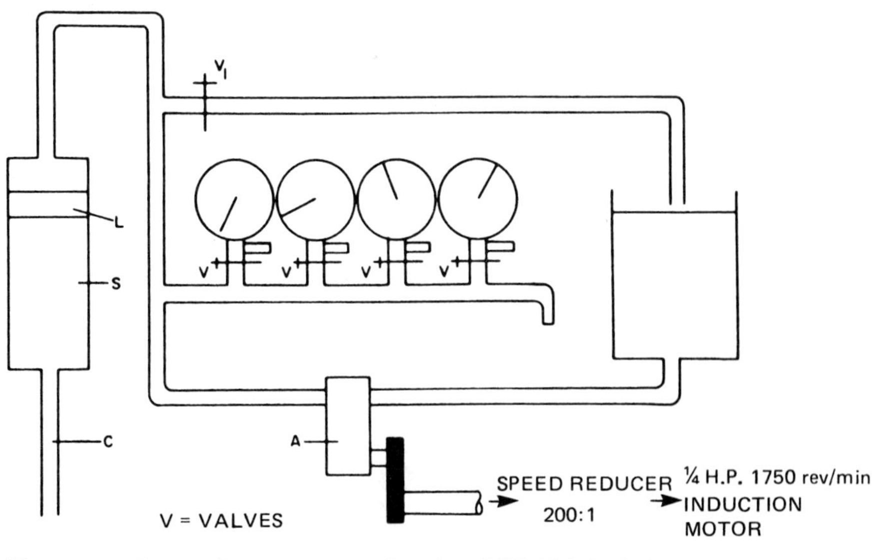

Figure 28 Grease viscometer system based on ASTM Method 1092.

Table 7 Summary of Major Viscometric Techniques Used Throughout Different Industries

Products	Type of viscometer	Operating principles—rangeability	Standard test methods	Suggested references
Lubricating oils	Glass capillary viscometers	Viscosity—shear rate measured in capillary tubes of different *L/D*, over temperature range. Figure 27 shows modified capillary where test fluid contained in reservoirs R1 and R2 and interconnecting tubing. Material separated from hydraulic fluid contained in I–IV. Adjusting press, between III and IV, translating piston drives test fluid thru capillary. Shear stress range is 30 Pa–5 MPa.	ISO Method 3104; IP Method 71; ASTM Method D445; ISO Viscosity Classification ISO 3448	Murphy, C. M. et al. *Trans. ASME, 71*, 561 (1949).
	Cold cranking simulator	Concentric cylinder viscometer with temperature control bath; measurement made at constant power, not constant shear stress or shear rate.	ASTM Method D2602	
	Bridgman falling-body rheometer	Measures viscosity at low shear rates and high pressures. Time measured for plummet of known weight to fall known distance vertically thru oil. Manual timing limits method to viscosity ranges of 10^{-3}–10^{3} Pa-s; accuracy >5%.		Irving, J. B., and A. J. Barlow, *J. Phys. E. Sci. Instr., 4*, 232 (1971).
	High-frequency oscillatory measurements	Measures shear modulus (G) and time delay from instant of application of high contact press, to attainment of equilibrium specific volume. Technique based on measurement of change in amplitude and phase of a pulsed train of mechanical waves on reflection at the interface between fuse quartz and oil. Waves are generated piezoelectrically.		Hutton, J. F., M. C. Phillips, J. Ellis, G. Powell, and E. W. Jones, in *Electrohydrodynamics and Related Topics*, C. M. Taylor, M. Gudet, and D. Berthe, eds., Mechanical Engr. Publ. Ltd., Bury St. Edmunds, 1979. Harrison, G., *The Dynamic Properties of Supercooled Liquids*, Academic Press, London, 1976. Matheson, *Molecular Acoustics*, Wiley, London, 1971.

	Optical techniques	Measurement depends on existence of Debye elastic waves caused by thermal motions of molecules. The waves display acoustic properties, containing Debye waves are propagating, hence Doppler shift of frequency of light probe source. Scattered light can be detected by an interferometer. Doppler frequency shift gives velocity of Debye waves. Oil sample must be near glassy state for this technique.		Alsaad, M. A., W. D. Winer, F. D. Medina, and D. C. O'Shea, *Trans. ASME (F) J. Lub. Tech.*, *100*, 418 (1978). Bair, S., and W. D. Winer, *Trans. ASME (F) J. Lub. Tech.*, *101*, 251 (1979). Dill, J. F., P. W. Drake, and T. A. Litovitz, *ASLE Trans.*, *18*, 202 (1975).
Lubricating greases	Weight penetration method	Measure of grease yield stress in terms of consistency. Basic test for consistency by dropping a specified cone from a given height into sample. After sinking time of 5 s the cone is stopped and depth of penetration measured in units of 10^{-4} meters. Greases classified according to penetration determined at 25°C.	Penetration of greases—IP Methods 50 and 310; ASTM Methods D217 and D1403.	
	Capillary viscometer, also Shell-de Limon viscometer	Measure of apparent viscosity. Data generates standard rheological flow curve (i.e., wall shear stress vs. shear rate). At low shear rates existence of yield stress observed. At high shear rates curve approaches line of zero slope indicative of the viscosity of the base oil from which grease is derived. Basic system described by ASTM Method D1092 is shown in Figure 28. Lowest shear rate range is $10\ s^{-1}$.	'Apparent Viscosity'—ASTM Method D1092. Flow Properties at High Temperatures—ASTM Method D3232.	Klamann, D., L. Endom, R. Rost, and A. Haak, *Erdol Kohle*, *20*, 219 (1967).
	Cone-and-plate viscometers	Operation described earlier in this chapter. To avoid slippage plates are usually grooved or roughened. Close spacing used between plates so that main portion of shearing boundary is seen by grease between the grooves.		Hutton, J. F. in *The Rheology of Lubricants*, T. C. Davenport, ed. Applied Science Publishers, Barking, 1973.

(continued)

Table 7 (Continued)

Products	Type of viscometer	Operating principles—rangeability	Standard test methods	Suggested references
	Torsional balance rheometer	Provides normal stress measurements. Operates by applying a known normal force to parallel plates between which sample is held. Plates are of fixed radii, of which one is free to move along axis joining centers. The rotation is applied and the equilibrium separation of plates denotes the equilibrium of the applied force with normal force generated by the grease sample at the measured shear rate.		
Organic coatings Paints Printing inks	Consistency cups	This is a single-point measurement instrument. Flowability measured by gravity-induced flow through an orifice. Paint standards require use of a conical bottom cup (Ford cup or ISO cups). Some designs employ a spherical bottom (Zahn and Shell cups). Efflux times provide measure of viscosity, which in part depends on the orifice geometry. High aspect ratios (orifice L/D) improve smoothness of flow and the linearity of the time–viscosity correlation. Viscosity measurement range can be extended by changing the orifice dimensions.	ASTM D1200-70; ISO 2431-1972	Westgate, M. W., *Off. Dig. Fed. Soc. Paint Technol.*, *32*, 616 (1960). McKelvie, A. N., *Progr. Org. Coatings*, *6*, 49 (1978). Walters, K. *Rheometry*, *Chapman and Hall*, London, 1975, p. 56. Patton, T. C., *Paint Flow and Pigment Dispersion*, Wiley, New York, 1964.
	Falling-sphere rheometer	The technique is best suited for materials that approach Newtonian behavior. The net velocity of a solid sphere submitted to buoyancy and viscous friction forces is measured. From Stokes law dynamic viscosity can be computed from the measured time of fall t of the sphere: $= (g/4.5)(s^{-1})r_2(l/t)$; where g = gravitational acceleration, l = distance		

	sphere falls, r = sphere radius. Equation ignores wall effects from the container. Corrections for wall effects can be made by a factor which accounts for tube-to-container radius: $= (g/4.5)(s^{-1})r_2(l/t)[l - 2.104(r/R) + 2.09(r/R)_3 \ldots]$ By using spheres of different radiis and/or densities, the range of viscosity measurements can be extended. The technique is not applicable for thixotropic fluids, which some paints are.		
Rising-bubble rheometer	The method is essentially the same as the falling-sphere method. Buoyancy works in the same direction as for spheres but with lower density than that of the test fluid. Hence, principle is based on time for bubble to rise through a column of fluid. Since the bubble is not truly spherical, the above equations do not apply rigorously. Relating bubble rise time to viscosity must therefore be based on calibration of the technique using fluids of known viscosity.	ASTM D1545-63	Patton, T. C., *Paint Flow and Pigment Dispersion*, Wiley, New York, 1964. Gardner, H. A., and G. G. Sward, *Physical and Chemical Examination of Paints*, Gardner Lab., Bethesda, MD, 1962.
Rolling-sphere test and sliding-plate test	Usually applied to measure viscosity changes in drying paint films. In situ measurements are made by either sliding, rotating, or rolling objects over the film surface and tracing the speed as a function of time. Film/substrate are usually inclined to some defined angle. These methods operate at given stress levels as opposed to a given shear rate. Generally, these provide a basis for a quality control test and/or for relative comparisons between samples.		Fischer, E. K., *J. Colloid Sci.*, *5*, 271 (1950). Goring, W., and N. Dingerdissen, *Farbe Lack*, *83*, 270 (1977). Kornum, L. O., *Fatipec XIV*, V. Takacs, ed. Budapest, 329 (1978). Kornum, L. O., *Rheol. Acta*, *18*, 178 (1979). Quach, A., and C. M. Hansen, *J. Paint Technol.*, *46*, 40 (1974).

(*continued*)

Table 7 (Continued)

Products	Type of viscometer	Operating principles—rangeability	Standard test methods	Suggested references
	Couette-type viscometers	Couette-type (coaxial cylinder geometry) viscometers are widely used for characterizing pigment dispersions. Since the sample for rheological testing is a dispersion, it's important that the tolerance between the cylinder be much larger than particle and aggregate dimensions. As a rule of thumb, the gap should be a factor of 10 larger. Various factors which can affect measurements are sedimentation of pigments, volatility of sample, pigment migration, wall slippage, entrapment of air with very viscous samples, and viscous heating. Pigment migration often translates into highly thixotropic behavior and it is therefore advisable to repeat measurements several times to ensure that thixotropy actually characterizes the product or if it is associated with the present state of the sample. One serious source of error can be introduced by using very large gaps between cylinders to accommodate dispersion—this leads to nonuniform shear over the gap which cannot be corrected for easily.		Middleman, S., *The Flow of High Polymers,* Interscience, New York, 1968. Mewis, J., *J. Non-Newtonian Fluid Mech., 6,* 1 (1979). van Wazer, J. R., J. W. Lyons, K. Y. Kim, and R. E. Colwell, *Viscosity and Flow Measurement,* Interscience, New York, 1963. Cheng, D. C-H., and B. R. Parker, Proc. VIIth Int. Congr. Rheol., C. Klason and J. Kubat, eds., Goethenburg, 518 (1976). Karnis, A., H. L. Goldsmith, and S. G. Mason, *J. Colloid Interface Sci., 22,* 531 (1966).
	Cone-and-plate viscometers	Advantages over Couette-type viscometers include: small sample size required, homogeneous shear rate, and possibility of immediate conversion of data to apparent viscosities. Similar problems of data inaccuracies as outlined for Couette-type instruments apply. Shear fracture of printing inks has been observed in cone-and-plate instruments. It is important that preliminary experiments be		Klijn, P-J., J. Ellenberger, and J. M. Fortuin, *Rheol. Acta, 18* 303 (1979). Cheng, D. C-H., *J. Appl. Phys., 17,* 253 (1966). Hadjistamov, D., and K. Degen, *Rheol. Acta, 18,* 168 (1979). Hussen, E., Ph.D. Dissertation, K. U. Leuven (1976).

	run to define at what shear rates this phenomenon occurs.	
Stormer–Krebs viscometer	Rotating-element viscometer which employs a two-bladed paddle. The device is most often used for product quality control testing and is operated at constant torque as opposed to constant speed as with other rotating-element viscometers. The constant torque is controlled by a weight *W* which causes the paddle to rotate at a given speed of 200 rpm, providing a measure of viscosity. The viscosity value is obtained from a weight–rotational speed diagram. Tests can be accelerated by the use of a stroboscopic instrument. Data are typically reported in arbitrary values of Krebs units (KU). The manufacturer provides a conversion table for W-KU. In the paint industry the device is normally applied as a single-point measurement device. It is however possible to generate a shear rate distribution by changing weights.	Epprecht, A. G., *Farbe Lack, 84*, 86 (1968). Camina, M., *Defazet, 28*, 117 (1974). Freier, H. J., *Farbe Lack, 69*, 87 (1963).
Falling-rod viscometer	As in the Couette geometry the test sample is held in an annular gap; however, one of the cylindrical walls (usually the outer one) moves under gravity while the other is stationary. This results in a telescopic flow field. The device has similar advantages to a cone-and-plate viscometer. The shear rate range can be extended by applying different weights to the rod. The actual measurement is the time for the rod to fall over a specified distance. Shear stresses can reach up to 10^4 N/m^2 with viscosities up to 100 Pa-s at shear rates as high as 10^3 s^{-1}.	Turner, T. A., *Brit. Ink Maker, 15*, 59 (1973).

(*continued*)

Table 7 (Continued)

Products	Type of viscometer	Operating principles—rangeability	Standard test methods	Suggested references
	Band viscometer	Based on the same principle as the falling-rod viscometer, except that a band moves through a slit filled with the test sample. Shear rate ranges can be extended by using different weights attached to the lower end of the band.		Taylor, J. H., and S. L. Cozzens, Natl. Printing Ink Res. Inst., Bull. No. 53, 1959. Patton, T. C., *Paint Flow and Pigment Dispersion,* Wiley, New York, 1964.
	Rotary tackmeters	Employed in the printing ink industry. Device consists of three rotating rollers, with the central roller driven by a motor. A second roller is positioned on top of the central roller, and the third roller provides reciprocating motion that smooths the liquid layer before it enters the nip between the other two rollers. The fluid sample transmits forces from the surface of one roller to the other. The external force required to maintain the roller in place provides a measure of the tack. This data can then be related to viscometric data.		Dobbels, F., and J. Mewis, *Chem. Eng. Sci., 33,* 493 (1978). Mill, C. C., *J. Oil Col. Chem. Assoc., 50,* 396 (1967).
Aqueous suspensions, e.g., suspensions of paper clays, mineral pigments for paper coating, ceramics coatings, drilling muds, pharmaceuticals (i.e., slurries for injections, emulsion-based suspensions), process waste streams.	Brookfield viscometer	Applied for low-shear viscosity quality control of products. The device has already been described in this chapter. In slurry viscometry work, a suspension should be prepared at as high a solids content as possible—single-point viscosity measurement in the paper industry is normally made at 100 rpm. The sample is then diluted by 2% in solids and a second measurement is made. The root reciprocal viscosity is plotted versus weight % solids and the amount of solids required for a viscosity of 5 poise is quoted. This is referred to as the *viscosity concentration* or	TAPPI Standard: T 648-sm-54	Clark, N. D., *Trans. Brit. Ceram. Soc., 49,* 409 (1950). Beazley, K. M., *J. Coll. Int. Sci., 41*(1), 105 (1972). Beazley, K. M., *TAPPI, 50*(3), 151 (1967).

	Clark viscosity. This terminology is normally used in paper coating technology and for calcium carbonate slurry preparation.	
Cone and plate	Already described above. One example of a commercial unit is the Ferranti–Shirley cone and plate instrument.	Huggenberger, L., W. Kogler, and M. Arnold, *TAPPI*, *62*, 37 (1979). Borruso, D., A. Croce, and A. Seves, *Ind. della Carta*, *7*, 100 (1969).
Gallenkamp torsion viscometer	This instrument is largely applied in the U.K. for characterizing ceramic casting slips. It employs a small metal cylinder which is suspended from a torsion wire and dips into the test suspension. The position and deflection of the measurement head is recorded by a pointer and scale. The head is immersed in the test fluid, then twisted through 360° and released. The extent of overswing provides a measure of the fluidity of the sample.	Moore, F., *Rheology of Ceramic Systems*, Institute of Ceramics, Stoke on Trent, 1965.
Vicat penetrometer	Often used for cement paste characterization, particularly in Europe and the U.K. Device uses a capped rod which moves vertically and carries a pointer over a scale at the same time. The lower end contains a needle which is used to penetrate the sample. The test sample is retained in a mold that sits on a nonporous plate. The needle is gently lowered into the sample and allowed to sink into it. The test is repeated several times until the needle no longer completely sinks into the sample. The time from the first experiment (i.e., when the cement paste is first prepared with water) to the point of no penetration is measured and used as a record of the initial setting time.	Jones, T. E. R., and S. Taylor, *Sil. Ind.*, *4/5*, 83 (1978).

consists of a concentric-cylinder viscometer. A constant-temperature bath is controlled to −18°C and the oil sample is injected into the viscometer volume via a syringe.

Several other viscometers specifically developed for lubrication research are illustrated in Figures 26 and 27. Brief descriptions of these instruments are provided in Table 7 along with techniques applied to rheological characterization of various products from other industries.

A number of other techniques have evolved in different industries that serve as both scientific devices for product characterization and as single-point measurement quality control techniques. Principal instruments are summarized in Table 7 according to industry preference. Key references describing the theory and application for that particular industry are included.

NOTATION

A	area
a	half-width
C_R	shear rate geometric parameter
C_1	integration constant
D_R	Mooney unit dial reading
E(t)	relaxation modulus
F	force
F'	normal thrust
f'	stress function
G'	dynamic rigidity function
$G(t)$	shear modulus
H	sample thickness
h	disk thickness
K	consistency index
L	length
l	length
l_{eff}	effective bob length
N	rotational velocity
N_f	Bagely correction factor
n	power law model exponent
n''	slope of torque versus rotational speed
P	pressure
P_{11}, P_{22}	normal stress differences
Q	volumetric flow rate
R, r	radius
R_b	bob radius
Re_{gen}	generalized Reynolds number
t	time
V	velocity

Greek Symbols

α'	angle
β	effective slip coefficient
Γ	Newtonian shear rate
γ_w	shear rate
δ	distance
η^*	complex dynamic viscosity
θ	time
μ	viscosity
φ	density
T	torque
τ_b	shearing stress at bob surface
τ_w	wall shear stress
ϕ	angle
ϕ'	deformation rate
ω	frequency

REFERENCES

Andrews, R. D., N. Hofman-Bang, and A. V. Tobolsky, *J. Polym. Sci., 3,* 669 (1948).

Aoki, Y. J., *Soc. Rheol. Japan, 7,* 20 (1979).

Bagley, E. B., *J. Appl. Phys., 28,* 624 (1954).

Bair, S., and W. O. Winer, *Trans. ASME (F) J. Lub. Technol., 101,* 251 (1979).

Ballman, R. L., *Rheol. Acta, 4,* 137 (1965).

Barnes, H. A., Chapt. 2, in *Rheometry: Industrial Applications,* K. Walters, ed. Wiley, New York, 1980, pp. 31–118.

Bird, R. B., W. E. Stewart, and E. N. Lightfoot, *Transport Phenomena,* Wiley, 1967.

Brizitsky, V. I., G. V. Vinogradov, A. I. Isaev, and Yu Ya Podolsky, *J. Appl. Polym. Sci., 22,* 665 (1978).

Calderbank, P. H., and M. B. Moo-Young, *Trans. Inst. Ch.E. (London) 37,* 26–33, (1959).

Chapman, F. M., and T. S. Lee, *SPE J., 26,* 37 (1970).

Chen, I. J., and D. C. Bogue, *Trans. Soc. Rheol., 16,* 59 (1972).

Cheremisinoff, N. P., *Fluid Flow Pocket Handbook,* Gulf Publishing Co., Houston, TX, 1984.

Cheremisinoff, N. P., *Instrumentation for Complex Fluid Flows,* Technomic Publishing Co., Lancaster, PA, 1986.

Cheremisinoff, N. P., and D. Azbel, *Fluid Mechanics and Unit Operations,* Ann Arbor Science Publishers, Ann Arbor, MI, 1983.

Cogswell, F. N., *Plastics and Polymers, 36,* 109 (1968).

Cogswell, F. N., *J. Non-Newt. Fluid Mech., 4,* 9 (1978).

Cotten, G. R., and J. L. Theile, *Rubber Chem. Technol., 57,* 749 (1979).

Dealy, J. M., paper presented at AICHE Meeting, Miami Beach, Nov. 1978.

Dexter, F. D., J. C. Miller, and N. Philippoff, *Trans. Soc. Rheol., 5,* 193 (1961).

Furuta, I., V. M. Lobe, and J. L. White, *J. Non-Newt. Fluid Mech., 1,* 207 (1976).

Goldstein, C., *Trans. Soc. Rheol., 18,* 357 (1974).

Gortemaker, F. H., and H. Janeschitz-Kriegle (1976).

Gortemaker, F. H., M. G. Hanson, B. de Cindio, H. M. Laun, and H. Janeschitz-Kriegel, *Rheol. Acta, 15,* 256 (1976).

Graessley, W. W., S. D. Glasscock, and R. L. Crawley, *Trans. Soc. Rheol., 14,* 519 (1970).

Han, C. D., *J. Appl. Polym. Sci., 15,* 2567, 2579, 2591 (1971).

Han, C. D., *Trans. Soc. Rheol., 18,* 103 (1974).

Han, C. D., *Rheology in Polymer Processing,* Academic Press, New York, 1976.

Han, C. D., and L. H. Drexel, *J. Appl. Polym. Sci., 17,* 3429 (1973).

Ide, Y., and J. L. White, *J. Appl. Polym. Sci., 22,* 1061 (1978).

King, R. G., *Rheol. Acta, 5,* 35 (1966).

Kreiger, I. M., and S. H. Maron, *J. Appl. Phys., 25,* 72 (1954).

Lee, B. L., and J. L. White, *Trans. Soc. Rheol., 18,* 467 (1974).

Lobe, V. M., and J. L. White, *Polym. Eng. Sci., 19,* 617 (1979).

Lodge, A. S., *Trans. Faraday Soc., 52,* 120 (1956).

Maxwell, B., and R. Chartoff, *Trans. Soc. Rheol., 9,* 41 (1965).

Meissner, J., *Rheol. Acta, 8,* 78 (1969).

Meissner, J., *J. Appl. Polym. Sci., 16,* 2877 (1972).
Metzner, A. B., and J. C. Reed, *AIChE J., 1,* 434 (1955).
Middleman, S., *Trans. Soc. Rheol., 13,* 123 (1969).
Mingawa, N., and J. L. White, *J. Appl. Polym. Sci., 20,* 501 (1976).
Mocosko, C. W., and J. M. Lornsten, *SPE Antec. Tech. Papers,* 461 (1973).
Mooney, M. J., *Rheology, 2,* 231 (1931).
Munstedt, H., *Proc. 7th Int. Rheol. Congr.,* 496 (1976).
Nakajima, N., and E. R. Harrel, *Rubber Chem. Technol., 52,* 9 (1979).
Novak, J. D., and W. O. Winer, *Trans. ASME (F) J. Lub. Tech., 90,* 580 (1968).
Philippoff, W., *Phys. Zeit, 35,* 884 (1934).
Pliskin, I., *Rubber Chem. Technol., 46,* 1218 (1973).
Skelland, A. H. P., *Non-Newtonian Flow and Heat Transfer,* Wiley, New York, 1967.
Stevenson, J. F., *AICHE J., 18,* 540 (1972).
Thomas, D. G., *Ind. Eng. Chem., 11*(18), 55 (1963).
Tobolsky, A. V., *Structure and Properties of Polymers,* Wiley, New York, 1960.
Treloar, L. R. G., *Trans. Faraday Soc., 36,* 538 (1940); *37,* 84 (1941); *43,* 284 (1947).
Vinogradov, G. V., A. Ya Malkin, E. P. Plotnikova, O. Yu Sabsai, and N. E. Nikolayeva, *Intl. J. Polym. Mat., 2,* 1 (1972).
Vinogradov, G. V., B. V. Radushkevich, and V. D. Fikham, *J. Polym. Sci., A-2,* 8 (1970).
Wales, J. L. S., *Rheol. Acta, 8,* 38 (1969).
Wales, J. L. S., and H. Janeschitz-Kriegel, *J. Polym. Sci., 5,* 781 (1967).
Wales, J. L. S., J. L. den Otter, and H. Janeschitz-Krief, *Rheol. Acta, 4,* 146 (1965).
Walters, K., *Rheometry,* Chapman & Hall, London, 1975.
Weissenberg, K., *Proc. 1st Intl. Rheol. Cong.* (1948).
White, J. L., and J. F. Roman, *J. Appl. Polym. Sci., 20,* 1005 (1976).
Zakharenko, N. V., F. S. Tolstukhina, and G. M. Bartenev, *Rubber Chem. Technol., 35,* 236 (1962).

3
Mixing Practices

INTRODUCTION

The mixing of non-Newtonian materials such as pastes, plastics, resins, and so on, involves factors distinctly different from those involved in the mixing of Newtonian fluids. A rotating propeller immersed in a pastelike material or a viscous liquid affects only a small portion of the total mass, leaving the greater part of it relatively undisturbed. Semisolids, viscous liquids, and heavy pastes require the mixing services of devices which are capable of affecting every portion of the material and achieving a homogeneous state by pushing one portion of the mass into another portion. Factors requiring special attention are the physical preparation of the materials, the behavior of the material under flow, and the effect of the order in which the ingredients are introduced into the mix. For highly viscous materials such as elastomers, equipment is often selected on the basis of its ability to shear material at low speed, or to wipe, smear, fold, stretch, or compress the mass being processed.

Of primary importance in the selection of a particular type of mixer for processing non-Newtonian materials is the behavior of that material while it is undergoing flow or shear. Many non-Newtonian materials exhibit different resistances to shear at different rates of shear. Some of these materials have greater resistance to shear, or flow, at high shear rates, while others show greater resistance at low rates of shear. The behavior of non-Newtonian materials in regard to these properties determines the type of equipment used for mixing and the amount of power that must be used to achieve a desired degree of mixing.

Since the properties of non-Newtonian materials play a major role in mixing equipment selection the general properties discussions of Chapter 1 are pertinent to this chapter. Some of these principles are reviewed and strengthened with additional discussions below.

GENERAL COMMENTS ON NON-NEWTONIAN BEHAVIOR

Recall that in Chapter 1 we considered the system of simple shear. Viewing this in another way, consider a thin layer of fluid between two parallel plates a distance dy apart as shown in Figure 1. One plate is stationary and a shearing force F is applied to the other. When conditions are steady, the force F will be balanced by an internal force in the fluid due to its viscosity. For a Newtonian fluid in laminar motion the shear stress is proportional to the velocity gradient, that is,

$$\frac{F}{A} = \tau \propto \frac{du}{dy}$$

This equation may be written as:

$$\tau = \mu \frac{du}{dy} = \mu\dot{\gamma} \tag{1}$$

where Newtonian viscosity μ is the tangential force per unit area exerted on layers of fluid a unit distance apart and has a unit velocity difference between them. The Newtonian viscosity depends only on temperature and pressure and is independent of the rate of shear. The diagram relating shear stress and rate of shear for Newtonian fluids, referred to as the *flow curve,* is therefore a straight line of slope μ (Figure 4A, Chapter 1). The single constant μ, in this case, completely characterizes the fluid. Newtonian behavior is exhibited in fluids (gases, liquids and various solutions) with comparatively small molecular mass. Liquids having a greater molecular mass show marked deviations from Newtonian behavior.

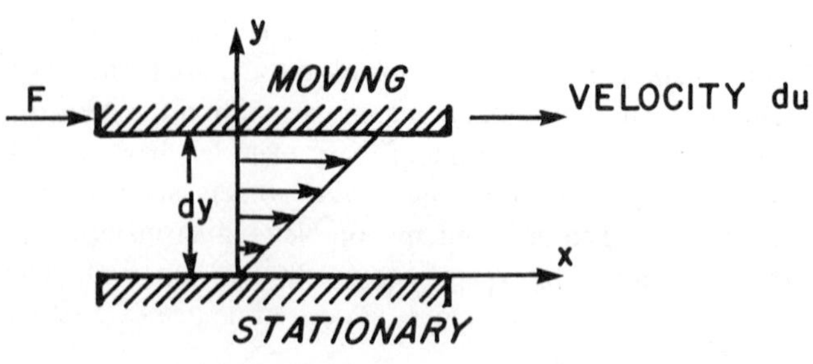

Figure 1 Fluid sheared between two plates.

In contrast, the flow curve of a non-Newtonian fluid is nonlinear. That is, the "viscosity" of a non-Newtonian fluid is not constant at a given temperature and pressure but depends on other factors such as the rate of shear in the fluid, the apparatus in which the fluid is contained, or even the previous history of the fluid.

These fluids may be classified into three broad types:

1. Fluids for which the rate of shear at any point is a function of the shearing stress at that point and depends on nothing else.
2. More complex systems for which the relation between shear stress and shear rate depends on the length of time the fluid has been sheared or on its previous history.
3. Systems which have characteristics of both solids and fluids and exhibit partial elastic recovery after deformation (so-called viscoelastic fluids).

Fluids of the first group, where properties are independent of time, are described by a rheological equation of the form:

$$\dot{\gamma} = f(\tau) \tag{2}$$

This equation implies that the rate of shear at any point in the fluid is a simple function of the shear stress at that point. These materials are termed *non-Newtonian viscous fluids* (or time-independent fluids).

Such fluids may be subdivided into three distinct types depending on the nature of the function in Eq. (2):

1. Bingham plastics
2. Pseudoplastic fluids
3. Dilatant fluids

Typical flow curves for these fluids were shown in Chapter 1 and compared with the linear relation typical of Newtonian fluids.

A *Bingham plastic* is an idealized fluid, although some materials can be approximated by this description. The flow curve is a line having an intercept τ_y on the shear-stress axis. The rheological equation for such a liquid may be written as

$$\tau = \tau_y + \eta_p \dot{\gamma}, \qquad \tau > \tau_y \tag{3}$$

where

τ = shear stress
$\dot{\gamma}$ = gradient of shear
η_p = plastic viscosity (i.e., the slope of the flow curve)

The properties of real liquids (e.g., slurries, concentrated suspensions, oil paints) can be approximated by this equation. The Bingham plastic at rest contains a three-dimensional structure of sufficient rigidity to resist any stress less than the yield stress value τ_y. Upon exceeding this stress the structure completely disintegrates and the system behaves as a Newtonian fluid under a shear stress $\tau - \tau_y$ (a straight-line relationship between shearing stress and rate of shear). When the shear stress falls below τ_y (i.e., $\tau < \tau_y$), the structure is said to be deformed.

Pseudoplastic fluids are frequently encountered in industrial practice. Among the most common materials are polymers. Their flow curve, as in the case of Newtonian liquids, starts at the origin. That is, these fluids have no yield stress. The typical flow curve for these materials reflects the ratio of shear stress to the rate of shear, which is termed the apparent viscosity, μ_a.

The curve for pseudoplastic fluids represents materials for which the required increase in shear stress to cause flow becomes smaller as the rate of shear increases. A decrease in the required shear stress is caused by the fact that the resistance to shear has decreased. Hence, the curve also indicates that the resistance to shear decreases as the shearing rate increases. A logarithmic plot of shear stress and shear rate is often found to be linear with a slope between zero and unity. As a result, an empirical functional relation known as the *power law* is widely used to characterize these fluids. This relation, which was originally proposed by Ostwald (1926) and has since been fully described by Reiner (1949), may be written as

$$\tau = k\dot{\gamma}^n \tag{4}$$

where k and n are constants with $n < 1$, for the particular fluid; k reflects a measure of the consistency of the fluid with the higher values representative of more viscous materials. Exponent n is a measure of the degree of non-Newtonian behavior, and the greater the departure from unity the more pronounced are the non-Newtonian properties of the fluid. It is important to remember that although n is nearly constant in many cases over wide ranges of shear rate, it is not a true constant for real fluids over all possible ranges of shear. This is not a great disadvantage since all that is needed is a rheological equation which describes the fluid over the particular range of shear rate encountered in a particular problem. Over such a range n may often be regarded as constant.

The dimensions of k depend on the index n and this fact has led to many objections to the use of the power law (Reiner, 1949). However, in most engineering calculations these objections are not serious. The apparent viscosity η_a for a power law fluid may be expressed in terms of n since

$$\eta_a = \frac{\tau}{\dot{\gamma}} \tag{5}$$

$$\eta_a = k\dot{\gamma}^{n-1}$$

Since $n < 1$ for pseudoplastics, the apparent viscosity decreases as the stress rate increases. This type of behavior is characteristic of suspensions of asymmetric particles or solutions of high polymers such as cellulose derivatives. This suggests that the physical interpretation of this phenomenon is probably that with increasing rates of shear the asymmetric particles or molecules are progressively aligned. That is, instead of the random intermingled state which exists when the fluid is at rest, the major axes are brought into line with the direction of flow. The apparent viscosity continues to decrease with increasing rate of shear until no further alignment along the streamlines is possible and the flow curve then becomes linear.

Since pseudoplastic fluids are time independent, this implies that the alignment of molecules suggested above occurs instantaneously as the rate of shear is increased or, at any rate, so quickly that the time effect cannot be detected using ordinary viscometric techniques.

Other empirical equations which have been used to describe pseudoplastic behavior are summarized below [see Wilkinson (1960) for a review of engineering models]:

Prandtl	$\tau = A \sin^{-1} (\dot{\gamma}/C)$	(6a)
Eyring	$\tau = \dot{\gamma}/B + C \sin (\tau/A)$	(6b)
Powell–Eyring	$\tau = A\dot{\gamma} + B \sinh^{-1} (C\dot{\gamma})$	(6c)
Williamson	$\tau = A\dot{\gamma}/(B + \dot{\gamma}) + \mu_\infty\dot{\gamma}$	(6d)

Coefficients A, B, and C are constants, specific to the particular fluid. These equations are considerably more difficult to apply than the simple power law expression.

Dilatant fluids are similar to pseudoplastics in that they show no yield stress. In contrast, the apparent viscosity for these fluids increases with increasing rates of shear. In the process industries dilatant fluids are much less common than pseudoplastics, but when the power law is applicable the treatment of both types is much the same. The dilatant curve represents materials (such as starch slurries, pigment–vehicle suspensions) for which the required increase in shear stress to cause flow increases as the rate of shear increases. This also indicates that the resistance of shear increases as the rate of shear increases.

The power law equation is again often applicable but in this case the index n has values greater than unity. The phenomenon of dilatancy may be explained according to Metzner (1956) as follows: The mixture consists of solid particles in intimate contact and of an amount of liquid just sufficient to fill up the voids between the particles. At low flow velocity, the internal friction in such a mixture is small as the liquid present acts as lubricant between the shifting solid particles. At higher flow velocity, the solid particles are no longer in close contact. The voids between them enlarge and cannot be filled up by the liquid, whereby the internal friction in the mixture increases.

Some general comments on time-dependent fluids are now necessary. Many real fluids cannot be described by a simple rheological expression such as Eq. (2). The apparent viscosity of more complex fluids depends on not only the rate of shear but also the time the shear has been applied. Recall that these fluids may be subdivided into two classes, thixotropic and rheopectic, depending on whether the shear stress increases or decreases with time when the fluid is sheared at a constant rate.

Thixotropic materials are those whose consistency depends on the duration of shear as well as on the rate of shear. They are encountered more often than rheopectic fluids. When a thixotropic material is sheared at a constant rate after a finite period of rest, the structure progressively disintegrates and the apparent viscosity decreases with time. The rate of structure breakdown during shearing at a given rate depends on the number of linkages available for breaking and must therefore decrease with time. The simultaneous rate of reformation of structure will increase with time as the number of possible new structural linkages increases. Eventually a state of dynamic equilibrium is achieved when the rate of buildup of structure equals the rate of breakdown. This equilibrium position depends on the rate of shear and moves toward greater breakdown at increasing shear rates.

A common example is material confined in a Couette viscometer. After the material has been resting for a long time one of the cylinders is rotated at a constant speed. The torque on the other cylinder then decreases with time as shown in Figure 2A. The rate of decrease and the final torque would both depend on the speed, that is, on the rate of shear. This behavior produces a hysteresis loop on the curve of shear stress plotted against rate of shear if the curve is plotted first for the rate of shear increasing at a constant rate and then for the rate of shear decreasing at a constant rate. This is illustrated in Figures 2B and C, where the curves A and B in Figure 2C are drawn for fluids of the Newtonian and pseudoplastic types which exhibit thixotropy.

Flow curves of increasing height can be obtained by applying shear for increasing lengths of time before making the return path. A single curve can be obtained by continuing the shearing process to equilibrium before returning. Pryce-Jones (1952) introduced the term *false body* to distinguish types of thixotropic behavior of Bingham plastics. True thixotropic materials break down completely under the influence of high stresses and behave like true liquids even after the stress has been removed, until such time as the structure has reformed. A false-bodied material, on the other hand, does not lose its solid properties entirely and can still exhibit a yield value even though this might be diminished. The original yield value is only regained after the material rests for a long time. Figure 3 shows the hysteresis loop on the flow curve for these two types of materials.

Consider a liquid in a vessel with a cylinder on a torsion wire immersed in it whereby the cylinder is deflected and the liquid stirred. If stirring is then stopped

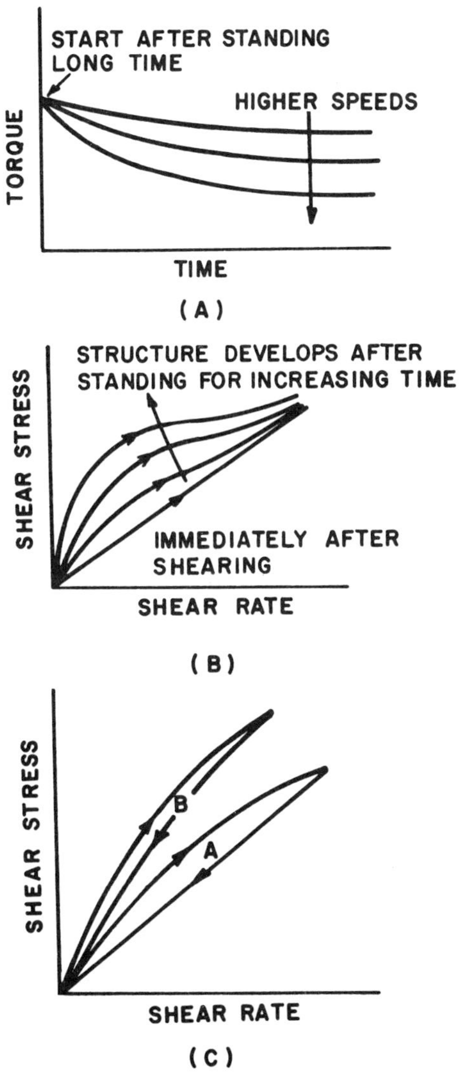

Figure 2 Development of hysteresis loops for thixotropic fluids.

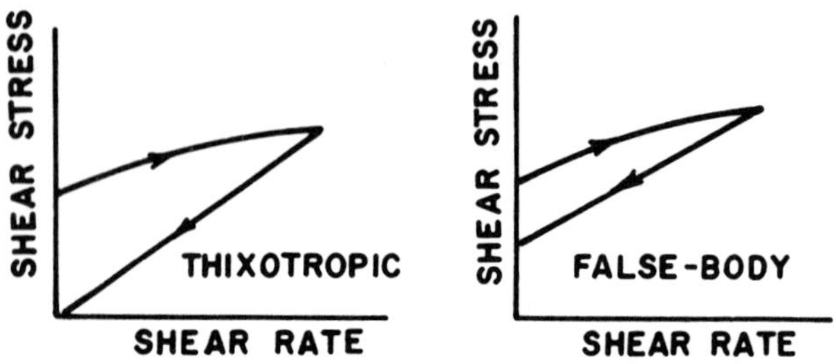

Figure 3 Types of hysteresis loops.

and the cylinder released, the torsion in the wire varies with time as in Figure 4 for the two types of material.

The false-bodied materials show a residual torsion in the wire indicating that the material can offer permanent resistance to shear immediately after stirring. That is, it retains a finite yield value. A true thixotropic material would show a residual torsion only if some time elapsed after stirring had ceased before the cylinder was released, to allow the structure to build up.

Similarly, false-bodied materials can retain elasticity. This would result in a recoil of the cylinder in the preceding experiment.

For *rheopectic fluids* the viscosity increase is a result of gradual formation of structure within the limits of low shear rates; reaching these rates the structure may be destroyed and the fluid will behave as a thixotropic fluid.

The two curves plotted in Figure 5 represent time-dependent non-Newtonian fluids. Curve A represents a rheopectic fluid whose shear stress increases as the time of shear (mixing time) increases. An increase in shear stress means that the shearing force *F* must be increased to produce flow. Since for any given rate of shear the consistency is the property that influences the amount of force required,

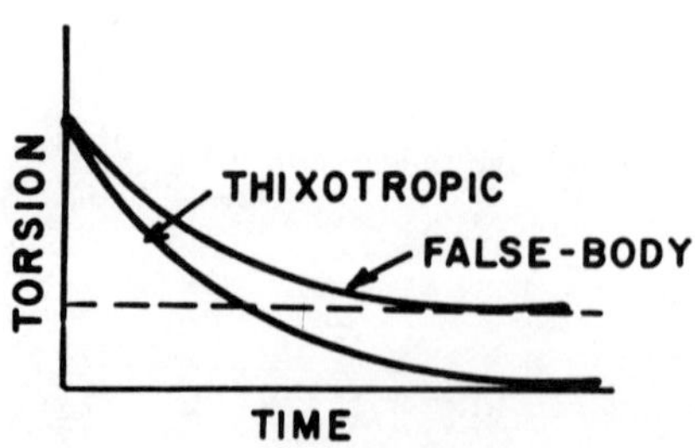

Figure 4 Comparison of true thixotropic and false-body behavior.

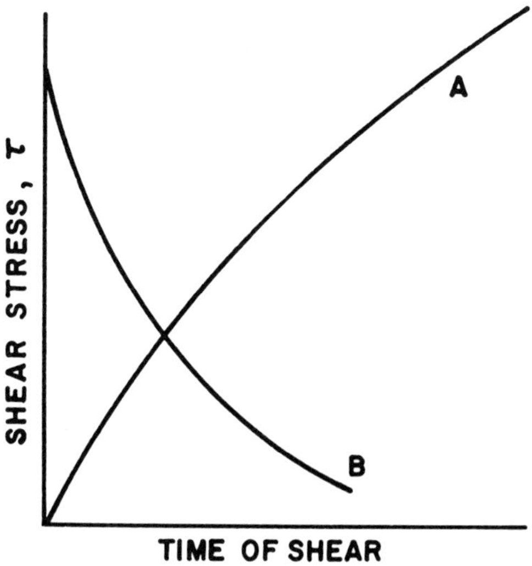

Figure 5 Plots of shear stress versus time of shear for non-Newtonian fluids.

an increase in shear stress must mean an increase in consistency. Hence, the material of curve A also exhibits an increase in consistency, or apparent viscosity, when the time of shear is increased.

Curve B is typical of a thixotropic liquid, exhibiting a decrease in shear stress with increase in time of shear. Hence, as the mixing time is prolonged the consistency or apparent viscosity of the fluid decreases. Steady-state flow permits the use of equations of pseudoplasticity for systems displaying "time effects." It should be noted that the "time effects" are reversible and after the stress is eliminated, the structure is completely reformed. "Time effects" have an important role in batch operations where as a result of long time action of shear after starting the machine there is a decrease in the apparent viscosity of materials to be mixed. For this reason the theoretical calculation of the energy balance is difficult, especially for the starting period.

Thixotropic behavior is similar to pseudoplasticity in that the time required for the alignment of particles is not negligible. Further, rheopectic fluids are superficially similar to their time-dependent counterparts (dilatant fluids) in that the time for structure buildup is negligibly small. The analogy in the latter case is not so close because with rheopectic fluids, structure buildup is brought about by small shearing rates only. That is, there is an upper limit to the shear rate beyond which the analogy is no longer valid.

Viscoelastic fluids are those that possess both elastic and viscous properties.

In other words, although the material may be viscous, it exhibits a certain elasticity of shape. An example of a viscoelastic fluid is pitch. For a simplified description, we consider Newton's law for the viscous component and Hooke's law for the elastic component. In steady-state flow under a shear stress τ the rate of shear will be τ/μ_0 where μ_0 is a constant Newtonian viscosity coefficient. If the shear stress is increased to $\tau + \delta\tau$ very rapidly, the material will be sheared through an additional angle $\delta\tau/G$, where G is the modulus of rigidity. There is therefore an additional rate of shear proportional to the rate of change of stress at any instant, and the total rate of shear is given by

$$\dot{\gamma} = \frac{\tau}{\mu_0} + \frac{\dot{\tau}}{G} \tag{7}$$

or in terms of a time constant λ_1:

$$\tau + \dot{\lambda}_1 \dot{\tau} = \mu_0 \dot{\gamma}, \qquad \text{where } \gamma_1 = \frac{\mu_0}{G} \tag{8}$$

Equation (8) is Maxwell's equation, and fluids described by it are usually referred to as Maxwell fluids.

The parameter λ_1 is the time constant of the exponential decay of stress at a constant strain. Observe from Eq. (8) that if the motion is stopped the stress will relax as $\exp(-t/\lambda_1)$. Consequently, λ_1 is known as the relaxation time. Schofield and Scott-Blair (1932) have applied the Maxwell equation to flour doughs.

Oldroyd (1953) investigated the elastic and viscous properties of emulsions and suspensions of one Newtonian liquid in another and derived theoretically the differential equation relating the shear stress τ and the rate of shear $\dot{\gamma}$ in the following form:

$$\tau + \lambda_1 \dot{\tau} = \mu_0(\dot{\gamma} + \lambda_2 \ddot{\gamma}) \tag{9}$$

The constants μ_0, λ_1, and λ_2 depend on the physical properties of the mixture. The elastic strain energy is stored during flow by virtue of the fact that interfacial tension provides the restoring force which makes the individual drops resist changes of shape. The same equation was also derived by Frölich and Sack (1946) for a dilute suspension of solid particles in a viscous liquid. Elastic strain energy is stored because the solid elastic particles are deformed by the flow of the surrounding liquid.

The constant μ_0 is the viscosity at low rates of shear in the steady state, that is, when $\dot{\tau} = \ddot{\gamma} = 0$. The constant λ_1 is a relaxation time and the physical significance of this is that if the motion is suddenly stopped, the shear stress will decay as $\exp(-\tau/\lambda_1)$; λ_2 is called a *retardation time* and reflects the fact that if all stresses are removed the rate of strain decays as $\exp(-\tau/\lambda_2)$.

Toms and Strawbridge (1953) found that the behavior of dilute solutions of polymethyl methacrylate in pyridine and some bitumens can be described by a similar expression. We conclude that the rheological equation for a viscoelastic fluid contains the time derivatives of both τ and γ in general. The general case is

$$f_1(D)\tau = f_2(D)\gamma \tag{10a}$$

or alternatively as a polynomial in D

$$\sum_{n=0}^{N} a_n D^n \tau = \sum_{m=0}^{M} \beta_m D^m \gamma \tag{10b}$$

where D is the differential operator d/dt.

GENERAL METHODS OF MIXING

Irrespective of the individual ingredients comprising a non-Newtonian material (i.e., liquid, solid particles in suspension, plastics), the objective of mixing is to decrease the concentration gradient to a desired minimum. In some cases this is accompanied by a decrease in the thermal gradient in processes involving heat transfer.

The minimum gradient of concentration (or temperature) is achieved when the infinitesimal samples of the mixture randomly extracted from any location within the mixing system, have the same composition. In some processes mixing is accompanied by dispersion, that is, by phenomena presenting a change in the physical characteristics of the components (e.g., intrusion of solid fillers into a resin mass, decrease in the particle size of a filler, etc.). In addition, mechanical mixing may be accompanied by physicochemical processes of plasticization.

The mixing methods and the design of mixing machines depend on the physical properties of components to be mixed. The mixing of non-Newtonian materials is basically carried out by different mechanical devices which both combine ingredients and provide intensive agitation. Common devices used are sigma-blade mixers, Banbury mixers, roller mills, anchor agitators, centrifugal disk-type mixers, and others.

Heavy pastes and materials with pronounced plastic properties require heavy machinery to achieve satisfactory mixing. Natural flow of materials and momentum effects are negligible and because of this the mixing must be forced. Sections of the mix must be separated from the whole mass and then forced back into a different part of the mixture volume by means of shearing, kneading, or wiping actions.

Highly viscous liquids and pastes can be mixed in either batch or continuous operations. A *batch mixing process,* as a rule, is performed in a closed volume of a mixing device. The principal parameter affecting mixing efficiency in this case is the minimum mixing time necessary for providing the desired homogeneity of the mixture. The capacity of a mixing device depends on the mixing time, the times of loading ingredients, starting, stopping, and unloading. Mixing time is established by not only the available volume of the mixing device, the design of the working elements, the rotational velocity, and the properties of the components to be mixed, but also, to a large degree, the initial orientation of components and the order in which they are introduced.

The rationally designed mixer must provide an equal mixing time independent of the initial location of components. Periodic reorientation of material during mixing will decrease the basic mixing time. In general, batch mixers are not very efficient at handling large capacities mainly because of difficulties encountered when attempting to automate operations. Over the course of a batch cycle, the properties of the mixture may change considerably.

During mixing and plastification of polymer materials the power consumption, as a rule, decreases. This often requires the installation of a motor drive of high capacity, that must be selected in accordance with the initial characteristics of the mixture. However, there is a need to use batch mixers in many cases, especially if required mixing times are short.

Continuous mixing operations involve the continuous loading and unloading of components. In this case there is a drastic variation in the concentration of components at the inlet. When properly performed, mixing decreases these variations at the exit to the desired minimum. Continuous operations have the advantage of providing a stable process. Also, the power consumption is both lower and more constant than in batch operations. The parameters of the mixture, as they move from inlet to outlet, vary even though the volume of the mixer remains constant. Despite the advantages, continuous mixing operations are difficult to justify in many cases because of the long mixing time requirements. Long mixing times are somewhat compensated for in continuous mixers by increasing process intensity (i.e., operating at higher rotational velocity). This may, however, result in uneconomical increases in the dissipation of mechanical energy and thermal destruction.

ESTIMATING POWER CONSUMPTION

The mixing action imposed by turbine, propeller, paddle, and anchor impellers results in complicated streamlines of fluid motion when applied to non-Newtonian materials. Recommended geometries for these types of mixers are summarized in Table 1 and detailed design practices are given by Azbel and Cheremisinoff (1983). In this process the power consumption depends not only on machine design, but also on the physical condition of the material.

Power consumption for mixing Newtonian materials cannot be determined on a purely theoretical basis even for the simplest of working devices. Therefore, the problems of scale-up and design must be addressed using similarity theory, which requires carefully planned experiments in model equipment. Non-Newtonian mixing is even more complex because of the interrelationships between fluid consistency, temperature, and shear. Fluid behavior of non-Newtonians during mixing is further complicated by the large dissipation of mechanical energy, which results in considerable variations in the temperature level of the process. Under actual process conditions there is little or no possibility of assessing the instantaneous viscosity of the material.

When a non-Newtonian fluid is processed in a mixing vessel, the variations in shear rate with distance from the impeller give rise to corresponding variations in fluid consistency. To allow for this, Metzner and Otto (1957) and Magnusson (1952) identified an average shear rate $(du/dr)_A$ in the vessel. The apparent viscosity μ_A corresponding to $(du/dr)_A$ is defined as the viscosity of that Newtonian fluid which would show exactly the same power consumption for agitation under identical conditions, at least in laminar flow. An empirical correlation that describes this is

$$\dot{\gamma}_A = \left(\frac{du}{dr}\right)_A = k_s N \tag{11}$$

Metzner and Taylor (1960) confirmed Eq. (11) independently. The experimental determination of parameter k_s for a given geometry is as follows:

1. N_p is measured for a particular N.
2. The corresponding Reynolds number is obtained from the appropriate plot for Newtonian fluids on a chart provided by Rushton (1950).
3. The apparent viscosity μ_A for these conditions is computed from the Reynolds number. (Laminar flow is used for maximum sensitivity to changes in μ_A.)
4. The shear rate $(du/dr)_A$ is obtained from the flow curve for the particular fluid under consideration, since $\mu_A = \tau/(du/dr)_A$.
5. The constant k_s is then evaluated from Eq. (11). The procedure is repeated for various N values to give an average k_s.

A compilation of experimental k_s values for a variety of geometries is given in Table 2 and Figure 6.

Table 1 Recommended Geometric Proportions of Common Mixing Devices and Range of Applications

<table>
<tr><th>Type of mixer</th><th>Size ratios</th><th>Sphere of application</th></tr>
<tr><td>Two Blades
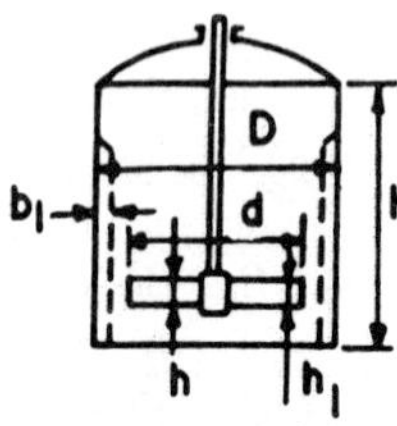
</td><td>$\frac{d}{D} = 0.5–0.7$
$\frac{h}{d} = 0.1–0.3$
$\frac{h_1}{d} = 0.14–0.2$
$\frac{b}{D} = 0.08$</td><td>For mixing of liquids with viscosities up to 15 N-s/m^2 and dissolution of solid particles, u = 1.5–3 m/s</td></tr>
<tr><td>Four Blades
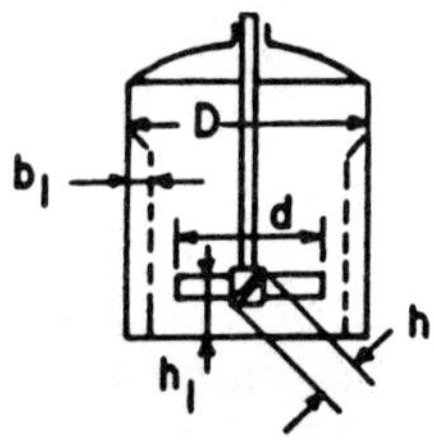
</td><td>$\frac{d}{D} = 0.2–0.4$
$\frac{h}{d} = 0.2–0.4$
$\frac{h_1}{d} = 0.4–0.8$
$\frac{b}{D} = 0.08$</td><td>For mixing of liquids of medium viscosity and dissolution of solid particles, u = 1.5–3 m/s</td></tr>
<tr><td>Anchor
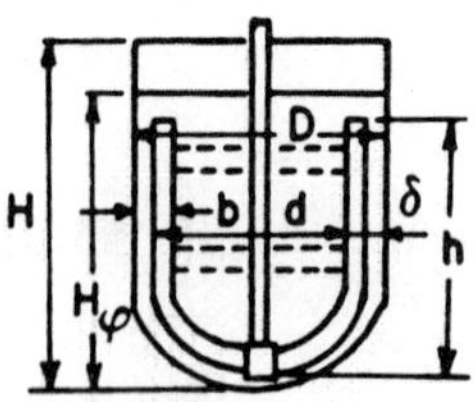
</td><td>$\frac{H_\varphi}{H} = 0.75–0.85$
$\frac{h}{H} = 0.56$
$\frac{b}{d} = 0.07$
δ = 25–40 mm</td><td>For mixing of liquids of viscosities up to 200 N-s/m^2 in tanks with heating elements, u = 1–3 m/s</td></tr>
<tr><td>Framed
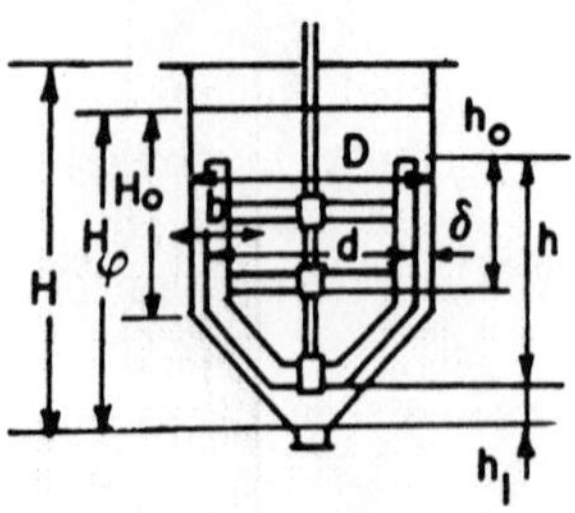
</td><td>$\frac{H_\varphi}{H} = 0.75–0.85$
$\frac{h}{H} = 0.6$
$\frac{h_0}{H_0} = 0.56$
h_1 = 190–275 mm
δ = 25–40 mm
$\frac{b}{d} = 0.07$</td><td>See anchor mixers</td></tr>
</table>

continued

Table 1 (Continued)

Type of mixer	Size ratios	Sphere of application
Turbine 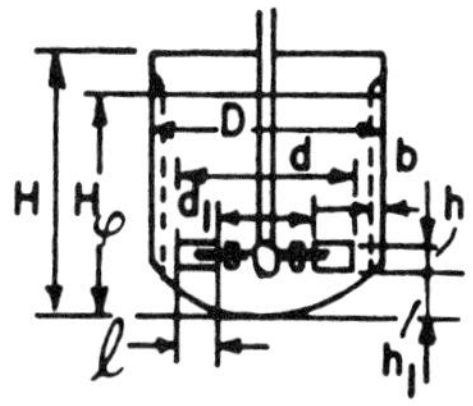 	$\frac{d}{D} = 0.3$ $\frac{H\varphi}{H} = 0.75–0.85$ $\frac{h}{d} = 0.2–0.3$ $\frac{l}{d} = 0.25$ $\frac{h_1}{d} = 0.5–1.5$ $\frac{d_1}{d} = 0.65$ $\frac{b}{D} = 0.08$	For intensive mixing of suspensions, dissolution and dispersion of liquids and gases May operate in the media with viscosities up to 25 N-s/m^2, u = 3–8 m/s
Shrouded Turbine 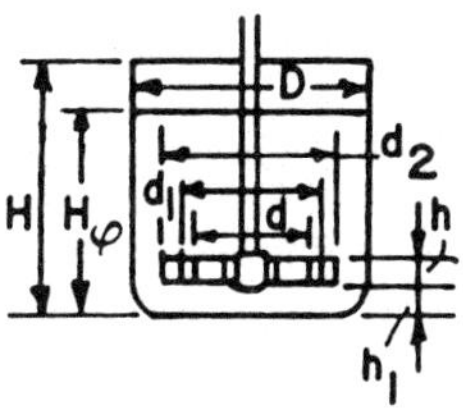 	$\frac{D}{d} = 2.4$ $\frac{h}{d} = 0.25$ $\frac{h_1}{d} = 0.85$ $\frac{d_1}{d} = 1.1$; $\frac{d_2}{d} = 1.6$ $\frac{H_\varphi}{d} 1.75$	
Propeller 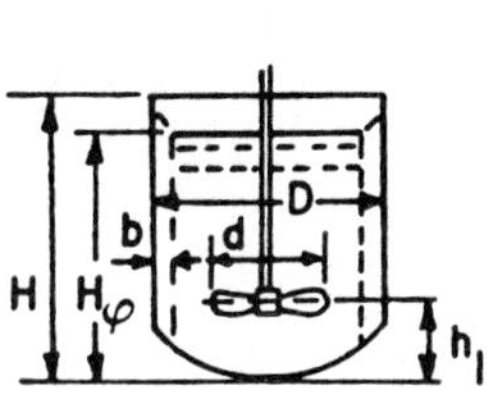 	$\frac{D}{d} = 2–4$ $\frac{H_\varphi}{d} = 2–4$ $\frac{h_1}{d} = 0.7–1.6$ $\frac{b}{D} = 0.08$ $\frac{t}{d} = 1–2$	For mixing of liquids with viscosities up to 4 N-s/m^2, dissolution and suspending of solid particles

Table 2 Description of Curves in Figure 6 as Recommended by Rushton et al. (1950)

Curve	Impeller	Baffles	D, ft	D_r/D	n	$k_s(n < 1)$
A–A	Single turbine with six flat blades	Four, $J/D_r = 0.1$	0.167–0.67	1.3–5.5	0.05–1.5	11.5 ± 1.5
A–A_1	Single turbine with six flat blades	None	0.167–0.67	1.3–5.5	0.18–0.54	11.5 ± 1.4
B–B	Two turbines, each with six flat blades and $D_r/2$ apart	Four, $J/D_r = 0.1$	—	3.5	0.14–0.72	11.5 ± 1.4
B–B_1	Two turbines, each with six flat blades and $D_r/2$ apart	Four, $J/D_r = 0.1$ or none	—	1.023–1.18	0.14–0.72	11.5 ± 1.4
C–C	Fan turbine with six blades at 45°	Four, $J/D_r = 0.1$ or none	0.33–0.67	1.33–3.0	0.21–0.26	13 ± 2
C–C_1	Fan turbine with six blades at 45°	Four, $J/D_r = 0.1$ or none	0.33–0.67	1.33–3.0	1.0–1.42	13 ± 2
D–D	Square-pitch marine propellers with three blades (downthrusting)	None, (i) shaft vertical at vessel axis, (ii) shaft 10° from vertical, displaced $R/3$ from center	0.417	2.2–4.8	0.16–0.40	10 ± 0.9
D–D_1	Same as for D–D but upthrusting	None, (i) shaft vertical at vessel axis, (ii) shaft 10° from vertical, displaced $R/3$ from center	0.417	2.2–4.8	0.16–0.40	10 ± 0.9
D–D_2	Same as for D–D	None, position (ii)	1.0	1.9–2.0	0.16–0.40	10 ± 0.9
D–D_3	Same as for D–D	None, position (i)	1.0	1.9–2.0	0.16–0.40	10 ± 0.9
E–E	Square-pitch marine propeller with three blades	Four, $J/D_r = 0.1$	0.5	1.67	0.16–0.60	10
F–F	Double-pitch marine propeller with three blades (downthrusting)	None, position (ii)	—	1.4–3.0	0.16–0.40	10 ± 0.9
F–F_1	Double-pitch marine propellers with three blades (downthrusting)	None, position (i)	—	1.4–3.0	0.16–0.40	10 ± 0.9
G–G	Square-pitch marine propeller with four blades	Four, $J/D_r = 0.1$	0.392	2.13	0.05–0.61	10
G–G_1	Square-pitch marine propeller with four blades	Four, $J/D_r = 0.1$	0.392	2.13	1.28–1.68	—
H–H	Two-bladed paddle	Four, $J/D_r = 0.1$	0.283–0.416	2–3	0.16–1.68	$10(n < 1)$
—	Anchor	None	0.927	1.02	0.34–1.0	11 ± 5
—	Cone impellers	None or four, $J/D_r = 0.08$	0.33–0.5	1.92–2.88	0.34–1.0	11 ± 5

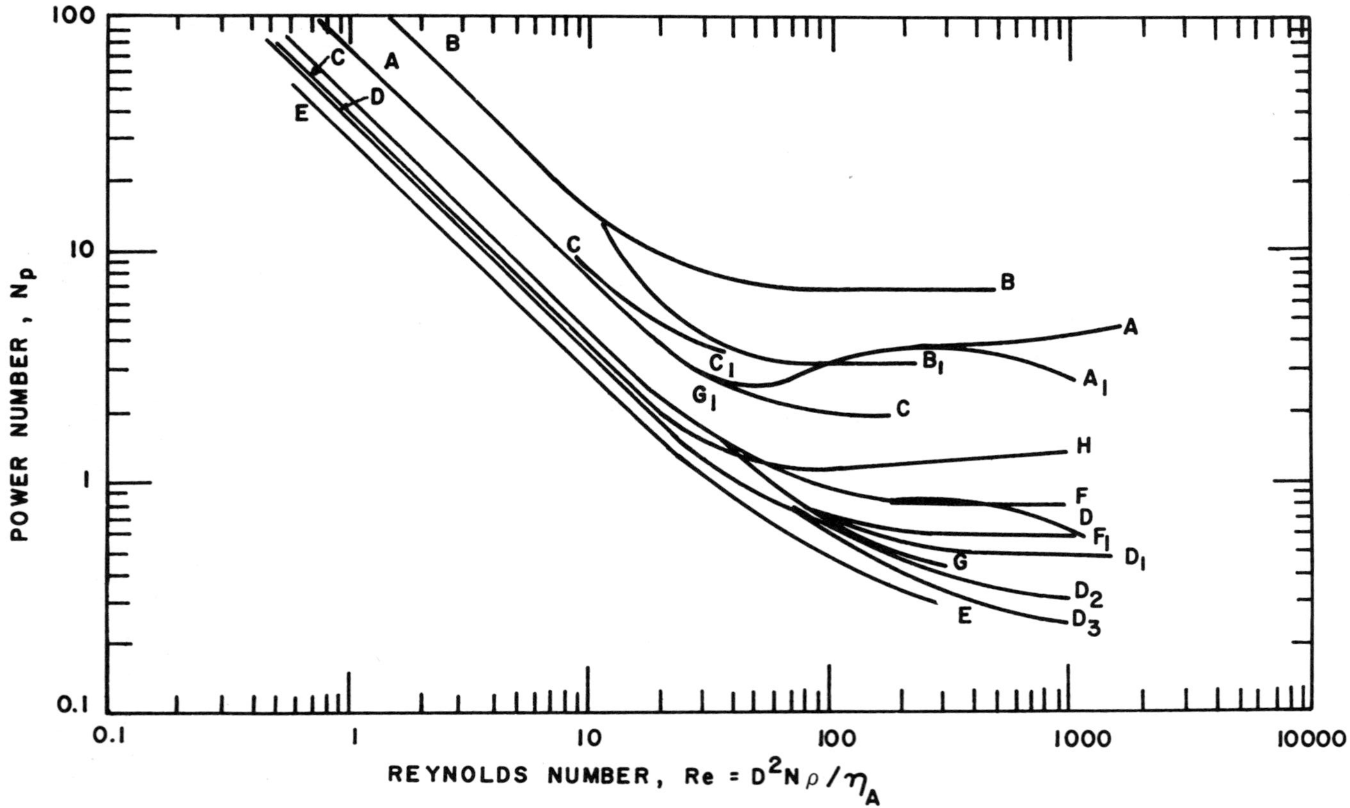

Figure 6 Correlation of the power number in the agitation of non-Newtonian fluids. (Based on data of Rushton et al., 1950.)

The prediction of power consumption in a particular non-Newtonian agitation system for a given impeller speed proceeds in the following manner:

1. Compute $(du/dr)_A$ from Eq. (11), using the appropriate k_s value from Table 2.
2. Evaluate the corresponding μ_A from the flow curve as $\tau/(du/dr)_A$.
3. Calculate the Reynolds number, $D^2N\rho/\mu_A$, and read the Power number, $P/D^5N^3\rho$, from the appropriate curve in Figure 6. P is then computed from N_p.

Calderbank and Moo-Young (1961) give the following relation for computing k_s for power law fluids:

$$k_s = B\left(\frac{4n}{3n + 1}\right)^{n/(1-n)} \tag{12}$$

In this expression n is the fluid's flow behavior index. Equation (12) may be used with Eq. (11) to compute $(du/dr)_A$. The following empirical relationships were recommended by the investigators for the constant B.

(a) For high-speed impellers:

(i) $n < 1, \quad \frac{D_T}{D} > 1.5, \quad R = 4, \quad \frac{J}{D_T} = 0.1,$

$$B = 11 \pm 10\% \tag{13}$$

(ii) $n > 1, \quad \frac{D_T}{D} < 3.0, \quad R = 4, \quad \frac{J}{D_T} = 0.1,$

$$B = \frac{22(D_T/D)^2}{(D_T/D)^2 - 1} \pm 15\% \tag{14}$$

(b) For anchor agitators:

$$n < 1, \quad D_T/D < 1.4, \quad B = 9.5 + \frac{9(D_T/D)^2}{(D_T/D)^2 - 1} \pm 10\% \tag{15}$$

The term involving n in Eq. (11) has the value $0.84 \pm 8.5\%$ for $0.05 < n < 1.68$ and may therefore be considered a constant for practical calculations. This means that a linear relationship between $(du/dr)_A$ and N is predicted by both Metzner and Otto and Calderbank and Moo-Young, with only minor dependence on the flow behavior index.

As already noted, the viscous behavior of numerous non-Newtonian fluids can be represented by a power law equation of the general form:

$$\tau = K(\dot{\gamma})^n \tag{16a}$$

This expression is applicable over a limited range of shear rates. Applying this to a viscoelastic material where

$$\eta_a = K(\dot{\gamma})^{n-1} \tag{16b}$$

with the Reynolds number defined as

$$\mathrm{Re} = \frac{D^2 N_p}{\eta_a(\dot{\gamma})} \tag{17}$$

then with Eq. (11), we have

$$\mathrm{Re} = \frac{D^2 N \rho}{K(\dot{\gamma}_A)^{n-1}} = \frac{D^2 N \rho}{K(k_s N)^{n-1}} = \frac{D^2 N^{2-n} \rho}{k_s^{n-1} K} \tag{18}$$

and from

$$\sigma_1 = h(\dot{\gamma})^{m-2} \tag{19}$$

and

$$N_{W_i} = \frac{\sigma_1(\dot{\gamma})}{\eta_a(\dot{\gamma})} N \tag{20}$$

and combining with Eqs. (16a) and (11), we obtain

$$N_{W_i} = \frac{h(\dot{\gamma}_A)^{m-2}}{K(\dot{\gamma}_A)^{n-1}} N = \frac{h}{K}(k_s N)^{m-n-1} N = \frac{h}{k_s K}(k_s N)^{m-n} \tag{21}$$

Perhaps the simplest non-Newtonian fluid to describe is the Bingham plastic, since it is a linear relationship. Adopting the procedure used for Newtonian fluids, the power for mixing a Bingham plastic is

$$\frac{P}{\rho N^3 D^5} = \phi\left(\frac{\mu_p}{\rho N D^5}, \frac{\tau_y}{\rho N^2 D^2}, \frac{N^2 D}{g}\right) \tag{22a}$$

Note that the second group is the ratio of yield stress to inertia forces. This can be expressed in terms of the Hedstrom number ($\mathrm{He} = \tau_y \rho D^2/\mu_p^2 = \mathrm{Re}^2 \tau_y/\rho N^2 D^2$). Thus

$$\frac{P}{\rho N^3 D^5} = \phi(\mathrm{Re}, \mathrm{He}, \mathrm{Fr}) \tag{22b}$$

Design procedures for estimating power requirements for mixing non-Newtonian materials are sparce. One procedure that appears useful for highly pseudoplastic materials is that of Metzner and Otto, applicable to flat-blade turbines. The procedure is as follows:

1. Find μ_a at a given N from the measured power coefficient and the appropriate correlation between power coefficient and Reynolds number used for Newtonian fluids; that is, measure $P/\rho N^3 D^5$ for the non-Newtonian system, and find $\rho ND^2/\mu_a$ from the correlation at this value of the power coefficient. This gives μ_a, knowing ρND^2 for the system, if μ_a is as defined above, that is, the viscosity of a Newtonian fluid which gives the same power coefficient in the same apparatus at the same speed.
2. Find the shear rate $\dot{\gamma}$, which gives the value of μ_a from the flow curve of the non-Newtonian fluid.
3. Find K by substituting the values of $\dot{\gamma}$ and the speed N in the equation

$$(\dot{\gamma})_{\text{av}} = KN \tag{23}$$

4. Repeat for different fluids at different speeds using various sizes of apparatus to determine the best value of K. All the experiments must be confined to the laminar flow region, since outside this region viscosity is not an important factor.

The value of K so obtained by Metzner and Otto was 13 where N is in rpm and $(\dot{\gamma})_{\text{av}}$ is in min^{-1}. Metzner and Otto's (1957) results are shown plotted in Figure 7. The difference between the two curves is that the laminar region for non-Newtonian pseudoplastic fluids extends further than that for Newtonian fluids. The reason for this is that μ_a increases with increasing distance from the impeller (regions of lower shear rate), which tends to depress eddies and delay the onset of turbulence.

Calderbank and Moo-Young (1961) extended this work for a wide variety of sizes and types of impellers and found that for Bingham and pseudoplastic fluids the value of K was 10 rather than 13. For dilatant fluids the relative size of the impeller and mixing vessel has an influence on the shear rate and therefore on the effective value of K. Calderbank and Moo-Young found in their work that for these fluids K could be represented by 12.8 $(D_1/D_T)^{1/2}$ where D_1 and D_T are the diameters of the impeller and tank, respectively.

HEAT TRANSFER

Discussion is now directed to the problem of turbulent heat transfer to non-Newtonian fluids. For slightly non-Newtonian materials such as dilute suspensions, numerous data on the heat transfer characteristics are reported in the literature [see Winding et al. (1944) and Orr et al. (1954) for examples]. For non-

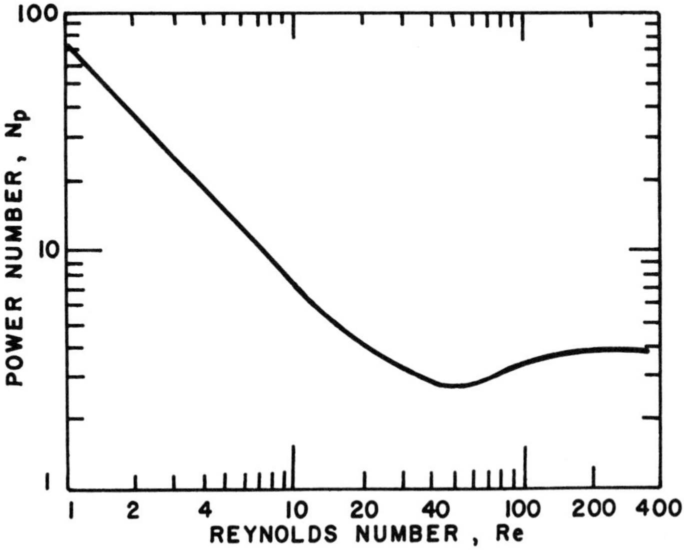

Figure 7 Power coefficient versus Reynolds number for non-Newtonian fluids based on the work of Metzner and Otto (1957).

Newtonian fluids in this category the flow behavior index n does not differ greatly from unity. The two most widely used correlations are the Dittus–Boelter equation for forced convection in circular tubes and that of Sieder and Tate, which incorporates a viscosity correction term:

$$\frac{hD}{k} = 0.023\left[\frac{\rho u_m D}{\mu}\right]^{0.8}\left[\frac{\mu C_p}{k}\right]^{x} \tag{24}$$

where exponent $x = 0.4 \sim 0.3$.

$$\frac{hD}{k} = 0.023\left(\frac{\rho u_m D}{\mu}\right)^{0.8}\left(\frac{\mu C_p}{k}\right)\left(\frac{\mu}{\mu_w}\right)^{0.14} \tag{25}$$

These equations are most often applied to heat exchanger calculations.

For heat transfer to highly non-Newtonian fluids, dimensional analysis has produced several correlations which have limited applicability. Metzner et al. (1957) suggest the following correlation:

$$\frac{\mu_a C_p}{k} = \frac{C_p m}{k}\left(\frac{u_m}{D}\right)^{n'-1}$$

where the generalized Reynolds number is defined as

$$\mathrm{Re}' = \frac{D^{n'} u_m^{2-n'} \rho}{m}, \qquad m = k' 8^{n'-1}$$

and the apparent viscosity μ_a is defined by relating the Newtonian Reynolds number to the generalized one:

$$\frac{\rho u_m D}{\mu_a} = \frac{D^{n'} u_m^{2-n'} \rho}{m}$$

or

$$\mu_a = m \left(\frac{u_m}{D} \right)^{n'-1}$$

Incorporating this into the Seider–Tate equation gives

$$\frac{hD}{k} = 0.023 \left(\frac{D^{n'} u_m^{2-n'} \rho}{m} \right)^{0.8} \left(\frac{mC_p}{k} \left(\frac{u_m}{D} \right)^{n'-1} \right)^{0.4} \left(\frac{m}{m_w} \right)^{0.14} \tag{26}$$

Equation (26) has been found useful for order-of-magnitude estimates of the turbulent heat transfer coefficient.

Skelland and Dimmick (1969) studied heat transfer between coils and pseudoplastic fluids with propeller agitation in baffled vessels. The pseudoplastic fluids studied obeyed the power law, with $0.528 \leq n \leq 0.910$. Various sizes of axially mounted marine-type propellers were used, each with three blades and square pitch. Various-sized coils were also tried, and both downthrusting and upthrusting agitation was employed. The correlation obtained from this study is

$$\frac{h_c D_e}{k} = 0.258 \left(\frac{D^2 N \rho}{\mu_A} \right)^{0.62} \left(\frac{c_p \mu_A}{k} \right)^{0.32} \left(\frac{\mu_A}{\mu_{AW}} \right)^{0.2} \left(\frac{D}{D_T} \right)^{0.1} \left(\frac{D_e}{D_T} \right)^{0.5} \tag{27}$$

where h_c is the individual heat transfer coefficient on the agitated side of the coil and D_e is the external diameter of the coil tube. In the case of a power law fluid,

$$\tau = K \left(\frac{du}{dr} \right)_A^n$$

$$\mu_A = \frac{\tau}{(du/dr)_A} = \frac{K(du/dr)_A^n}{(du/dr)_A} = \frac{K}{(du/dr)_A^{1-n}}$$

Substituting Eq. (11) for $(du/dr)_A$,

$$\mu_A = \frac{K}{(k_s N)^{1-n}} \tag{28}$$

Thus, for power law fluids:

$$\frac{h_c D_e}{k} = 0.258 \left(\frac{D^2 N^{2-n}}{k_s^{n-1} K}\right)^{0.62} \left(\frac{c_p K}{k(k_s N)^{1-n}}\right)^{0.32} \left(\frac{K}{K_W}\right)^{0.2} \left(\frac{D}{D_T}\right)^{0.1} \left(\frac{D_e}{D_T}\right)^{0.5} \tag{29}$$

The form of Eq. (27) is not restricted to power law materials, but applies in general to any time-independent non-Newtonian fluid over the limits $332 \leq D^2 N\rho/\mu_A \leq 2.6 \times 10^5$. Furthermore, it reduces exactly to the recommended Newtonian correlation with regard to the effects of the six key variables (D, N, D_T, D_e, μ, μ/μ_w).

Since the temperature of the fluid varies during the heating or cooling process the temperature at which to evaluate μ_A in Eq. (27) (or K in Eq. (29)) becomes an issue of concern. Skelland and Dimmick (1969) showed that evaluation of μ_A at the mean run temperature, $(T_1 + T_2)/2$, was adequate throughout their work. An additional study worth noting is that of Hagedorn and Salamone (1967) for heat transfer in the simpler geometry of a jacketed vessel. Their mildly non-Newtonian results ($0.69 \leq n \leq 1.0$) were correlated with an average deviation of ±14%, whereas for their highly non-Newtonian data ($0.36 \leq n \leq 0.69$) the average deviation from their correlation increased to ±20%.

A practical design problem is that of *scale-up in coil-heated, baffled vessels with propeller agitation.* Skelland (1983) presented the example where the following criteria are required to duplicate some product characteristic when scaling up using power law fluids:

(A) Equal rates of heat transfer per unit surface of coil are to be maintained in the model and prototype equipment.
(B) Equal volumes of heat transfer per unit volume of vessel contents are to be maintained.

For power law fluids Eq. (29) may be expressed as follows:

$$h_c = \phi D_e^{-0.5} D^{1.34} D_T^{-0.6} N^{0.92-0.3n} \tag{30}$$

where parameter ϕ is defined as

$$\phi = 0.258 k^{0.68} \rho^{0.62} C_p^{0.32} (K/K_w)^{0.2} k_s^{0.3(1-n)} K^{-0.3} \tag{31}$$

From similarity, the ratio of impeller speeds in the two vessels is written for criterion A:

$$\frac{N_1}{N_2} = \left[\left(\frac{D_{e_1}}{D_{e_2}}\right)^{0.5}\left(\frac{D_2}{D_1}\right)^{1.34}\left(\frac{D_{T_1}}{D_{T_2}}\right)^{0.6}\frac{\Delta T_{1m_2}}{\Delta T_{1m_1}}\right]^{1/(0.92-0.3n)} \tag{32}$$

Equation (32) assumes that the same fluid is handled over temperature ranges with the same midpoint and that the tube wall temperatures have the same average value in the two different vessels. That is:

$$\left.\begin{aligned} \phi_{\text{av},1} &= \phi_{\text{av},2} \\ h_{c_1}\,\Delta T_{1m_1} &= h_{c_2}\,\Delta T_{1m_2} \end{aligned}\right\} \tag{33}$$

Subscripts 1 and 2 refer to the different-sized vessels. For criterion B we write

$$\begin{aligned} \frac{q_1}{V_1} &= \frac{q_2}{V_2} \\ \frac{h_{c_1}\pi D_{e_1}L_{c_1}\Delta T_{1m_1}}{H_1(\pi D_{T_1}^2/4)} &= \frac{h_{c_2}\pi D_{e_2}L_{c_2}\Delta T_{1m_2}}{H_2(\pi D_{T_2}^2/4)} \end{aligned} \tag{34}$$

where L_c is the length of submerged coil and H is the height of the liquid in the vessel.

Substituting Eq. (30) for h_c and solving for the ratio of shaft speeds required to meet criterion B:

$$\frac{N_1}{N_2} = \left[\frac{H_1}{H_2}\left(\frac{D_{T_1}}{D_{T_2}}\right)^{2.6}\left(\frac{D_{e_2}}{D_{e_1}}\right)^{0.5}\frac{L_{c_1}}{L_{c_2}}\left(\frac{D_2}{D_1}\right)^{1.34}\frac{\Delta T_{1m_2}}{\Delta T_{1m_1}}\right]^{1/(0.92-0.3n)} \tag{35}$$

To simultaneously duplicate both criteria, Eqs. (32) and (35) give the following:

$$\frac{H_1}{H_2} = \frac{D_{e_1}}{D_{e_2}}\left[\frac{D_{T_2}}{D_{T_1}}\right]^2\frac{L_{c_1}}{L_{c_2}} \tag{36}$$

Note that Eq. (36) cannot be satisfied when strict geometrical similarity prevails in the two vessels of different size. Consequently, criteria A and B cannot both be achieved in this example.

Jackets are frequently used instead of coils for heating or cooling. Baker and Walter (1979) give one correlation predicting the heat transfer coefficients inside jackets fitted to provide jetting action of the heating and/or cooling medium.

Helical ribbon impellers are widely used in promoting heat transfer to non-Newtonian fluids of high consistency, when operation is confined to the laminar and transitional flow regimes. For heat transfer in jacketed vessels agitated by double helical ribbon impellers, and when $D^2N\rho/\mu_A$ is below 1000, Nagata (1975) recommends the following correlation:

$$\frac{h_j D_T}{k} = 1.75 \left(\frac{D^2 N\rho}{\mu_A}\right)^{1/3} \left(\frac{c_p \mu_A}{k}\right)^{1/3} \left(\frac{\mu_A}{\mu_{AW}}\right)^{0.2} \left(\frac{D_T - D}{D_T}\right)^{-1/3} \tag{37}$$

where μ_A has the same definition as given earlier for non-Newtonian power consumption and μ_{AW} is the viscosity at the wall temperature. h_j is the individual agitated-side heat transfer coefficient to or from the jacket.

Scraped surface heat exchangers are widely used for processing non-Newtonian fluids of high consistency which may tend to adhere to the heat transfer surface. Such devices are comprised of rotating blades which continually scrape material from the surface and are exemplified by the Votator (Skelland, 1958), the Dopp Kettle (Huggins, 1931), and the helical ribbon impeller fitted with rubber scrapers (Nagata, 1975).

To analyze the heat transfer in such units, consider the fluid in the vicinity of the heat transfer surface between scrapings by two consecutive blades. It can be assumed that

1. Plug flow exists in the fluid over the depth of heat penetration.
2. The fluid thickness exceeds the depth of heat penetration.
3. There is complete mixing of the fluid displaced by the scraper blade with the bulk fluid after each displacement from the vicinity of the heat transfer surface.

A standard approach to analyzing these systems is to first develop the unsteady-state three-dimensional conduction equation, and then reduce it to that for one-dimensional thermal transport.

Consider a fluid volume element in the vicinity of the heat transfer surface, awaiting removal by the next scraper blade, as shown in Figure 8. Unsteady-state

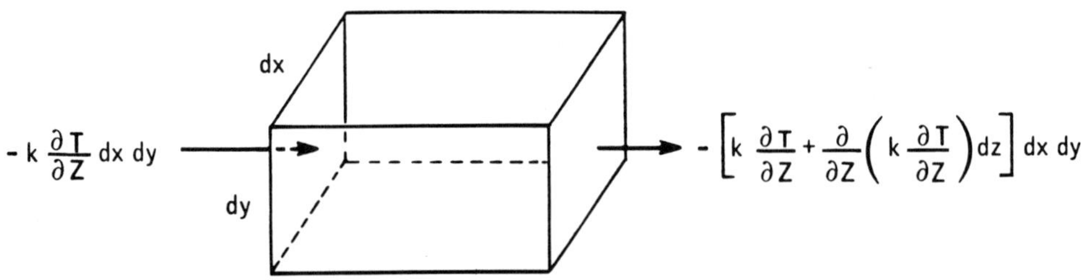

Figure 8 Analysis of heat transfer problem.

conduction may occur through all six faces of the element, depending on the symmetry of the temperature field.

Fourier's law describes the rate at which heat enters through the left face of the element by conduction:

$$-k \frac{\partial T}{\partial z} \, dx \, dy$$

The rate at which heat leaves by conduction through the right face is

$$-\left(k \frac{\partial T}{\partial z} + \frac{\partial}{\partial z}\left(k \frac{\partial T}{\partial z}\right) dz\right) dx \, dy$$

The rate of accumulation of heat within the volume element from these two sources is therefore

$$\frac{\partial}{\partial z}\left(k \frac{\partial T}{\partial z}\right) dx \, dy \, dz$$

From the top and bottom faces the accumulation rate is

$$\frac{\partial}{\partial y}\left(k \frac{\partial T}{\partial y}\right) dx \, dy \, dz$$

and from the front and rear faces it is

$$\frac{\partial}{\partial x}\left(k \frac{\partial T}{\partial x}\right) dx \, dy \, dz$$

Another expression for the rate of accumulation of heat in the volume element is

$$dx \, dy \, dz \, c_p \rho \frac{\partial T}{\partial t}$$

Thus, for constant thermal conductivity k,

$$\frac{\partial T}{\partial t} = \alpha\left(\frac{\partial^2 T}{\partial x^2} + \frac{\partial^2 T}{\partial y^2} + \frac{\partial^2 T}{\partial z^2}\right) \tag{38}$$

where α is the thermal diffusivity $k/\rho c_p$.

For one-dimensional conduction, in the z-direction for example, this equation reduces to

$$\frac{\partial T}{\partial t} = \alpha \frac{\partial^2 T}{\partial z^2} \tag{39}$$

Equation (39) is the basic equation for unsteady-state conduction in the radial (or z) direction.

The boundary conditions for this system are as follows:

$$\begin{aligned} T(z,0) &= T_\infty \\ T(0,t) &= T^* \\ T(\infty,t) &= T_\infty \end{aligned} \tag{40}$$

The temperature at the heat transfer surface ($z = 0$) is constant at T^*, as is the temperature T_∞ at z effectively equal to infinity. Equation (39) may be expressed as follows:

$$\frac{\partial T'}{\partial t} = \alpha \frac{\partial^2 T'}{\partial z^2} \tag{41}$$

where

$$\begin{aligned} T' &= T - T_\infty \\ T^{*\prime} &= T^* - T_\infty \end{aligned}$$

The boundary conditions then become:

$$\begin{aligned} T'(z,0) &= 0 \\ T'(0,t) &= T^{*\prime} \\ T'(\infty,t) &= 0 \end{aligned} \tag{42}$$

The Laplace transform of both sides is taken with respect to time t,

$$\int_0^\infty e^{-\rho t} \frac{\partial^2 T'}{\partial z^2} dt = \frac{1}{\alpha} \int_0^\infty e^{-\rho t} \frac{\partial T'}{\partial t} dt$$

$$\frac{d^2 F}{dz^2} = \frac{1}{\alpha} \rho F - T'(z,0) = \frac{\rho}{\alpha} F$$

where

$$F(z,\rho) = L_t[T'(z,t)]$$

whence

$$\frac{d^2F}{dz^2} - \frac{\rho}{\alpha} F = 0 \tag{43}$$

The second boundary condition transforms to $F(0,\rho) = T^{*\prime}/\rho$. The solution of the ordinary differential equation for F, subject to $F(0,p) = T^{*\prime}/\rho$ and F remains finite as $z \rightarrow \infty$, is

$$F(z,p) = \frac{T^{*\prime}}{p} e^{-\sqrt{\rho z^2/\alpha}}$$

Skelland (1983) obtained the function having this as its Laplace transform, thus providing an expression that gives the temperature in the fluid near the heat transfer surface as a function of position z and time t. The instantaneous heat flux from the surface is

$$\frac{q}{A} = -\alpha c_p \left(\frac{\partial T}{\partial z}\right)_{z=0} \tag{44}$$

Evaluating $(\partial T/\partial z)_{z=0}$ from Skelland's solution:

$$\frac{q}{A} = (T^* - T_\infty)c_p\rho \left(\frac{\alpha}{\pi t}\right)^{1/2}$$

The total thermal energy penetrating the fluid in an exposure time t_e is therefore

$$\int_0^{t_e} \frac{q}{A}\, dt = (T^* - T_\infty)c_p\rho\left(\frac{\alpha}{\pi}\right)^{1/2} \int_0^{t_e} t^{-1/2}\, dt$$

$$= 2(T^* - T_\infty)c_p\rho\left(\frac{\alpha t_e}{\eta}\right)^{1/2} \tag{45}$$

and the average rate of transfer during exposure is obtained by dividing the last expression by t_e,

$$\left(\frac{q}{A}\right)_{\text{av}} = 2(T^* - T_\infty)c_p\rho\left(\frac{\alpha}{\pi t_e}\right)^{1/2} \tag{46a}$$

or

$$\frac{(q/A)_{\text{av}}}{T^* - T_\infty} = h = 2\left(\frac{c_p\rho k}{\pi t_e}\right)^{1/2} = 1.1284\left(\frac{c_p\rho k}{t_e}\right)^{1/2} \tag{46b}$$

The time of exposure t_e is the time between passage of two consecutive blades. For n_b scraper blades rotating at N rps, $t_e = 1/n_bN$ where the blades are uniformly distributed around the shaft. Then the scraped surface heat transfer coefficient h is given by

$$h = 1.1284(c_p\rho k n_b N)^{1/2} \tag{47}$$

And in dimensionless form:

$$\frac{hD}{k} = 1.1284\left(\frac{D^2N\rho}{\mu_A} \cdot \frac{c_p\mu_A}{k}\, n_b\right)^{1/2} \tag{48}$$

Equation (48) shows that the heat transfer coefficient varies with rotary shaft speed N raised to the ½ power, which has been confirmed by Skelland (1958) using water data in the Votator obtained by Houlton (1944). Other investigators (Harriott, 1959; Skelland et al., 1962; Bott and Azoory, 1969) have studied fluids of low to moderate consistency or viscosity and obtained an exponent close to 0.5. With high consistency or viscosity fluids Skelland (1958), Bott and Azoory (1969), and Ramdas et al. (1980) found lower values for the exponent on N (0.113 ~ 0.18).

For *helical ribbon impellers* fitted with two rubber scrapers, Nagata (1975) gives the following correlation:

$$\frac{h_j D_T}{k} = 5.4\left(\frac{D^2N\rho}{\mu_A}\right)^{1/3}\left(\frac{c_p\mu_A}{k}\right)^{1/3}\left(\frac{\mu_A}{\mu_{AW}}\right)^{0.2} \tag{49}$$

This was developed from data obtained in a vessel having a diameter of 30 cm.

MASS TRANSFER

Mass transfer of *gas–liquid* non-Newtonian systems is reasonably well correlated by the following empirical equation:

$$\frac{k_L a D^2}{D_L} = 0.060\left(\frac{D^2N\rho}{\mu_A}\right)^{1.5}\left(\frac{DN^2}{g}\right)^{0.19}\left(\frac{\mu_A}{\rho D_L}\right)^{0.5}\left(\frac{\mu_A V_s}{\sigma}\right)^{0.6}\left(\frac{ND}{V_s}\right)^{0.32} \times [1 + 2(N'_{Wi})^{0.5}]^{-0.67} \tag{50}$$

where N'_{Wi} is the Weissenberg number

$$N'_{Wi} = \frac{\text{characteristic material time}}{\text{characteristic process time}}$$

k_La is the liquid phase mass transfer volumetric or capacity coefficient, a is the interfacial area per unit volume, V_s is the superficial gas velocity (length/time), and μ_A is the average apparent viscosity, obtained from the flow curve as described earlier:

$$\mu_A = \frac{\tau}{(du/dr)_A} = \frac{\tau}{k_s N} \tag{51}$$

For Newtonian fluids, μ_A becomes μ and the Weissenberg number equals zero. The numerator of the Weissenberg number is essentially the reciprocal of the shear rate at which the reduced complex viscosity, that is, the ratio of the apparent viscosity μ'_A to the zero-shear viscosity μ_0, is 0.67.

For *liquid–liquid systems,* Skelland and Seksaria (1978) report visual observations revealing the minimum impeller speed N_m at which one phase was completely dispersed as droplets in the other. This study produced the following empirical correlation:

$$\frac{D^{1/2}N_m}{g^{1/2}} = C_1\left(\frac{D_T}{D}\right)^{a_1}\left(\frac{\mu_c}{\mu_d}\right)^{1/9}\left(\frac{\Delta\rho}{\rho_c}\right)^{0.25}\left(\frac{\sigma}{D^2\rho_c g}\right)^{0.3} \tag{52}$$

The values of C_1 and a_1 are tabulated by the investigators for different impeller types and location. Depending on the type impeller and its position, the average percent deviation between predicted and observed N_m ranged from 4.28% to 18.15%. However, almost all the average deviations were below 12% for nearly 200 data points.

For agitation of two-phase Newtonian fluids, Treybal (1963) showed that the Power number–Reynolds number plot coincides with that for single-phase liquids when the Power and Reynolds numbers are defined as follows:

$$\rho_M = \rho_c\phi_c + \rho_d\phi_D \tag{53}$$

$$\mu_M = \frac{\mu_c}{\phi_c}\left(1 + \frac{1.5\ \mu_d\phi_D}{\mu_d + \mu_c}\right) \tag{54}$$

where ϕ_c and ϕ_D are volume fractions of the continuous and disperse phases.

The effects of a non-Newtonian disperse phase have not been extensively studied, although if the drops behave as rigid particles (high surface tension σ, low $\Delta\rho$, small diameter), then the effects of μ_d may not be important. In this regard, Harriott (1962) and Calderbank (1967) experimentally, and Azbel (1981) theoretically, state that small drops and bubbles behave as rigid spheres when suspended in agitated liquids.

CIRCULATION CAPACITY, CIRCULATION TIME, AND MIXING TIME

The pumping capacity of an impeller is primarily responsible for the overall flow around a vessel, generating fresh interface even within a single-phase liquid. The action of mixing is thus promoted by molecular and eddy diffusions. The pumping or circulation capacity may be measured by velocity traverses across the stream ejected by the impeller. More often, and with less accuracy, it is estimated by determining the average time for some suitable small particle, of neutral buoyancy, to complete one circulation loop within the vessel, as established by visual observation and stopwatch measurements.

In either case the results are correlated as the dimensionless circulation number N_{Ci}, defined as:

$$N_{\mathrm{Ci}} = \frac{q}{ND^3} \doteq \frac{V}{\theta_c ND^3} \tag{55}$$

where q is the volumetric discharge rate of the impeller, V is the volume of liquid in the vessel, and θ_c is the time for one circulation loop to be completed.

Many non-Newtonian mixing systems are confined to the laminar or transitional flow regimes. In these cases forms of helical ribbon or helical screw and draught tube devices are most often used.

Carreau et al. (1976) measured both circulation times (θ_c) and mixing times (θ_m) in high consistency Newtonian, pseudoplastic, and viscoelastic fluids stirred with helical ribbon impellers. The ratio θ_m/θ_c was about 3.5 for the Newtonian fluids, increasing by a factor of 2 or 3 for slightly elastic solutions, and increasing markedly for the highly elastic fluids (up to sevenfold).

Carreau and co-workers (1976) showed that the mixing time θ_m increased by factors of 3 to 7 as the elastic characteristic time increased. The pumping capacity of the helical ribbon was correspondingly reduced by increasing elasticity. Furthermore, in the absence of significant stagnation zones, it was found that the mixing effectiveness (defined as $V/ND^3\theta_m$ = 1/dimensionless mixing time), is directly related to the axial pumping capacity of the ribbon impeller.

For a *helical screw inside a draught tube,* Chavan and Ulbrecht (1973) examined axial discharge rates, circulation times, and axial velocities for Newtonian, pseudoplastic, and viscoelastic fluids. Again, the circulation number N_{Ci} was found to be independent of impeller speed, Reynolds number, and the rheology in the laminar region for Newtonian and pseudoplastic fluids ($D^2N^{2-n}\rho/K \leq 10$ or 100, depending on geometry), where power law behavior was observed. Elastic effects, however, reduced the circulation number by up to 50%, in a way that also depended on geometry. A straightforward correlation is provided by Chavan and Ulbrecht for the circulation number in terms of the geometry of the screw and draught tube unit. The empirical nature of the expression requires its application within the tabulated ranges of a large number of geometrical variables used in the study.

Ford and Ulbrecht (1975) obtained an empirical correlation for the dimensionless blending time, $N\Theta_B$, using a helical screw in a draught tube. The end point for blending was assessed by optical means in systems comprised of blends of equal amounts of polymer solutions having different rheological properties. The fluids included Newtonian, pseudoplastic, and viscoelastic materials, and the correlation takes the form of a plot of $N\Theta_B(\mu_{OM}/\mu_{OL})^{-z}(1 + 0.52N_{\text{Wi}})^{-1.25}$ versus $\text{Re} = D^2N\rho/\mu_a'$ with

$$\mu_a' = \mu_0 - mN^{b-1} \tag{56}$$

where μ_0, m, and b are parameters of the flow curve of the *final* mix.

The exponent z on the ratio (μ_{OM}/μ_{OL}) in the ordinate is related to the initial orientation of the two components to be blended, with a value $z = 0.059$ when the more viscous liquid is initially in the upstream position in the draught tube, $z = 0.17$ when the more viscous liquid is initially in the downstream position in the draught tube. Also, μ_{OM} and μ_{OL} are the zero-shear viscosity of the more and less viscous components, respectively. Skelland (1983) provides further discussion.

Chevan et al. (1975) examined the influence of pseudoplasticity and viscoelasticity on mixing. In the laminar region ($D^2N^{2-n}\rho/K < 10$) they found that $N\Theta_c$ and $N\Theta_M$ were constants for a given geometry, unaffected by the shear thinning viscosity behavior. Mixing time was observed to be a factor of 2 for viscoelastic materials. The interaction between shearing and normal stresses in viscoelastic fluids caused substantial reduction in the axial velocity component, with substantial increase in the angular component of velocity.

Ribbon–screw combinations were studied by Coyle et al. (1973), who found the mixing time of nonelastic liquids to be independent of the apparent viscosity, μ_A, above 15 Poise, with $N\Theta_M$ being constant for a given geometry in laminar flow. Measurements of both circulation and mixing times showed that $\Theta_M \doteq 3\Theta_C$.

In some cases the range of applicability of fan turbines and of helical impellers overlap, depending upon the fluid consistency involved and the degree of mixing considered satisfactory. Coyle et al. (1970) have compared the economics of alternative forms of agitation when satisfactory results are obtainable

Table 3 Relative Cost Comparison between Axial Flow Turbines and Helical Impellers for Equal Blend Times and Equal Heat Transfer Characteristics

Impeller type	Relative values			
	Hp	Speed	Torque	Initial cost
Axial flow turbine	1	1	1	1
Helical impeller	1/3–1/10	1/15–1/6	1.5–3.0	2.5–3.5

from either source. The relative cost data, based on 1970 dollars, are summarized in Table 3.

The flow patterns generated in Newtonian and purely viscous non-Newtonian fluids by axial and radial flow impellers (e.g., propellers and turbines) are qualitatively well known, consisting of recirculation loops described in standard mixing textbooks [see Azbel and Cheremisinoff (1983)]. With viscoelastic fluids, however, secondary flows not encountered with purely viscous fluids may arise. This is particularly true for rotating turbines [Kelkar et al. (1972) has simulated this behavior using rotating spheres]. The competing influence of shearing and normal stresses may cause a toroidal vortex to form above and below the rotating body, the vortices becoming segregated at a particular numerical combination of Weissenberg and Reynolds numbers [see, for examples, Chavan and Mashelkar (1980) and Ulbrecht (1974)]. Fluid trapped in these vortices fails to mix with the bulk of the material in the vessel, thereby reducing overall mixing efficiency and diffusion of ingredients.

In contrast, no such segregated vortices have been observed with propeller agitators. In this case the screw propulsion process apparently is enhanced by the viscoelasticity, thus accentuating the usual axial recirculation pattern as described by Ulbrecht (1974). This constitutes an advantage for axially thrusting propellers over turbines and paddles when agitating viscoelastic materials.

We shall now turn attention to the hardware of mixing non-Newtonian fluids.

BLADE MIXERS

Principles of Operation

Blade mixers are used to promote kneading and mixing action accompanied with heating or cooling of different deformable or plastic solids and high consistency pastes. Their operation involves compressing the fluid mass flat, folding it over on itself, and then compressing it again. The material is usually torn apart and high shear is produced between the moving and stationary fluid elements. The mixing is usually performed by two Z-shaped heavy blades rotating in opposite directions at different speeds on parallel horizontal shafts.

Typical blade designs are illustrated in Figure 9. Figure 9A shows the widely used sigma-blade employed for general-purpose kneading. There is great variation in commercially available blade designs, ranging from lightweight to heavyweight constructions. Selection of the specific design depends on the consistency of the mix. The size of the mixer is limited primarily by considerations of power input, weight, speed of mixing, materials of construction, and methods of mix discharge. Many machines are equipped with a tilting mechanism which facilitates the discharge of the batch while the blades are turning. Figure 10 shows a double-arm mixer in operation. For special mixing requirements, the sigma-blade mixer may incorporate special construction features, such as jackets for heating or cooling the troughs, dust-tight or vacuum-tight lids, and linings of special alloys for the mixing of corrosive or abrasive materials. Capacities of the equipment for handling heavy-consistency materials range from laboratory-sized batches to several hundred gallons and capacities for lighter-consistency material

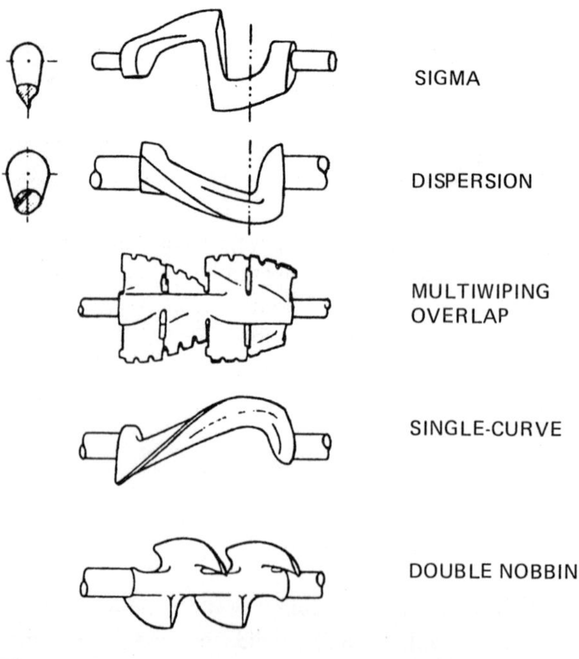

Figure 9 Different agitator blades for double-arm kneaders.

are much larger. In general, these mixers require very heavy drive mechanisms and motors.

The agitator blades in the sigma-blade mixer may be mounted so that their paths are either tangential or overlapping. In the tangential mounting arrangement the two blades rotate side by side with their circular paths of rotation not quite touching. In the overlapping arrangement, the paths of rotation overlap; and consequently the blades must be designed and the speed of rotation adjusted so that the blades always clear each other. In the tangential type the blades can have any relative speed because their paths of rotation do not overlap. Many modifications may be made to the shape of blades to meet specific mixing requirements. Many manufacturers provide double-arm mixers with interchangeable blades.

Examples of materials handled by double-arm kneaders of the sigma-blade type are resins, putty, adhesives, baker's dough, and cellulose additives. Sigma-type agitator blades are best for general-purpose use.

Figure 10 A 300-gal double-arm mixer designed for vacuum operation. The unit employs a direct drive through a gear reducer coupled to the front agitator shaft. Hydraulic tilt. (Courtesy of J. H. Day Co., Cincinnati, OH.)

The single-curve (or spiral agitator) blade, shown in Figure 9D, provides the same smooth blending and folding action as the dispersion blade (Fig. 9B) but has been designed for larger overlapping-type mixers. The dispersion agitator blades provide excellent mixing for fiber-reinforced products. They do not have the same heavy shearing action as other types, and therefore produce a smooth

Figure 11 A 50-gal double-arm mixer. Direct drive to front agitator shaft through gear reducer. Master hydraulic system operates both compression ram cover and bowl tilting mechanism. (Courtesy of J. H. Day Co., Cincinnati, OH.)

homogeneous mix with no fiber breakdown. An example of a single-curve (spiral agitator) double-arm mixer in operation is shown in Figure 11. The unit provides direct drive to the front agitator shaft through a gear reducer. The master hydraulic system operates both the compression ram cover and the bowl tilting mechanism.

The double nobbin agitator blade produces the extra shearing action required on certain materials. This agitator design resists blade distortion from stresses in the mixing load. Commercial designs are available in both overlapping and tangential action mixers. These units are most often used for heavy plastic materials. Figure 12 shows an example of this type of mixer.

Parametric Calculations

By optimizing important process parameters, any mixing system has the potential of providing high-quality mixing at minimum power consumption per unit volume of materials production. A proper mixer should also be of a low weight, compact, and durable. To optimize a design we first require a system of equations that will enable parametric calculations or sensitivity studies to be made.

Following the analysis of Pikovsky et al. (1960), consider a particle of material or fluid positioned on a mixer blade which is located at angle α to the

Figure 12 A 300-gal double-arm mixer with tank in tilted position. Unit has hydraulic tilt mechanism, antifriction bearings, and double nobbin agitator blades available in overlapping or tangential construction. (Courtesy of J. H. Day Co., Cincinnati, OH.)

plane of rotation. This fluid particle is under action of gravity force G and centrifugal force P,

$$P = \frac{G}{g} \omega^2 x \tag{57}$$

where x is the particle's distance from the axis of rotation (m), g is the gravitational acceleration (m/s^2), and P has units of kg$_f$. Under the action of these two forces there arise in the plane of particle motion the following forces.

A friction force of the opposing material against the bottom of the trough:

$$W_1 = \left(G \sin \varphi + \frac{G}{g} \omega^2 x\right) f, \quad \text{kg} \tag{58}$$

A component of the gravity force:

$$W_2 = G \cos \psi, \quad \text{kg}$$

A friction force against the blade:

$$W_3 = W_1 f \cos \alpha, \quad \text{kg}$$

A friction force due to the gravity force:

$$W_4 = W_2 f \cos \psi \cot \alpha, \quad \text{kg}$$

where

ψ = angle of blade rotation, rad
ω = blade angular velocity, rad/s
f = friction coefficient of particle against the metal

The total resistance is thus

$$W = W_1 + W_2 + W_3 + W_4$$

$$= G(1 + f \cot \alpha) \left(\frac{w^2 fx}{g} + f \sin \psi - \cos \psi\right), \quad \text{kg} \tag{59}$$

Figure 13 summarizes the forces acting on the particle.

An elemental amount of work dA spent for passing over the total resistance at the blade's rotation at an angle $d\psi$ is

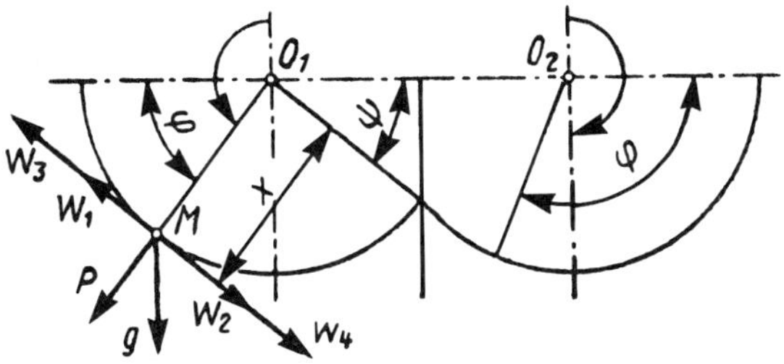

Figure 13 Forces acting on particle.

$$dA = (W_1 + W_3)R\, d\psi - (W_2 + W_4)x\, d\psi, \quad \text{kg-m} \tag{60}$$

The total work at the blade rotation from $\psi = 0$ to $\psi = \pi - \psi$ is

$$A = G(1 + f \cot \alpha)[fR(0.24\omega^2 x + 1.71) - 0.71x] \tag{61}$$

where

R = blade radius, m
ψ = extension angle of the blade from the bottom zone, rad

To compute the power, with respect to the dimensions of the blade, the following formula is recommended:

$$N = 0.105nzb\gamma' \sin \alpha(1 + f \cot \alpha)\,(CR^3 - r^3)x\,(0.0009fRn^2 - 0.24) + 0.86fR(R^2 - r^2) \tag{62}$$

where

N = power, kg-m/s
r = initial radius of the blade, m
b = blade width, m
n = rotational velocity, rot/min
z = number of blades
γ' = specific gravity of material, kg/m^3

The motion of a blade during mixing can be visualized by the analogy of a slab moving through a viscous liquid. Consider an elemental area dF (shown in Figure 14), where $dF = b\,dx$, m^2; b is the length of the elementary area, and dx

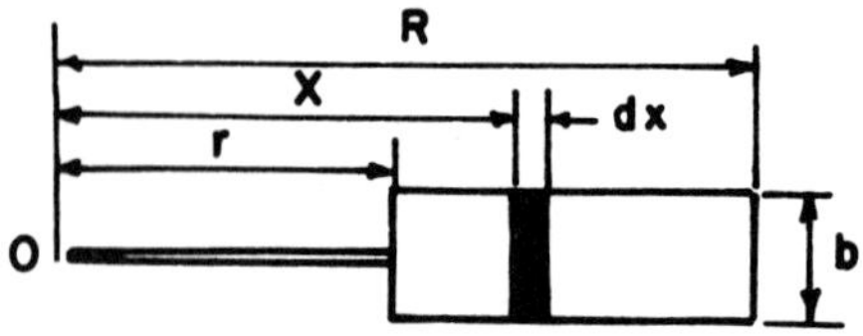

Figure 14 Slab moving through a viscous fluid.

is the width. According to Newton's law, the resistance acting on the elementary plane is

$$dP = C \frac{\gamma'}{g} dF v^2, \quad \text{kg}$$

where

γ' = liquid specific weight, kg/m^3
$v = x\omega$, rotational velocity of the plate, m/s
C = drag coefficient

Then

$$dP = C \frac{\gamma'}{g} b\omega^2 x^2\, dx, \quad \text{kg}_\text{f}$$

and after integration

$$P = C \frac{\gamma'}{g} b\omega^2 \int_r^R x^2\, dx = C \frac{\gamma'}{3g} b\omega^2 (R^3 - r^3) \tag{63}$$

The power required to rotate the blade is

$$N = \int_r^R dPv$$

or

$$N = \int_r^R C \frac{\gamma'}{g} b\omega^3 x^3\, dx = C \frac{\gamma'}{4g} b\omega^3 (R^4 - r^4), \quad \text{kg-m/s}$$

If the number of blades is $Z = 2$, then for a drive efficiency η_{eff} the required horsepower is

$$N = C \frac{\gamma'}{g} \left[\frac{\omega^3 (R^4 - r^4)}{410 \ \eta_{\text{eff}}} \right] \tag{64}$$

Kharkhuta et al. (1968) have experimentally determined that the coefficient C for asphalt–concrete mixtures is a function of the blade velocity n. At $n = 60$ to 70 rpm, $C = 6$; at $n = 70$ to 80 rpm, $C = 5$.

From dimensional analysis the following formula was derived for highway mixers:

$$N = 150 d^{4.56} n_0^{2.78} \left(\frac{\gamma'}{g} \right)^{0.78} \mu_a^{0.22}, \quad \text{kg-m/s} \tag{65}$$

where

d = circle diameter traced out by the blade
n_0 = rotational blade velocity, rot/s
μ_a = apparent mixture viscosity, kg-s/m^2

For approximate calculations of power, Kharkhuta et al. (1968) recommend the following formulas:

$$\text{For } Q_M < 1400 \text{ kg}, \qquad N = 0.035Q, \quad \text{kW} \tag{66a}$$

$$\text{For } Q_M > 1400 \text{ kg}, \qquad N = 30 + 0.01Q, \quad \text{kW} \tag{66b}$$

where Q_M is the mass of the mix in kilograms.

The capacity of a batch mixer is

$$G_M = \frac{Q_M}{T} \tag{67}$$

where T is the time for a single mix.

The capacity of a continuous mixer is

$$G_c = \frac{Q_M}{t_m} \tag{68}$$

where $Q_M = V\rho$, the mass of material in the mixer (kg), determined by the mixer volume V and fluid density ρ; t_m = mixing time.

Because the capacity is equal to the product of the cross-sectional area of the mixer F and its length l, the capacity expression may be restated as follows:

$$G_c = FS_d\rho, \quad \text{kg/hr} \tag{69}$$

where

F = cross-sectional area of material in the mixer, m^2
S_d = material displacement, m/hr
ρ = density of the mixture, kg/m^3

Capacity is the main parameter determining the basic elements of the mixer. From the capacity, the mass of the mix (Eqs. (67) and (68)) can be computed. Further, from the mass of the mix Q_M the volume of mixer trough can then be determined. For batch mixers the following formulas are used for computing the unit's dimensions of length L, width B, and height H:

$$\begin{aligned} L &= 750 + 0.67\,Q_M, \quad \text{mm} \\ B &= 830 + 0.37\,Q_M, \quad \text{mm} \\ H &= 448 + 0.385\,Q_M, \quad \text{mm} \end{aligned} \tag{70}$$

The blade radius as a function of mix mass is determined from the following equation:

$$R = \sqrt[3]{\frac{Q_M}{12.6\rho K_f}} \tag{71}$$

where

ρ = density of mixture, kg/m^3
K_f = filling coefficient
Q_M = mix mass, kg

The length of the trough can be computed from

$$L = 4.25R, \quad \text{m}$$

The width b_b and height h_b of the blades can be computed from

$$b_b = 0.42R, \quad \text{m}; \qquad h_b = 0.5R, \quad \text{m}$$

The blade pitch is normally established by the manufacturer, and a typical value is

$$t = 0.75, \quad \text{m}$$

The blade velocity is

$$n_b = \frac{45.3}{\sqrt{R}}, \quad \text{rpm}$$

The size of a continuous mixer as a function of the mix mass is

$$R = \sqrt[3]{\frac{Q_M}{(1.21 \sim 1.41)\gamma' z k_c}}, \quad \text{m} \tag{72}$$

Figure 15 shows the recommended dimensions of a blade mixer where the distance between two blade shafts is set at $l = 1.5R$, the trough width is generally $B = l + 2R$, the height of the part of the trough over the blade is $h = 1.35R$, and the height of the trough is $H = h + R = 2.35R$.

The length of the trough can be computed from the following formula:

$$L = \frac{Q_M}{\rho K_d F}, \quad \text{m} \tag{73}$$

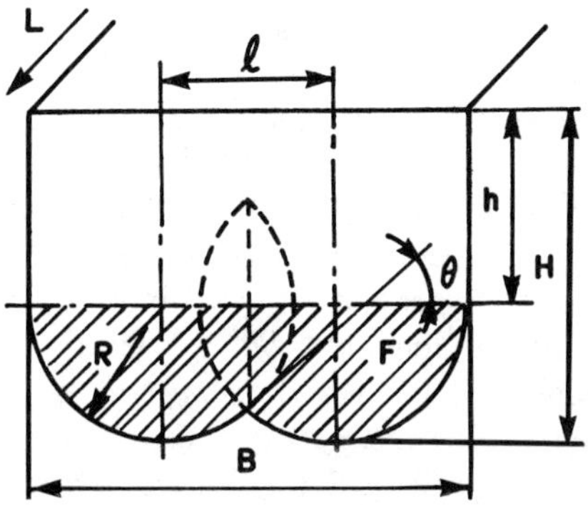

Figure 15 Recommended dimensions of a blade mixer.

where the area of mixing chamber is

$$F = R^2 \left[\pi - 0.5 \left(\frac{\pi\psi}{90°} - \sin 2\psi\right)\right], \quad m^2 \tag{74}$$

The shafts of mixers are designed for combined deformation from bending and torque moments. Increasing the rotational speed n_0 is practical only up to a certain limit; beyond this limit any increase in n_0 does not provide appreciable increases in mixing efficiency. The critical rpm is typically 180 to 200 for blade mixers. The approximate rpm may be estimated from the following relationships developed by Kharkhuta et al. (1968):

For a mix with 40 to 160 kg

$$n_0 = \left(\frac{100}{m} + 54.5\right), \quad \text{rpm}$$

For a mix with 160 to 360 kg

$$n_0 = \left(\frac{180}{m} + 46\right), \quad \text{rpm}$$

For any size mixer

$$n_0 = \left(\frac{9600}{m} + 52\right), \quad \text{rpm}$$

where m = mass of the mix, kg.

The rotational speed and the distance from the blade tip to the axis R_m establishes the velocity of the blade tip in the mixing material, which thus influences mixing efficiency, where R_{max} is typically 270 to 300 mm.

The area of one blade is S = 225 to 275 cm^2 for batch mixers and S = 150 cm^2 for continuous mixers.

The value of ψ and the distance between blades along the shaft determines the character of the helical surface and the rate of material displacement along the mixer. The smaller the value of ψ, the faster the material will pass through the mixer. For materials requiring substantial mixing time, $\psi = 90°$.

The formulas presented in this subsection can be used to establish preliminary dimensions for blade mixers. Parametric calculations can be made to establish a range of mixer sizes that will provide high-quality mixing at acceptable or minimum power consumption.

BANBURY MIXERS

Principles of Operation

There are a variety of kneading machines where the trough or container is open, so the contents may be unloaded by tilting. Some designs, however, have an enclosed mixing chamber, in which case the units are known as *internal mixers*. This type of mixer is often used for dissolving rubber and for dispersing rubber in liquids or in rubber compounding applications. The Banbury mixer illustrated in Figure 16 is a typical example. It is a heavy-duty double-arm mixer in which the agitators are in the form of interrupted spirals.

The Banbury mixer resembles a double-arm kneader in that it has two adjoining cylindrical sections which meet tangentially. In the Banbury design, however, the rotating agitators are much more rugged in construction and are able to apply considerably more power per gallon of working volume. Following Figure 16, two rotors (1) turn slowly toward each other at slightly different speeds. Each rotor has a blade which extends along the length of the rotor, roughly in the form of a spiral. Each rotor is cored to permit cooling or heating by the passage of cooling water or an appropriate heating agent. The mixing chamber (2) may also be cooled or heated, by means of the sprays (3). The materials to be mixed are fed into the mixing chamber through the feed hopper (4). A floating weight (5) is operated by compressed air and rests on top of the feed, serving to confine the material to the mixing space and to exert pressure on it. The mixed batch is discharged by a power-operated sliding discharge mechanism (6). The saddle (7) between the two rotors, which is attached to the sliding mechanism, slides out from under the rotors, leaving an opening through which the mix discharges.

Because of the high degree of agitation a Banbury mixer can complete a mix in a relatively short time. It is capable of reducing to a matter of minutes mixing operations requiring hours in other types of mixers. Examples of applications include the mixing of wood flour and linseed oil in linoleum manufacture, and the incorporation of fillers, softeners, and accelerators into rubber stock. The mixing of viscoelastic polymeric materials such as rubber with different ingredients is considered a mechanical process where under different shear deformations the dispersion of ingredients takes place. In addition to the mechanical processes of mixing, physicochemical processes take place between the material, fillers, and other ingredients.

The mixing process is normally carried out under laminar conditions of material flow, created by the rotating parts of the machine. During this action the non-Newtonian material is subjected to different stresses and strains. The selection of the mixing regime depends on several parameters, such as the amounts of materials loaded, the order of loading, and the loading times, the time of a mix,

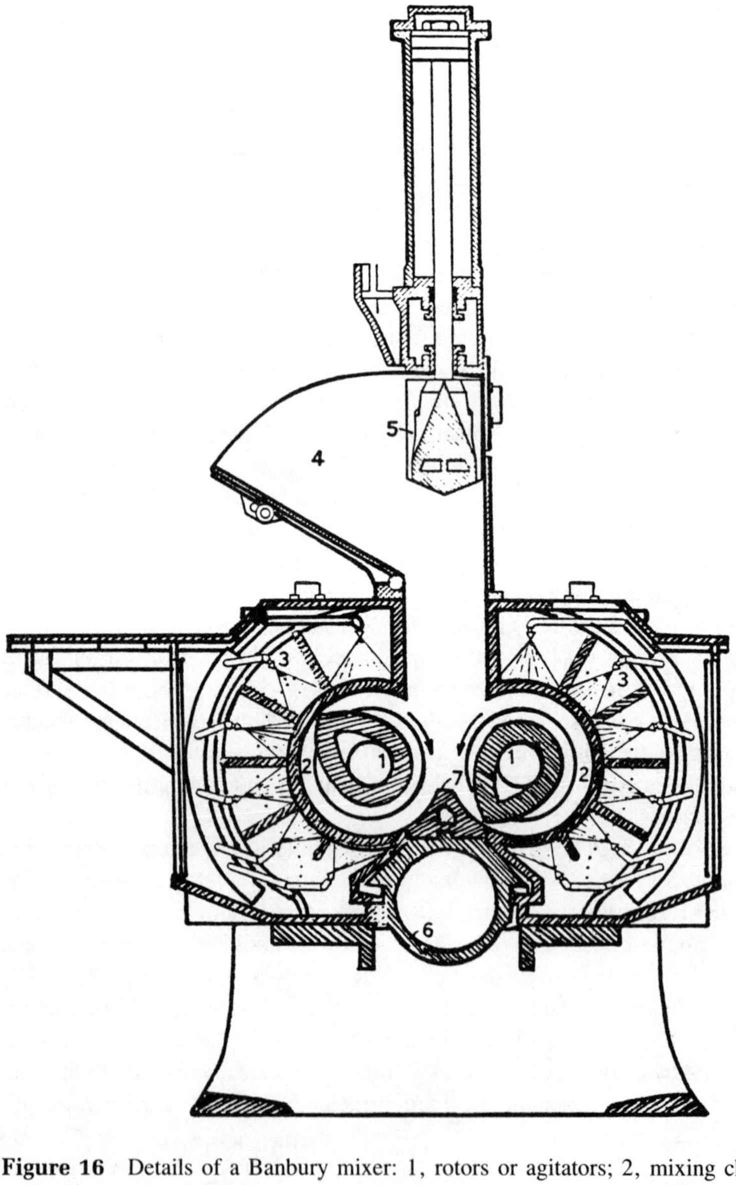

Figure 16 Details of a Banbury mixer: 1, rotors or agitators; 2, mixing chamber; 3, cooling sprays; 4, feed hopper; 5, floating weight; 6, sliding discharge; 7, saddle discharge opening.

the rotational speed of the blades, and the pressure in the chamber. By varying these parameters the process may be considerably intensified. However, the basic source of intensification is an increase of blade velocity and an increase of the pressure in the mixing chamber. An increase in the rotational velocity of the blades increases the shear velocity, which leads to more frequent dividing of the mass and renewal of its surface. This promotes rapid distribution of ingredients in the mass. An increase in rotational velocity leads to temperature rises in the mixture as well as increase in the pressure exerted against the walls of the chamber.

The largest amounts of mixing energy are consumed in applications involving viscous components and when filler materials are added in the form of powders to viscous solutions. The energy consumption in these cases is variable. It depends on various technological factors as well as the physicochemical properties of components. Figure 17 shows a typical time plot of mixer power consumption. The area limited by the power consumption curve and x-axis is equal to the energy consumption per mixing cycle. A plot such as Figure 17 provides an evaluation of the intensity of the process as well as an indication of the distribution of the ingredients in the paste. The energy consumption for a given amount of mixture of specified composition and temperature is constant and independent of the rotational velocity of the blades.

For Newtonian fluids a good approximation is to assume that the drive horsepower is proportional to the rotational velocity. For non-Newtonian materials, however, this is not a valid assumption because nonlinear relationships exist between shear stress and shear rate. Furthermore, at increased speeds the rate of shear deformation also increases, which promotes temperature rises due to energy dissipation. Heat transfer promoting mixture cooling is generally poor; thus an increase in temperature of the mixture results in a lowering of the fluid's viscosity. This, in turn, decreases the energy consumption of mixing. It should be noted that the power of the drive and the energy consumption depend on the

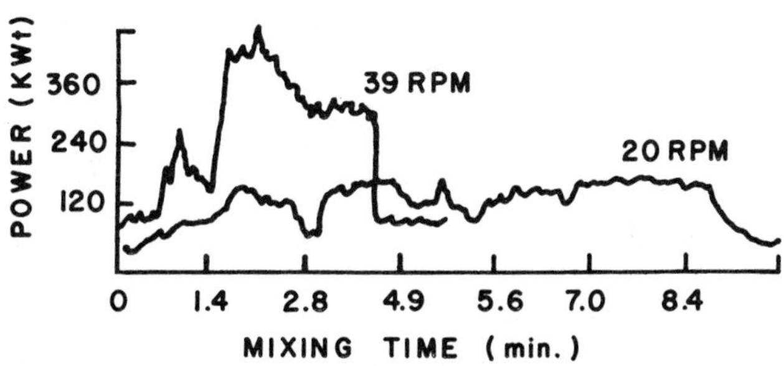

Figure 17 Time plot of mixer power consumption.

order of loading mixture components. For materials loadings in succession, a decrease in the time intervals between loadings of discrete portions demands power increases due to a shortening of the mixing cycle. The maximum power of the drive occurs when all materials are loaded simultaneously. Pressure increases over the mixture influence the energy consumption when the chamber is totally filled. This is especially true when loading hard mixtures. At the same time, energy consumption increases but not proportionally with pressure increases. Pressure rises cause temperature increases in the mixture, resulting in delaying of the power consumption. During this action the specific energy consumption does not undergo significant changes but, rather, gradually decreases with mixing time.

Parametric Calculations

Mixing theory of non-Newtonian materials is complex and not completely understood. Consequently, a calculation procedure for establishing the drive horsepower cannot be recommended. As a rule, this parameter must be established on the basis of experimental data and information obtained from mixers of similar designs.

Attempts at establishing guidelines for the power consumption in closed mixers were made by Bernhardt (1962), who considered the deformation process of viscous liquids when shear takes place between the tip of a blade and the wall of the chamber. The material was assumed to be of constant viscosity, the radius of the chamber was large in comparison to the gap, and the process was isothermal.

These assumptions, particularly constant viscosity and temperature, prohibit an accurate evaluation of the power consumption. They do, however, enable the system evaluation to be made qualitatively.

In the case where the gap h between the blade tip and chamber wall is constant, the velocity distribution along the gap is

$$\nu = u\left(1 - \frac{u}{h}\right) \tag{75}$$

where

u = peripheral velocity of the blade tip
y = coordinate

The velocity varies linearly from $\nu = 0$ at $y = h$ (at the wall of the chamber) to $\nu = u$ at $y = 0$ (at the blade surface). In this case the shear stress is

$$\tau = \mu \frac{dv}{dy} = -\mu \frac{u}{h} = 2\mu \frac{Q}{h^2} \tag{76}$$

where Q is the volumetric rate related to a unit length of blade in the axial direction. As h is a constant, the shear stress is also constant along the channel. The power per unit length is the product of the resistance force to the blade motion and velocity:

$$N = u \int_0^L \tau \, dx = \mu u^2 \frac{L}{h} \tag{77}$$

where L is the blade width.

Thus, the power for a rectilinear channel per unit length of blade is

$$N = 4\mu Q^2 \frac{L}{h^3} \tag{78}$$

Assuming the gap varies linearly along the length $L(x)$, then

$$h = h_0(1 - cx) \tag{79}$$

The shear stress will be

$$\tau = 2\mu \frac{Q}{h_0^2} \left(\frac{1}{1 - \beta\rho}\right)^2 \tag{80}$$

where

$$\beta = \frac{\alpha - 1}{\alpha}, \qquad \rho = \frac{x}{L}; \qquad \alpha = \frac{h_L}{h_0}$$

and the power is

$$N = 15.5\mu Q^2 \frac{L}{h_0^3} \tag{81}$$

Thus, the shear stress and power consumption are considerably higher in an inclined channel.

Equation (78) may be rewritten as

$$N = 4\mu u^2 \frac{L}{h_0} \tag{82}$$

It is assumed in this case that the channel has a constant depth, from which the shear stress is

$$\tau = \mu \frac{u}{h_0^3} \tag{83}$$

The product μu may be assumed to be a constant. Using the preceding equations Bernhardt et al. (1962) assumed that the shear stress in the closed mixer depends only on the gap size between the blade and the wall. However, the assumption that μu is a constant is valid only as a first approximation. For an alternative approach for estimating the power consumption Guber (1966) recommends a power law for the shear stress

$$\tau = K\dot{\gamma}^n \tag{84}$$

where

τ = shear stress, kg/cm²
K and n = rheological constants
$\dot{\gamma}$ = shear rate, s^{-1}

Equation (84) has been found to represent rubber mixtures over moderate ranges of shear rate reasonably well. Keeping in mind that the shear rate in the mixer is proportional to the rotational rate of the blades and an average stress is proportional to the power consumed, the following analysis can be applied to estimating the power consumption of the mixer.

The specific energy spent in the deformation of a unit volume of material is

$$N_{sp} = \tau\dot{\gamma} \tag{85}$$

The total power of the drive is

$$N = N_{sp} \times V \tag{86}$$

And from Eqs. (84), (85), and (86) we obtain

$$N = k\dot{\gamma}^{n+1}V \tag{87}$$

where

N = horsepower, kg-cm/s
V = volume of mixture, cm^3
$\dot{\gamma}$ = average shear rate, s^{-1}

The average shear rate of the material in the gap is

$$\dot{\gamma} = \frac{u}{h} \tag{88}$$

where

u = peripheral velocity of the blade's ridge, cm/s
h = gap between the blade tip and chamber wall, cm

Guber (1966) has verified this relation experimentally. Guber also developed the following set of equations for Banbury mixers that can be used for parametric calculations. The section of material in the gap is divided into a series of elements; then the power is determined for each element and summed as follows:

$$N = 2k\left(\frac{\pi n_{av}}{60}\right)^{n+1} \{[l_1(0.5^{n+1} + 0.865^{n+1}) + 2\, l_2 \times 0.705^{n+1}]F' + (D_r/h_0)^{n+1}h_0\delta\,(l_1 + l_2)\}, \quad \text{kg-cm/s} \tag{89}$$

where

n_{av} = average velocity of rear and front blades, rpm
l_1 = length of the long blade ridge, cm
l_2 = length of the short blade ridge, cm
D_r = diameter of blade ridge, cm
h_0 = gap between the blade ridge and the chamber, cm
δ = width of the blade ridge

Parameter F' is given by

$$F' = \sum_{i=1}^{j}\left(\frac{D_k}{h_i} - 2\right)^{n+1} f_i \tag{90}$$

where

i = number of elements which the deforming volume is divided into, or the width of deformation zone of the material in the gap between the blade and the chamber
D_k = chamber inside diameter, cm
h_i = gap between the blade ridge and the chamber for an element, cm
f_i = cross-sectional area of the ith element, cm^2

Guber gives values of F' as a function of the deformation zone width for different rheological constants n. For geometrically similar equipment, the following relationship applies:

$$\frac{N''}{N'} = \frac{k''\left(\frac{\pi n''_{av}}{60}\right)^{n''+1} u^3\alpha}{k'\left(\frac{\pi n'_{av}}{60}\right)^{n'+1}} \tag{91}$$

where

u = geometrical scale
α = F''/F' coefficient considering the rate of filling (space factor)

The prime (′) denotes the model and (″) denotes the prototype. When both mixers are filled with the same mixture and operated at the same temperature, Eq. (91) simplifies to the following:

$$\frac{N''}{N'} = \left(\frac{n''_{av}}{n'_{av}}\right)^{n+1} \alpha u^3 \tag{92}$$

Experiments have shown that the pressure in the mixer develops between the gap and that the maximum mixture pressure arises for the minimum gap. At any point along the gap the pressure is approximately three times as great as the shear stress at the corresponding point. The following equations were obtained for determining these variables:

$$\begin{aligned} P_{av} &= 3\tau_{av} = 3K\dot{\gamma}^n_{av} \\ \tau_{av} &= \frac{2M_t}{3V} = \frac{10N}{\pi n_{av} V} \\ P_{av} &= \frac{2M_t}{V} = \frac{30N}{\pi n_{av} V} \end{aligned} \tag{93}$$

where

P_{av} = average mixture pressure in the chamber, kg/cm²
τ_{av} = average shear stress of deforming mixture, kg/cm²
M_t = blade torque moment, kg-cm
V = volume of deforming mixture, cm³
n_{av} = average blade velocity, rpm

For rubber mixtures of different compositions the horsepower versus velocity and floating weight is approximately linear, as shown by Zakharkin et al. (1966) and Starov (1961). However, in power–velocity coordinates the slope of the approximating curve exceeds that in the coordinates of power–pressure. Thus, mixing intensity shows a greater dependency on velocity than pressure. From a mixture containing 30% carbon black and 30% chalk as fillers, the following empirical relation was obtained:

$$W = A + Bp \qquad \text{and} \qquad W = C + Dn_0 \tag{94}$$

where

W = horsepower, kW
p = pressure on the mixture, kg/cm^2
n_0 = rotational velocity of the rear blade, rpm

The values of the coefficients in Eq. (94) are given in Table 4.

From a series of experiments, the horsepower calculation of any similar type of mixer, where it is necessary to change the operating regime, may be made by using the following relationship:

$$N = N_0 + \frac{\Delta P_g}{\Delta P_d} \Delta N_p + \frac{\Delta n_g}{\Delta n_d} \Delta N_d \tag{95}$$

where

N = horsepower at the new operating regime, kW
N_0 = horsepower at the existing operating regime
ΔN_d = increase in horsepower in the model
ΔP_g, Δn_g = the prototypes degree of increase in pressure and rpm, %
ΔP_d, Δn_d = the model's degree of increase in pressure and rpm, %

Table 4 Values of Coefficients in Eq. (94)

Rear blade velocity (rpm)	Velocity limits (rpm)	Specific pressure on mixture (km/cm^2)	Limits of specific pressure (km/cm^2)	A	B	C	D
32	32–100	0.66	0.66–1.32	4.05	0.57	1.25	0.1
50	32–100	0.88	0.88–1.32	5.26	1.41	1.8	0.1
80	32–100	1.32	0.88–1.32	7.76	2.41	1.25	0.12

The following example problem illustrates some of the principles outlined above.

Example 1 The blade speed of a 140-liter mixer is 30 rpm, and the pressure in the mixture is 5 kg_f/cm^2. The horsepower is 630 kW. Determine the horsepower required to operate in a different mixing regime, where the speed is 40 rpm and the pressure is 6.65 kg/cm^2.

Solution. From the graph developed by Starov (1961) of $N = f(P)$ for a model mixer, we determine the horsepower increase (ΔN) and pressure increase (ΔP), in %:

$$\Delta P = \frac{P_2 - P_1}{P_1} 100 = \frac{1.32 - 0.88}{0.88} \times 100 = 50\%$$

$$\Delta N = \frac{N_2 - N_1}{N_1} 100 = \frac{4.8 - 4.5}{4.5} \times 100 = 6.67\%$$

The pressure coefficient is

$$K_p = \frac{\Delta N}{\Delta P} = \frac{6.67}{50} = 0.133$$

From a plot of $N = f(n)$ for the model:

$$\Delta n = \frac{n_2 - n_1}{n_1} \times 100 = \frac{40 - 30}{30} \times 100 = 33.3\%$$

$$\Delta N' = \frac{N_2' - N_1'}{N'} \times 100 = \frac{8 - 6}{6} \times 100 = 33.3\%$$

The velocity coefficient is

$$K_n = \frac{\Delta N'}{\Delta n} = \frac{33.3}{33.3} = 1$$

The change in pressure for the new regime is

$$\Delta P_d = \frac{6.65 - 5}{5} \times 100 = 33\%$$

The required horsepower of the prototype for this increase in pressure is

$$\Delta N_d = \Delta P_d K_d = 33 \times 0.133 = 4.4\%$$

that is, the required increase in horsepower for an increase in pressure of 33% is 28 kW.

The change in rotational velocity in the prototype is

$$\Delta n_d = \frac{40 - 30}{30} \times 100 = 33.3\%$$

The increase in horsepower corresponding to this increase in velocity is

$$\Delta N_d = \Delta n_d \times k_n = 33.3 \times 1.0 = 33.3\%$$

Thus, the horsepower for the new regime is

$$N = 630 + 28 + 210 = 868 \text{ kW}$$

This is an approximate calculation; a more accurate estimate of the horsepower can be made by using Eq. (95). To intensify mixing at the expense of greater velocity and pressure requires more powerful drives and, consequently, greater power consumption. With large capacity machines the main portion of energy introduced is dissipated, and provisions must therefore be made for the evacuation of large amounts of heat. Thus, calculation of a thermal regime is of great importance. The primary reason for a thermal design is to ensure that the mixture temperature does not exceed a prespecified level that may harm the process or the products.

The largest portion of heat generated during mixing arises from the intense shearing action imposed on the viscous material. It is often assumed that heat generation is constant during the mixing cycle for a fixed rotational velocity. This assumption must, however, be applied with reservations because as mixing proceeds the properties of the viscous mass change. Kapitonov (1968) reports that the temperature of the mixture first increases sharply, then stabilizes because of an equilibrium established between heat generation and evacuation by an appropriate cooling medium (e.g., water).

Most of today's commercial mixers operate in an intensive regime, and thus the mixing cycle is shorter than the time needed to arrive at a stable or equilibrium-state regime. Calculations must therefore be based on the principles of unsteady heat transfer. For design purposes a relationship between the mixture temperature t_m, heat transfer coefficient h, and cooling water flow rate G_w is needed. The temperature interval can be assumed to be so small that the mean arithmetic temperature can be applied under the condition that

$$\frac{\Delta t_{\text{fin}}}{\Delta t_{\text{in}}} < 2$$

where Δt_{in} and Δt_{fin} are initial and final temperatures of the mixture and cooling medium. The temperature of the mixture at the end of the ith time interval is

$$t_{m_i} = \frac{Q_g - Q_{\text{ev}_i}}{M_m C_m} + t_{m_{(i-1)}} \tag{96}$$

The amount of heat evacuated by the cooling medium from the mixture in time $\Delta\tau_i$ is

$$Q_{\text{ev}} = KF\, \Delta\tau_i(t_{m_i} - t_{w_i}) \tag{97}$$

After some transformations the following expressions are obtained:

$$Q_{\text{ev}} = \frac{\Delta\tau_i h\, FM_m C_w\, (t_{m_i} + t_{m_{(i-1)}} - 2t_{w_i})}{2M_w C_w + \Delta\tau_i h\, F} \tag{98}$$

and

$$t_{m_i} = \frac{Q_g\, (2M_w C_w + \Delta\tau_i h\, F) + 2\Delta\tau_i h\, FM_w C_w t_{w_i}}{M_m C_m(2M_w C_w + \Delta t_i h\, F) + \Delta t_i h\, M_w C_w} + \frac{M_m C_m(2M_w C_w + \Delta\tau_i h\, F) - \Delta\tau_i h\, M_w C_w}{M_m C_m(2M_w C_w + \Delta\tau_i h\, F) + \Delta\tau_i h\, FM_w C_w}\, t_{m_{(i-1)}} \tag{99}$$

where

Q_g = amount of generated heat in the mixture for $\Delta\tau_i$ time interval
h = average heat transfer coefficient from the mixture to the water
F = total heat transfer area
M_w = water flow rate for $\Delta\tau_i$
C_w = specific heat of water
t_{w_i} = initial water temperature

Denoting the polynomials in Eq. (99) as A and B, respectively, the equation for the first time interval is simply

$$t_{m_1} = \mathrm{A} + Bt_{\text{min}} \tag{100a}$$

For the second time interval $t_{m_{(i-1)}} = t_m$:

$$t_{m_2} = A + AB + B^2 t_{\text{min}} \tag{100b}$$

and for the nth time interval:

$$t_{m_n} = A(B^0 + B^1 + B^2 + \cdots + B^{n+1}) + B^n t_{m_i} \tag{100c}$$

From Eq. (99), the maximum mixture temperature at steady state, considering the given cooling regime, can be determined as follows:

$$t_{m_{\max}} = Q_g \left(\frac{1}{\Delta\tau_i h\, F} + \frac{1}{2M_w C_w} \right) t_{w_i} \tag{101}$$

The output of a Banbury mixer is determined by the mass of loaded mixture, which in turn depends on the mixing chamber volume:

$$M = 60\, \frac{V\rho}{\tau} \psi, \quad \text{kg/hr} \tag{102}$$

where

V = volume of loading, m^3 (typically 55–60% of the chamber volume)
ρ = density of the mixture, kg/m^3
ψ = operation factor (assumed to be equal to 0.8–0.9)
τ = cycle duration, min

The cycle duration is

$$\tau = \tau_1 + \tau_2 + \tau_3 + \tau_4 + \tau_5 + \tau_6 + \tau_7 \tag{103}$$

where

τ_1 = loading time
τ_2 = time for closing feed hopper
τ_3 = time for opening of hopper gate
τ_4 = mixing time
τ_5 = time for opening of discharge mechanism
τ_6 = time for closing of discharge mechanism
τ_7 = unloading time

Mixing time τ_4 is the main component and depends on the process technology and its intensity. All other components of the mixing cycle are considered to be auxiliary operations. The smaller the mixing time, the more severe are the demands for auxiliary operations. If the mixing time is 3 min, the auxiliary operations should not exceed 30 to 40 s.

To increase loading operations, mechanical transportation and automatic techniques for dosing ingredients are employed. These mixers are provided with pneumo- or hydrodrives.

Consider the design of a *hydraulic drive gate*. The equation describing the gate motion is

$$m \frac{dv}{dt} = P_d - P_{\text{r.c.}} - P_r \tag{104}$$

or

$$m \frac{d^2y}{dt^2} = P_d - P_{\text{r.c.}} - E' \tag{105}$$

where

m = mass of moving parts
$v = dy/dt = y'$ − gate velocity
y = gate path
d^2y/dt^2 = gate acceleration
P_d = driving force from the pressure of the working liquid on the piston of the hydrocylinder
$P_{\text{r.c.}}$ = constant component of resistance forces of the gate
P_r = variable component of resistance forces of the gate

The driving force is

$$P_d = P_1 \frac{\pi}{4} (D^2 - d_1^2) \tag{106}$$

where

P_1 = pressure of working liquid in the cylinder
D = inside diameter of the cylinder
d_1 = piston rod outside diameter

Pressure d_1 is determined from Bernoulli's equation, which is applied over two sections: the section at the pump discharge at nominal pressure P and the section at the outline in the gate–cylinder space at P_1; then

$$P_d = A' + \nu_1^2 C' \tag{107}$$

where A' and C' = coefficients from Kapitonov (1968); ν_1 = speed of the cylinder. The constant component of forces which resist the gate motion $P_{r.c.}$ arises from friction forces in the guides, and the friction force of the mixture against the sliding gate is

$$P_{r.c.} = \mu_1 P_0(l - y)s + \mu_2[G + P_0(l - y)s] = B \tag{108}$$

where

P_0 = pressure of the mixture in the chamber
μ_1 = mixture friction against the gate
μ_2 = friction against the guides
s = width of sliding gate
G = weight of gate moving parts
y = length of the path covered by the gate

Despite variation in mixture loadings [which is equal to $P_0(l - y)s$], it is common practice to assume the load to be constant both during mixing and mixture unloading.

During unloading operations, there are no sharp reductions in mixture level, and the amount of the mixture to be unloaded is proportional to the increasing area of the discharge opening. The variable component of the resistance force P_2 is determined by hydraulic resistances:

$$P_2 = E' + F'u_1^2 \tag{109}$$

where E' and F' = coefficients. Taking into account Eq. (109) we rewrite Eq. (105) as follows:

$$\frac{G}{g} y'' = (C' - F')(y')^2 = A' - B' - E' \tag{110}$$

or

$$y'' + A'(y')^2 = B' \tag{111}$$

where

$$A' = \frac{g}{G}(F' - C') = \text{constant}$$

and

$$B' = \frac{g}{G}(A' - B' - E')$$

Solving Eq. (111) relative to time, we obtain:

$$\tau = \sqrt{\frac{A'}{B'}}\,L, \quad \text{s} \tag{112}$$

and the gate velocity becomes

$$y' = \nu_1 = \sqrt{\frac{B'}{A'}} = \text{constant} \tag{113}$$

To decrease the time of auxiliary operations (in this case the time of gate work) an increase in the pressure of the working liquid, a decrease in the hydraulic resistances, friction factor of the mixture against the gate, friction factor of the guides as well as the weight of the moving parts of the gate are recommended.

MIXING ROLLS

Principles of Operation

Mixing rolls subject pastes and deformable solids to intense shear by passing them between smooth or corrugated metal rolls that revolve at different speeds. These machines are widely used in the rubber and plastics industries. The principal design consists of two horizontal rolls or cylinders, arranged side by side and rotating toward each other at different speeds. The ratio of the peripheral speeds of the rolls, known as the friction ratio, ranges from 1 to 2 but is usually around 1.2. The higher friction ratios leads to greater heat generation. Friction, speed, and the sizes of rolls influence the cooling of the material mass and the intensity of its treatment. As such, data obtained from a unit is specific to that machine and cannot be applied to a different one.

Figure 18 illustrates the basic mixing operation. The material enters the mixing rolls in the form of lumps, powder, or friable laminated materials. As a result of rotation, adhesion, and friction, the material is entrained into the gap between the rolls, and upon discharge it sticks to one of the rolls, depending on their temperature difference and velocities. The rolls are temperature controlled. The rolling process is also influenced by the gap between the rolls. In batch mixing rolls the mass after loading passes through the gap between the rolls several times and the mixing action is due to the different speeds of the rolls

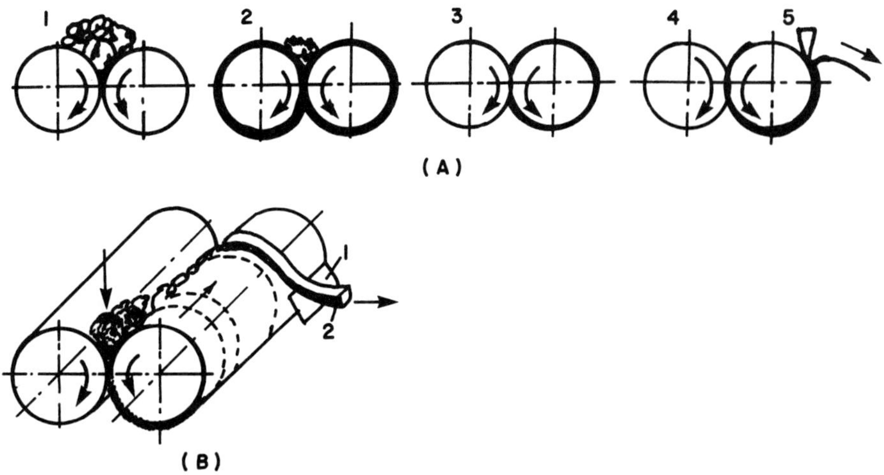

Figure 18 Operation of mixing rolls: (A) Batch operation (1, loading; 2, rolling; 3, end of rolling; 4, mass shear; 5, knife). (B) Continuous operation (1, knife; 2, continuous discharge of mass).

(Figure 18A). In continuous mixing rolls the mass enters continuously from one side of the machine and passes between the rolls in rotational and forward motion along the unit's axis. The mix is continuously discharged in narrow strips (refer to Figure 18B). Both the shearing action and entrainment of material into the gap are very important to the mixing process and in transporting the material through the unit.

Parametric Calculations

Shearing strain is required for initial mixing of regular or isolated systems consisting of two components. The objective of this strain is to mix the system as a determined statistical variation of any of the properties of a series of samples to a minimum. According to mixing theory, three basic principles may be formulated:

1. The total component surface contact area should increase during mixing.
2. The elements of surface contact area should be evenly distributed in the mixing mass.
3. Mixture components must be distributed such that for any element of volume the relative content of components is the same as in the total system.

Parametric calculations for mixing rolls make it possible to estimate the basic parameters of the machine, namely, its productivity, the thrust forces between rolls, torque, and horsepower. The calculation of these parameters marks the starting point of the structural design of mixing rolls. The questions of time and quality of mixing, the required number of passes of the mixture through the rolls, as well as other mixing characteristics, must be established from industrial and laboratory experiments.

The issues outlined below are related to deformation and flow of polymeric non-Newtonian materials between the gap of the mixer rolls. An examination of the flow behavior in this region will provide insight into computing the basic parametric values which characterizes the operation of mixing rolls. Figure 19 shows the basic scheme of the rolling process. The rolls act on the material by compressing and pushing it along the machine's horizontal axis. We define the rolls as having radius R, as being located at a distance $2h_0$ from each other, and as rotating at velocity ν. The material wedge thickness at the gap entrance is $2h_2$, and at discharge, $2h_1$. The hydrodynamic theory of this process is covered by Bernhardt (1962), Soroka et al. (1965), Bekin and Nemytkov (1966), and Lukach et al. (1967), where it is shown that the streamline equation is obtained by integration of the following stream function:

$$\psi = \int \nu_x \, dy \tag{114}$$

where

ψ = streamline (ψ = const)
ν_x = velocity component in the x-axis direction
ν_y = velocity component in the y-axis direction determined from the continuity equation:

$$\frac{\partial \nu_x}{\partial x} + \frac{\partial \nu_y}{\partial y} = 0 \tag{115}$$

Defining the following dimensionless parameters [from Bernhardt (1962) and Lukach et al. (1967)]:

$$\rho = \frac{x}{\sqrt{2Rh_0}} = \pm\sqrt{\frac{h}{h_0} - 1} \tag{116}$$

and

$$\eta = \frac{y}{\sqrt{2Rh_0}}$$

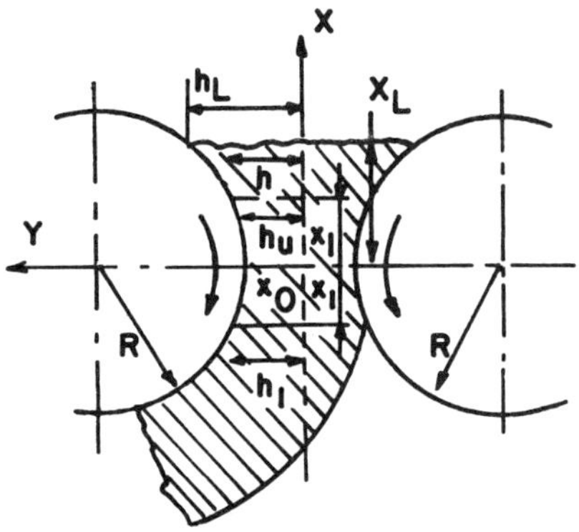

Figure 19 Mechanisms governing rolling process.

The velocity in the x-axis direction is

$$\nu_x = \frac{3\nu}{2}\left\{\left[\frac{\rho^2 - \rho_1^2}{(1 + \rho^2)^3}\right]\frac{4h^2}{C^2} + \frac{2 - \rho^2 - 3\rho_1^2}{3(1 + d^2)}\right\} \tag{117}$$

where ρ_1 is the value of ρ at the gap outlet and $C = \sqrt{2h_0R}$ = constant. Substituting the value of ν_x from Eq. (117) into Eq. (114) and integrating using the boundary conditions ($\psi = 0$ at $\nu_x = y = 0$), we obtain:

$$\psi = \frac{\nu h_0 \mu}{C}\left[\frac{4h^2\,(\rho^2 - \rho_1^2)}{C^2\,(1 + \rho^2)} + \frac{2 + 3\rho_1^2 - \rho^2}{1 + \rho^2}\right] \tag{118}$$

The results of this expression are presented in Figure 20 for a series of ρ and η values. As shown in Figure 20, near the roll surface the material flows parallel and the roll surface itself is the streamline. At the central zone located at the wedge entrance, countercurrent flow is observed. For a wedge of a final size, the countercurrent pattern takes the form of two closed eddies. This is the amount of material which may be observed in rolls of equal peripheral velocities. The material passing out of the gap sticks to one of the rolls rotating with it, and again touching with the portion of the wedge where rotational motion takes place in order to enter in the gap.

The contact area increases very slowly for part of the material. As shown by Figure 20A the streamline $\psi = 0$ forms a closed contour which passes through the countercurrent zone where the velocity gradient is very small. Beyond the stagnation point the velocity gradient for $\psi = 0$ is equal to zero. Thus, for the case

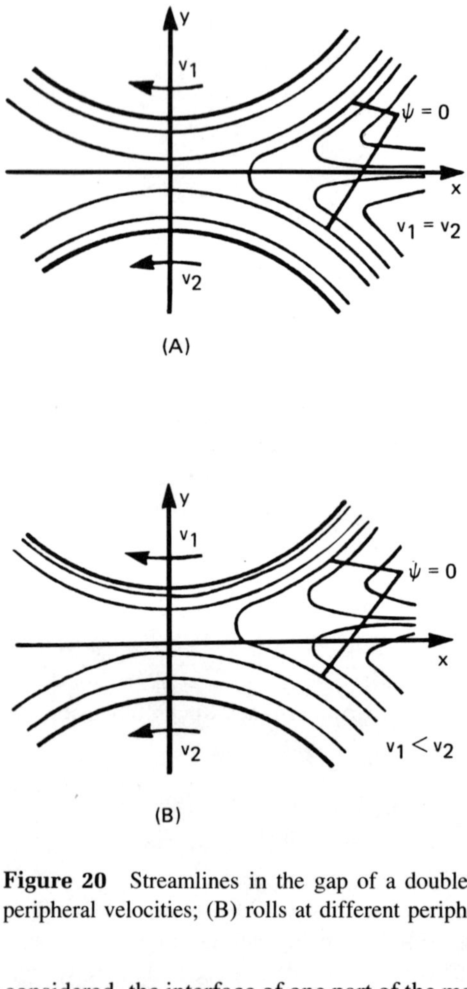

Figure 20 Streamlines in the gap of a double-roll type mixer: (A) rolls at the same peripheral velocities; (B) rolls at different peripheral velocities.

considered, the interface of one part of the material increased considerably slower than the one on the streamlines which pass through the zones of high-velocity gradients. The elements of the interface are not evenly distributed in the mass. Since only the interface is oriented along the streamlines, the material stops to change its orientation because there is no motion across the streamlines. The existence of closed streamlines and the absence of a distinct motion along the z-axis parallel to the roll generatrix are the reasons why the elements are not distributed evenly in the mixing mass. Even if it is assumed that the components are evenly distributed along the roll axis, there will always be some nonbroken interfaces. Thus, the symmetrical rolling process does not provide satisfactory distribution of the ingredients.

The mixing process may be intensified by imposing different velocities and temperatures on the rolls. The equations describing the flow in the gap between the rolls may be obtained by essentially the same method described by Lukach et al. (1967). In this latter situation, a double integration of Eq. (117) is made over the limits of $v_x(h) = v_1 > v_x(-h) = v_2$, where v_1 and v_2 are the peripheral velocities of the front and rear rolls. The equation of streamlines for rolling with friction is

$$\psi = \frac{v_0 h_0 v}{C}\left[\frac{4\eta(\rho^2 - \rho_1^2)}{C^2(1 + \rho^2)} + \frac{2 + 3\rho_1^2 - \rho^2}{1 + \rho^2} + \frac{2\lambda\eta}{3C(1 + \rho^2)}\right] \tag{119}$$

where

$$v_0 = \frac{v_1 + v_2}{2}, \qquad \lambda = \frac{v_1 - v_0}{2v_0}$$

As follows from Eq. (119), there is a closed streamline $\psi = 0$ at $\eta = 0$ (Figure 20B). In this case the countercurrent zone is displaced to the roll with a lower peripheral velocity. It is important in this case that beyond the stagnation point the streamlines pass through the zone of definite shear strain, and in comparison to the symmetrical rolling, the shear of material increases. However, even for this case the streamlines are closed and the rolling with friction does not satisfy the requirements of perfect mixing, because the interfaces oriented along the streamlines maintain their positions.

To eliminate these drawbacks, the operator uses the technique of material undercut, and then introduces it again to the gap from the other side of the roll. The closeness of streamlines is violated and the material is displaced along the axial direction. The material undercut is a necessary part of the process in two-roll mixers, where for a continuous process they are provided with a special automatic device.

Determination of Thrust Forces The methods for evaluating thrust forces are based on the hydrodynamics of rolling, the theories of elastic and plastic deformation, as well as the theory of similarity. The Soviet literature gives detailed analytical treatment of this subject [see Bekin (1966) and Karpachev et al. (1964)].

Polymer materials mixed by rolling undergo both combined and total deformation:

$$\epsilon = \epsilon_{el} + \epsilon_{h.el} + \epsilon_{pl} \tag{120}$$

where

ϵ = total deformation
ϵ_{el} = elastic deformation
$\epsilon_{h.el}$ = high elastic deformation
ϵ_{pl} = plastic deformation

The plastic deformation of a polymer involves viscous flow with irreversible mutual displacement of molecules. The plastic deformation is highly sensitive to temperature.

It should be noted that the relationships between the temperatures of different components and total deformation play different roles. From these points of view there are three different amorphous states of polymers: glassy—in the zone of lowest temperatures (in this case elastic deformation prevails), high elastic—the zone of medium temperatures (in this case high elastic deformation prevails), and the viscous-flowing state—the zone of high temperature (in this case the deciding factor in plastic deformation). The lower regions of elastic and high elastic deformation are best suited for processing polymers by the rolling method.

The law of viscous flow depends on the properties of the materials and may be written as follows:

For a Newtonian fluid: $$\tau = \mu_a \frac{dv}{dx} \tag{121a}$$

For a Bingham material: $$\tau = \tau_0 + \mu_a \frac{dv}{dx} \tag{121b}$$

For a non-Newtonian fluid: $$\tau = K\left(\frac{dv}{dx}\right)^n \tag{121c}$$

where

τ = shear stress
μ_a = apparent viscosity
dv/dx = velocity gradient
τ_0 = limiting shear stress
K and n = rheological constants of the flow (n − tangent of linearized section of the curve)

To determine the flow curve of a non-Newtonian material, the rheological parameters K and n must be known. Flow curves may appear linear on logarithmic coordinates over a portion of the shear rates. This means that the material may be more viscous over a portion of the shear rates and less viscous over others. These properties are accounted for through the coefficient of apparent viscosity:

$$\eta = \mu_a = \frac{\tau}{\dfrac{dv}{dy}} = K\left(\frac{dv}{dy}\right)^{n-1} \tag{122}$$

We now consider the method based on the *hydrodynamic theory of rolling*. The behavior of a polymer material approaches that of a non-Newtonian (quasi-viscous) fluid; therefore, polymers may be considered as a viscous fluid entrained in the gap by rotating rolls. In deriving design equations the following assumptions are applied:

1. The flow regime in the gap is laminar.
2. The material sticks to the roll surface; that is, there is no slip between the relative motion of the polymer and the walls, which limits the flow (the layers of material adjacent to the rolls are in motion at the same velocity as the rolls).
3. The gravitational and inertial forces are small compared to surface forces, and thus may be neglected.
4. The flow of material is assumed to be one-dimensional; that is, the material passes mostly through the gap [$v_z = 0$, $v_x \geq v_y$, and $(\partial v_x/\partial y) \leq (\partial v_y/\partial y)$] as illustrated in Figure 21.
5. Hydrodynamic pressure at the polymer gap entrance and at the outlet from the rolls is zero. The pressure in the plane's parallel to the roll axis does not change in the y- and z-directions; that is,

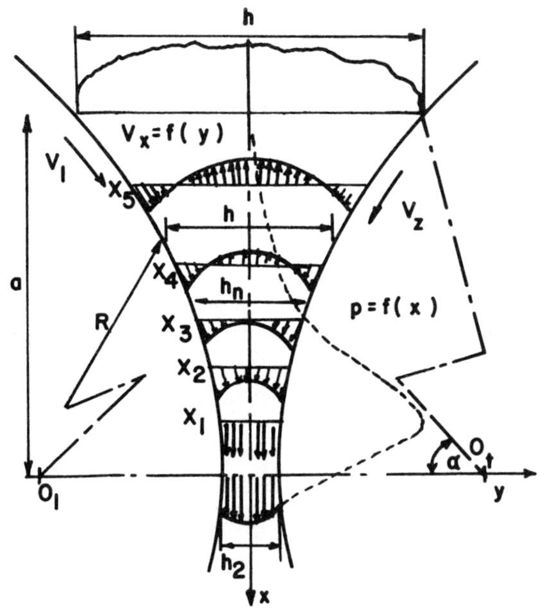

Figure 21 Pressure and velocity distributions of material in the roll section.

$$\frac{\partial P}{\partial y} = \frac{\partial P}{\partial z} = 0$$

The Navier–Stokes equations are the starting point in the analysis. Considering the system to consist of a steady plane-parallel flow of viscous liquid in the gap between the rolls and applying the above assumptions, we write:

$$\frac{\partial^2 \nu_x}{\partial y^2} = \frac{1}{\mu_a} \times \frac{\partial P}{\partial x}$$

$$\frac{\partial P}{\partial y} = 0 \tag{123}$$

where p = pressure.

In the zone of maximum pressure $(\partial P/\partial x) = 0$ [i.e., $(\partial \nu/\partial y) = \text{const}$]. In this zone the flow rate of polymer changes uniformly from ν_1 to ν_2 (taking into account that material sticks to the roll surface).

Upon integration of Eq. (123), we obtain

$$\nu_x = \frac{1}{\mu_a} \times \frac{\partial P}{\partial x} \times \frac{y^2}{2} + C_1 y + C_2 \tag{124}$$

Substituting the friction value $f = \nu_2/\nu_1$ into Eq. (124) and applying the boundary conditions ($\nu_{-h/2} = \nu_1$; $\nu_{h/2} = \nu_2 = f\nu_1$), we obtain (after determining and substituting C_1 and C_2):

$$\nu_x = \frac{1+f}{2}\nu_1 + \frac{1}{2\mu_a} \times \frac{\partial P}{\partial x}\left(y^2 - \frac{h^2}{4}\right) + (1+f)\frac{\nu_1}{h} \tag{125}$$

where h is a moving coordinate (the distance between the rolls).

Using Eq. (125) and determining the flow rate in the gap, the differential equation of pressure becomes

$$\frac{\partial P}{\partial x} = \pm \frac{1+f}{2} 6\mu_a \nu_1 \sqrt{2} \frac{h - h_1}{h^3\sqrt{h - h_k}} \tag{126}$$

After integration, we obtain a formula for determining the specific thrust force:

$$P = \frac{1+f}{2} 2.22 \frac{\mu_a \nu_1 R}{h_k}, \quad \text{kg/cm} \tag{127}$$

where

μ_a = apparent dynamic viscosity, kg-s/cm^2
v_1 = peripheral velocity of the roll, cm/s
R = roll radius, cm
h_k = minimum gap between rolls, cm

The thrust force acting between the rolls is

$$P = pL \tag{128}$$

where

p = specific thrust force from Eq. (127)
L = length of the roll

The coefficient of apparent viscosity μ_a may be approximately determined from the velocity gradient formula:

$$\frac{\partial v}{dy} = \frac{2v_1}{h_k} \tag{129}$$

From either literature data or rheological testing a flow curve of $\mu_a = f(\partial v/\partial y; t°C)$ can be constructed and a value for μ_a obtained.

From the laws of plastic deformation of material between the rolls, the average specific pressure on the roll is:

$$P_{av} = K\sigma_T \frac{2h_{n.l.}}{(\delta - 1)\,\Delta h}\left[\left(\frac{h_{n.l.}}{h_k}\right)^{\delta} - 1\right] \tag{130}$$

where

K = coefficient, ($\simeq$1.15)
σ_T = yield limit of rolling material, kg/cm^2
$h_{n.l.}$ = neutral layer thickness
δ = $\mu/\tan \alpha/2$, coefficient
Δh = $2R(1 - \cos \alpha)$, linear rolling reduction
α = angle of material seizure

The thrust force is

$$P_{\text{th}} = P_{\text{av}} B_{\text{in}} l_s, \quad \text{kg} \tag{131}$$

where $l_s = \sqrt{R\,\Delta h}$, the arch of seizure (refer to Figure 22A).

According to the laws of elastic deformation of materials the thrust force between rolls is

$$P_{\text{th}} = \frac{4ERb_H}{h_H} \sin\alpha(1 - \cos\alpha), \quad \text{kg} \tag{132}$$

where E = elastic modulus of material, kg/cm²; b_H, h_H (from Figure 22B).

From similarity theory the following empirical equations, which may be used for evaluating the thrust forces in rubber mixtures, were obtained by Lukach et al. (1967):

$$\left.\begin{aligned} \frac{P}{\gamma D^3} &= C_1 B^{x_1} \left(\frac{h_k}{D}\right)^{y_1} \left(\frac{L}{D}\right)^{z_1} \left(\frac{D_1}{D}\right)^{k_1} \\ \frac{P}{\gamma D^3} &= C_2 M^{x_2} \left(\frac{h_k}{D}\right)^{y_2} \left(\frac{L}{D}\right)^{z_2} \left(\frac{D_1}{D}\right)^{k_2} \end{aligned}\right. \tag{133}$$

where

B and M = reducibility and softness of mixture
δ = specific gravity of mixture
D = diameter of the roll
L = length of material band on rolls
h_k = minimum gap between rolls

Coefficients C_1, C_2, x_1, x_2, y_1, y_2, z_1, z_2, k_1, k_2 are determined from experiments. Design equations for some synthetic and natural mixtures are also reported by Lukach et al. (1967). For example, for butadiene acrylonitrile blends:

$$\left.\begin{aligned} P &= C_1 \gamma D^{1.4} L^{0.7} h_k^{0.1} B^{-0.4} \\ P &= C_2 \gamma D^{1.4} L^{0.7} h_k^{0.1} M^{-1.8} \end{aligned}\right\} \tag{134}$$

The values for C_1 and C_2 are as follows:

	C_1	C_2
CSN-40	260	71
CSN-26	180	54

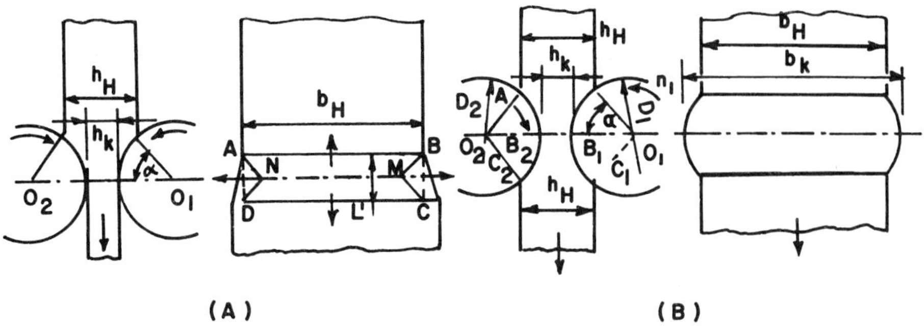

Figure 22 Deformation of material between rolls: (A) plastic deformation; (B) elastic deformation.

Determination of Horsepower As in estimating thrust forces, different methods of calculating torque and horsepower are available. The principal methods are those based on the theory of plastic or elastic deformation, on the hydrodynamic theory of rolling, and on the theory of similarity. The power consumed by mixing rolls depends on several factors: the properties and temperatures of materials, the peripheral velocities of the rolls, the rate of friction, the gap size, and others.

The influence of some of these factors on the energy consumption of the mixing rolls can be understood through the theory of plastic deformation. An increase in the peripheral velocity leads to an increase in the volume of material that can be deformed between the rolls. This results in an increase in the work of deformation. Changes in the peripheral velocity also influence the rate of deformation. Elastic deformation, which does not depend on time, does not play an important role. For highly elastic and particularly for plastic deformation, a decrease in the time of deformation results in an increase in the necessary load on the rolls, which in turn leads to higher horsepower requirements.

Greater friction results in an increase in the volume of material undergoing deformation as well as the rate of deformation and consequently the horsepower consumed.

Enlarging the gap between rolls leads to an increase in horsepower requirements; however, the horsepower increase falls behind the gap increase. This can be explained by the fact that for larger gaps, an increase in the volume of material passing through requires more work of deformation. On the other hand, as the gap is enlarged, decrease in the linear rolling occurs, which reduces the work of deformation. The opposite may also occur whereby the role of relative deformation becomes determinant.

The energy consumption in batch operations may change significantly with time because of variation in the material's plastic properties. The resistance torque to the rotation of the rolls may be assumed to consist of two components:

$$M = M_p + M_f, \quad \text{kg}_\text{f}\text{-m} \tag{135}$$

where M_p is the torque required for overcoming the resistance of the material deformation as determined by thrust forces, and M_f is the torque for overcoming friction in the bearings with respect to gravity and thrust forces.

The torque for overcoming thrust forces is

$$M_p = PD \sin \frac{\alpha}{2}, \quad \text{kg-cm} \tag{136}$$

where

P = thrust force determined from Eq. (128)
D = roll diameter
α = angle of seizure

The torque attributed to friction is

$$M_f = \mu(P + G)d, \quad \text{kg-cm} \tag{137}$$

where

μ = friction coefficient in bearings
G = roll gravity force
d = diameter of journal of roll

The horsepower can be computed from the following formula:

$$N = \frac{(M_p + M_f)}{9700\eta}, \quad \text{kW} \tag{138}$$

where

n = rotational velocity of the roll, rpm
η = efficiency of the drive

The electric power of mixing rolls reaches its maximum value only for a short time (within the first 2 to 3 min). Therefore, the horsepower of an electric motor may be 1.5 to 2.0 times less than the value computed from Eq. (138).

Based on the hydrodynamic theory of rolling, the horsepower consumed by the mixing rolls is established by the shear stresses arising in the material. The change of shear stress in the direction of deformation for a roll which rotates slowly is (Bekin and Nemytkov, 1966):

$$\tau_{+h_k/2} = \mu_a \left[\frac{3(1-f)(\delta - \delta_o)}{\delta^2 h_k} + \frac{(1-f)}{\delta h_k} \right] \tag{139}$$

For rapidly rotating rolls:

$$\tau_{-h_k/2} = -\mu_a \left[\frac{3(1-f)(\delta - \delta_o)}{\delta^2 h_k} + \frac{(1-f)}{\delta h_k} \right] \tag{140}$$

where

μ_{ap} = apparent dynamic viscosity
v_1 = peripheral velocity of slow rotating roll
f = friction
$\delta = \Delta = h/h_k$, dimensionless parameter
$\delta_0 = h_1/h_k$, dimensionless parameter
h = moving width of the gap
h_k = minimum width of gap (see Figure 20)

The corresponding section modulus from the forces of viscous friction is

$$M_{+h/2} = \frac{3C_1 K \alpha_{ap}^n (1+f) R\sqrt{R}\, L v_1^n}{\alpha_{ap_1}\sqrt{h_k}}$$

$$M_{-h/2} = \frac{3C_2 K \alpha_{ap}^n (1+f) R\sqrt{R}\, L v_1^n}{\alpha_{ap_2}\sqrt{h_k}} \tag{141}$$

where α_{ap} = coefficient determined from the following formulas:

$$\alpha_{ap_1} = \frac{(1+f)\sqrt{R}\, T_1}{l\sqrt{h_k}}$$

$$\alpha_{ap_2} = \frac{(1+f)\sqrt{R}\, T_2}{l\sqrt{h_k}} \tag{142}$$

$$\alpha_{ap} = \frac{3(1+f)\sqrt{R}\, T}{2l\sqrt{h_k}}$$

where C_1, C_2, T_1, T_2 are the experimental constants and l is the length of the roll.

The total torque is

$$M = \frac{6CK\alpha_{ap}^{n}\nu_1^{n}(1 + f)^2\sqrt{R}\,L}{\alpha_{ap}\sqrt{h_k}} \tag{143}$$

The total technological horsepower is

$$N = \frac{3EK\alpha_{ap}^{n}\nu_1^{n}(1 + f)^2\sqrt{R}\,L}{\alpha_{ap}\sqrt{h_k}} \tag{144}$$

where

E = constant
L = length of the roll

The third method available for establishing horsepower requirements is based on similarity theory. The following equations were obtained for basic marks of butadiene acrilonitrile caoutchoucs:

$$\left.\begin{aligned} N &= K_1\gamma\omega D^2L^{0.6}h_k^{0.1}f^{-0.2}B^{-0.7} \\ N &= K_2\gamma\omega D^2L^{0.6}h_k^{0.1}f^{-0.2}B^{-2.5} \end{aligned}\right\} \tag{145}$$

And for styrene-butadiene rubbers:

$$\left.\begin{aligned} N &= K_1\gamma\omega D^{2.3}L^{0.6}h_k^{0.1}f^{-0.2}B^{-0.7} \\ N &= K_2\gamma\omega D^{2.3}L^{0.6}h_k^{0.1}f^{-0.2}M^{-2.5} \end{aligned}\right\} \tag{146}$$

where values of the parameters K_1 and K_2 are

	K_1	K_2
Caoutchouc AN-40	0.11	0.0192
Caoutchouc AN-26	0.1	0.02
Styrene-butadiene rubber SBR-30	0.115	0.0145
Styrene-butadiene rubber SBR-10	0.084	0.014

Equations (145) and (146) were obtained on the basis of experiments where parameters were varied over the following limits: h_k = 0.6–2.5 mm; ν_1 = 6.28–18 m/min; f = 1–3; L = 150–1050 mm; D = 200–400 mm; initial caoutchouc plasticity η_a = 0.06–0.15; with the temperature of the rolls and caoutchouc kept constant over all plastification regimes (rolls were 45 ± 5°C; caoutchouc 80 ± 5°C).

Considering rolls as a double energy contour, the friction gear will transfer power that is slightly greater than that obtained by the driving roll. Part of the horsepower may be transferred because of friction from a slower moving roll to the faster one. As a result, the horsepower required for the rapid moving roll may exceed that consumed by the drive.

Torques exerted on the rolls are not equal. The total torque may be estimated from the following formula:

$$M = M_{d_2}\left(1 \pm \frac{K_{ci_2}}{f}\right) + M_f\left(1 + \frac{1}{f}\right) \tag{147}$$

M = total torque
M_{d_2} = torque on driving roll
M_f = torque of friction on the bearing of one roll
K_{ci_2} = coefficient of circulation determined from the graph developed by Lukach et al. (1967).

In addition to the above, the influence of circulation patterns must also be included in the structural design of mixing rolls.

Productivity Considerations

Another area important to parametric evaluations is the roll productivity. Rolls may operate in either batch (multiple passing of material through rolls) or continuous processes (single-pass of material through the gap of the rolls). The productivity design of mixing rolls should be done correspondingly.

Batch Mixing Rolls The productivity of batch mixing rolls depends on a single-time volume loading of the machine, the processing cycle, the size of the rolls, and the operational factor. Productivity may be calculated from the following formula:

$$q = \frac{60 q_1 \rho \eta_e}{\tau}, \quad \text{kg/hr} \tag{148}$$

where

q_1 = volume of mixture for one-time loading, dm^3
ρ = mixture density, kg/dm^3 (ρ = 0.9–1.1 kg/dm^3)
η_e = efficiency factor (η_e = 0.8–0.9)
τ = time cycle, min

The one-time volume q_1 is determined from experiments for each material and roll size. Preliminary calculations can be made using the following correlation:

$$q_1 = (0.0065 \sim 0.0085)DL, \text{ dm}^3 \tag{149}$$

where D is the rear roll diameter, cm.

Continuous Mixing Rolls Productivity may be determined from

$$Q = 60\pi Dnhb\rho\nu_e, \quad \text{kg/hr} \tag{150}$$

where

D = rear roll diameter, dm
ν_e = rear roll velocity, rpm
h = thickness of material band leaving the gap
b = width of material band leaving the gap

The average velocity of the material is determined from:

$$\nu_m = \frac{Q}{bh\rho}, \quad \text{m/min} \tag{151}$$

where

Q = productivity of rolls, kg/min
b and h = width and height of band from the gap, dm

Heat Balances

Complicated heat processes occur during the treatment of plastic materials in mixing rolls. Depending on the operating regime and material properties, heat flow may be either introduced into or evacuated from the machine. The requirements for cooling or heating are determined by a heat balance over the machine:

$$G_m i_{\text{in}} + Q_N + G_c C_c t_{w.\text{in}} = G_m i_{\text{out}} + G_{w.\text{in}} C_w t_{w.\text{out}} + Q_l \tag{152}$$

where

G_m = amount of material to be treated per unit time
i_{in} = enthalpy of material at initial temperature
i_{out} = enthalpy of material at final temperature

Q_N = amount of heat obtained by transformation of mechanical energy into thermal energy
G_c = amount of cooling water entering the rolls
C_c, $t_{w.\text{in}}$ = specific heat and temperature of water entering rolls
$t_{w.\text{out}}$ = temperature of water leaving rolls
Q_l = heat losses of rolls

From Eq. (152) the amount of heat absorbed by the cooling water is established.

$$Q_w = G_c C_c (t_{\text{in}} - t_{\text{out}}) = Q_N - Q_l - G_m(i_{\text{in}} - i_{\text{out}}) \tag{153}$$

The amount of heat due to energy dissipation is:

$$Q_N = 860 N \eta_e \tag{154}$$

where

N = average horsepower consumed by the rolls
η_e = drive efficiency

The heat losses consist of losses by convection and radiation:

$$Q_l = (q_c + q_R)F \tag{155}$$

$q_c = \bar{h}_c(t_w - t_a)$, the specific heat flow by convection
$\bar{h}_c$ = heat transfer coefficient from a roll to air
t_w = temperature of wall of roll (assumed to be equal to the temperature of the material to be treated)
t_a = temperature of ambient air
q_R = specific heat flow by radiation

$$q^R = \epsilon C_0 \left(\frac{T_w}{100}\right)^4 \tag{156}$$

where

ϵ = emittance, and accounts for the fact that one roll is covered with material
T_w = absolute temperature of the wall of the roll
C_0 = radiation constant of an absolute black body

According to calculations, depending on material properties, the amount of heat spent in raising the temperature of the mass from t_{in} to t_{out}, including heat losses, can be as high as 10 to 25% of the total heat generation Q_N. The amount of heat absorbed by the cooling water is determined from a heat balance equation:

$$Q_w = UL\ \Delta t_{av} \tag{157}$$

where U is the heat transfer coefficient through the cylindrical wall.

$$U = \frac{1}{1/(2\pi K)\ \ln(d_{out}/d_{in}) + (1/\bar{h}\pi d_{in})} \tag{158}$$

where

K = conductivity of roll material
d_{out} = outside roll diameter
d_{in} = inside roll diameter (diameter of roll hollow or average diameter of a circle passing through the center's peripheral channels)
$\bar{h}$ = heat transfer coefficient from internal roll space to the moving water (determined by one of the equations of forced convection, depending on the flow regime)
L = roll length
Δt_{av} = average temperature difference defined as:

$$\Delta t_{av} = t_w - \frac{t_{w.out} - t_{w.in}}{2} \tag{159}$$

The method of heat calculations of rolls amounts to the following procedure. First, determine Q_l and G_c (a temperature difference can be assumed; typically $t_{w.out} - t_{w.in} = 5–7°C$). Knowing the size of the roll hollow (or the sizes of peripheral channels) it is then possible to determine the velocity of the cooling water in order to estimate the heat transfer coefficient U.

From Q_w, U, and L, determine Δt_{av}, and from Eq. (159), calculate $t_{w.out}$ or $t_{w.in}$. For specified $t_{w.out}$ and $t_{w.in}$, apply Eq. (157) to establish whether the length is sufficient for evacuation of the given amount of heat, Q_w.

RIBBON BLENDERS AND PLANETARY MIXERS

Few design correlations are reported in the open literature for ribbon blenders despite their wide use in the manufacturing of foods, drugs, plastics, paper, pigments, adhesives, chemicals, and cleansers. There are a number of design variations, with the differences primarily in the agitator configuration. Figure 23 shows four configurations of ribbon agitators applicable for mixing materials ranging from dry powders to moderately viscous pastes. The continuous ribbon agitator (Figure 23A) is a standard construction with inner and outer ribbons which may be arranged for center or end discharge. The continuous agitator construction (Figure 23A) is a standard construction with inner and outer ribbons which may be arranged for center or end discharge. The cut-out agitator construction (Figure 23B) is used when materials to be mixed are heavier than normal. Basically the same as the continuous type, the cut-out agitator has alternate sections of the outer ribbon removed. The cut-in agitator construction (Figure 23C) is used for cutting oils, fats, and shortenings into flours and powders. It is essentially a continuous agitator with cutting bars added to the inner ribbon, and cutting wires mounted through the ribbon arms. The plow-type construction (Figure 23D) is used for mixing extremely heavy crystalline or abrasive materials.

Figure 24 shows examples of ribbon blenders: a small batch and a large continuous unit. The standard drive for ribbon blenders consists of a heavy-duty base-mounted gear motor which drives the agitator through two sprockets and roller chain. On larger units, the drive utilizes a drive motor, gear reducer, and controlled torque coupling, with sprockets and chains. All drive motors can be started under load. For cases where products of different consistency during blending are mixed, variable speed drives are available.

Ribbon blenders are widely used in the food and pharmaceutical industries as sanitary blenders. Figure 25 shows one such installation. These units are designed for fast breakdown, to facilitate cleaning. They are generally constructed of stainless steel. Designs may also be equipped with partial or full jackets, with a range of 150 psig and higher available for either heating or cooling purposes.

Some general information concerning the types of discharge valves and stuffing boxes used on ribbon blenders is given in Figure 26. Also, Table 5 can serve as a rough guide to sizing ribbon blenders. Materials to be mixed should always be tested first in a unit to establish product quality achievable in a particular mixer type or design.

Note that the agitator of the ribbon blender consists of outer and inner spiral ribbons supported on a shaft which runs longitudinally through the center of the blade's tank. This arrangement provides a triple mixing action. First, as the agitator rotates, material is tumbled and transported radially around the inside of the blender. Second, the pitch of the outer ribbon moves the material toward the unit's discharge.

The widths of the inner and outer ribbons must be balanced to provide even movement of material in both directions and to prevent the load from building up at any one point in the blender. Close tolerances are also maintained between the outer ribbon and blender vessel. On heavy-duty machines, where the outer

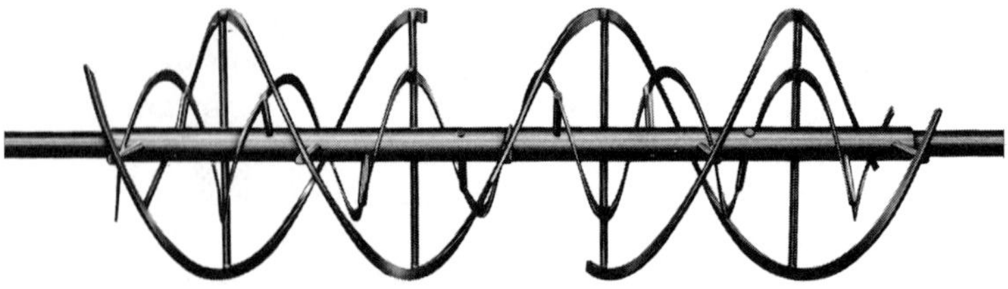

Continuous Ribbon agitator: standard construction for all Day Ribbon Blenders, has continuous inner and outer ribbons, and may be arranged for center or end discharge.

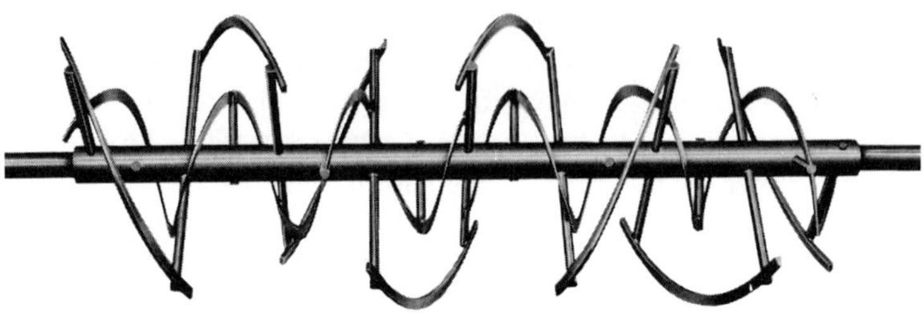

Cut-out agitator construction is used when products to be mixed are heavier than normal. Basically the same as the continuous type, the cut-out agitator has alternate sections of the outer ribbon removed, efficiently mixes heavier batches.

Figure 23 Different ribbon agitator designs. (Courtesy of Day Mixing Co., Nautamix Europa, Cincinnati, OH.)

ribbon may be as much as 2.5 cm thick, the trailing edge of the ribbon is relieved to eliminate any high bearing load between the ribbon and tank walls.

Ribbon blenders are essentially self-contained mixers best suited for batch or semicontinuous mixing. An example of a unit for laboratory scale work with a variable drive is shown in Figure 27.

Planetary mixers are also well suited for a wide range of liquid and solid mixing applications. Double planetary mixers provide a mix cycle consisting of two regularly shaped stirrer blades revolving around the tank on a central axis. Simultaneously, each blade revolves on its own axis at approximately the speed of the central rotation. With each revolution on its own axis each blade advances along the tank wall. This movement ensures that good homogeneity of the material being mixed will be achieved within a relatively short time frame. Mixing does not depend on the flow characteristics of the mix.

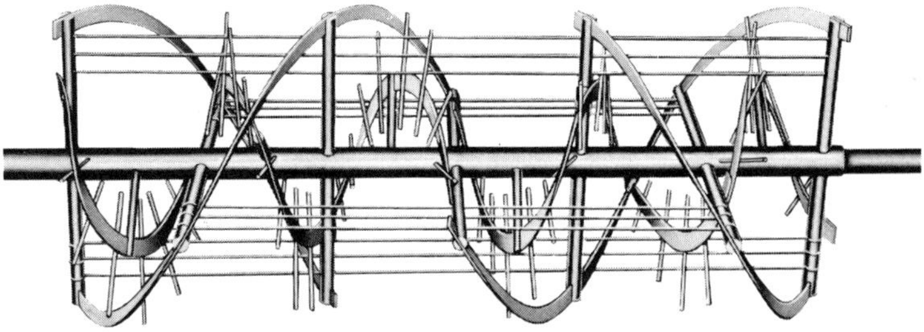

Cut-it-in agitator construction is used for cutting fats, oils and shortenings into flours and powders. Essentially a continuous agitator with cutting bars added to inner ribbon, and cutting wires mounted through ribbon arms. Leather or Tygon wipers are furnished.

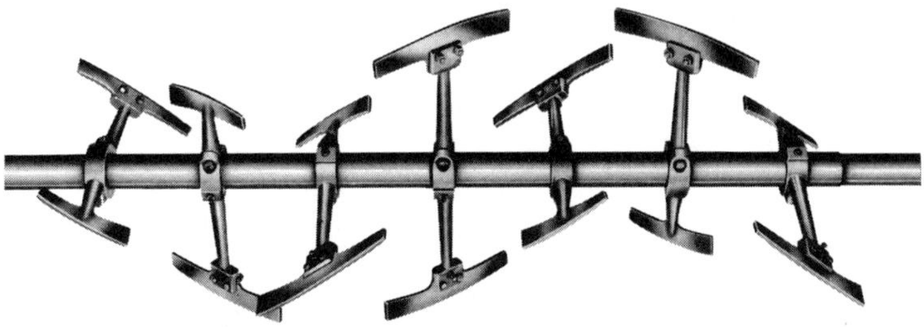

Dry Color agitator or plow type construction is used for mixing extremely heavy crystalline or abrasive materials. A series of 'T' shaped paddles are spaced 90° apart on the shaft. All except center assembly have one long and one short arm and paddle, effectively circulate material at outer and inner areas of tank; center assembly consists of two long arms to facilitate discharge.

Double planetary mixers have no packing glands or bearings in the product zone. Hence, cleaning between batches is minimized. Figure 28 shows a planetary mixer unit. The machine is equipped with a hydraulic lift which makes it possible to lower and raise the stirrer blades into and out of the mixing tank. The unit shown is capable of handling applications ranging from low-viscosity fluids to very high-viscosity pastes and doughlike materials. Typical capacities of these units range from 10 to over 300 gallons.

Figure 29 illustrates orbital revolutions of the agitator blade on a horizontal plane of fluid. The stirrers essentially cover every point within the mixer can. These machines are operated strictly in a batch mode. Designs feature rugged construction for a wide range of mixing applications ranging from low to very

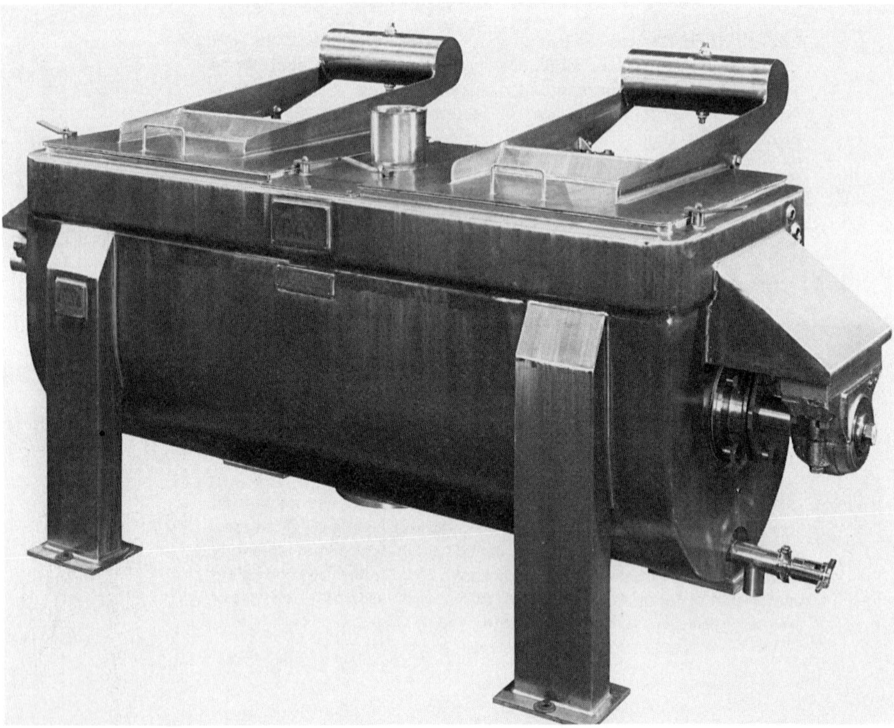

Figure 24 Ribbon blender units. (Courtesy of Day Mixing Co., Nautamix Europa, Cincinnati, OH.)

Figure 25 Sanitary ribbon blender in operation in the finishing department of a pharmaceutical firm. (Courtesy of Day Mixing Co., Nautamix Europa, Cincinnati, OH.)

DISCHARGE VALVES

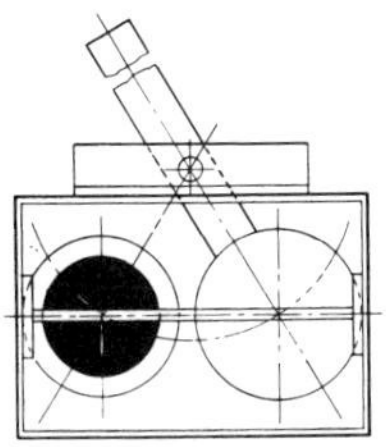

Standard Slide Paddle Valve
This valve is the most common choice for the majority of applications. Available for manual or pneumatic operation. Provides positive solids shut off.

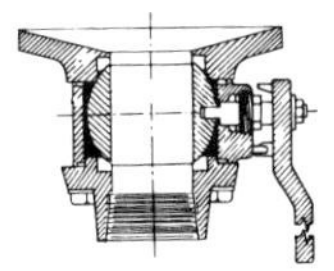

Ball Valve
Used for vacuum service or for retention of liquids.

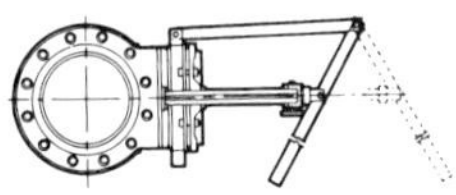

Optional Slide Gate Valve
Used for dust tight operation or where liquids are present in the blend.

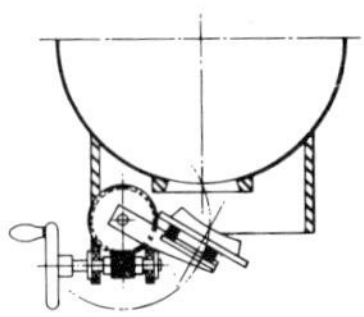

Flush Plug Valve
Eliminates dead spots in the mix trough. Equipped for either manual of pneumatic operation.

STUFFING BOXES

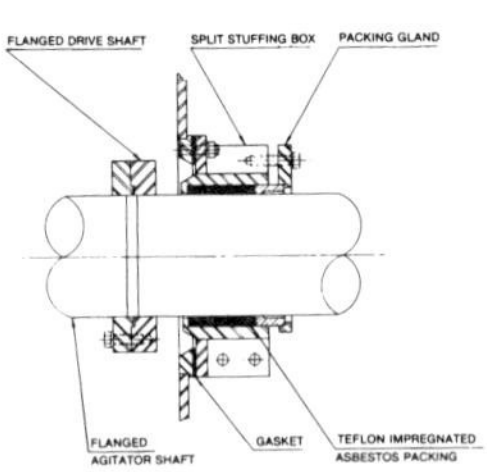

Standard Type
The split housing is readily disassembled for repacking. Teflon-impregnated, braided asbestos packing is normally used.

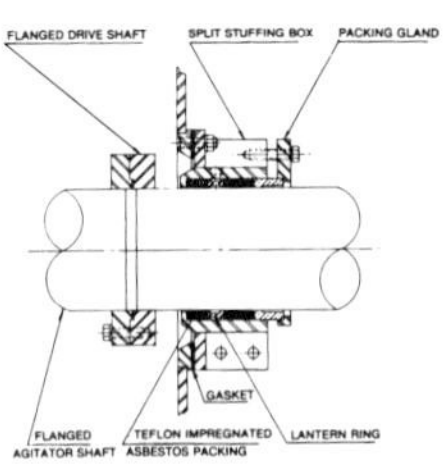

Air or Liquid Seal
Designed to prevent migration of liquids or solids into the packing area. Either air or a liquid enter the blender continuously under slight pressure through a lantern ring in the stuffing box.

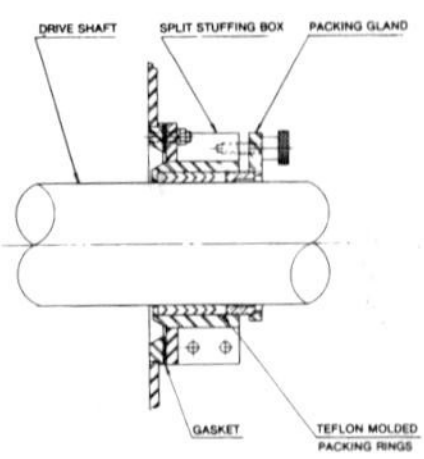

Sanitary Type
When thorough cleaning must be achieved between batches, this design is used. Stuffing box is easily disassembled with simple hand tools and includes special molded Chevron-Teflon packing rings.

Figure 26 General information on discharge valves and stuffing boxes used on ribbon blenders. (Courtesy of Charles Ross and Son Co., Hauppauge, NY.)

Table 5 Typical Dimensions and Specifications of Ribbon Blenders

	Working capacity		Full capacity		Tank size			Agitator speed (rpm)		Motor HP[a]		Approx. ship. wt. lb.		Overall dimensions			
														Length			
Model	Cu. ft.	Gal.	Cu. ft.	Gal.	Dia.	Lgth.	Depth	Stdrd.	Heavy duty	Stdrd.	Heavy duty	Stdrd.	Heavy duty	Standard A[1]	Heavy duty A[2]	Width[b] B	Height[c] C
0	1.3	10	1.6	12	13″	16″	15″	60		1		550		3′ 4″		1′10″	3′11½″
B	5.0	37	6.0	45	17″	38″	18½″	60		1½		825		5′ 2″		2′	4′ 3″
C	10.8	80	12.5	93	21″	48″	24″	60		3		1160		6′ 2″		2′ 4″	4′ 9″
D	17.5	130	19.5	145	24″	60″	26½″	60		5		1475		7′ 3″		2′ 7″	4′11″
D-8	23.0	170	25.5	190	26″	66″	28½″	50		7½		2080		8′ 3″		2′10″	5′ 1″
D-10	36.0	270	40.0	300	30″	78″	33″	50		10		2300		9′ 3″		3′ 4″	5′ 6″
F	54.0	400	59.0	440	34″	90″	37¼″	45		10		3150		10′ 6″		3′ 7½″	5′10″
G	62.5	465	69.0	515	36″	96″	39¼″	40		15		3720		10′11″		3′ 9½″	6′
40×8	80	600	93	700	40″	8′	47″	25	20	25-50	60-100	5720	6220	11′ 4″	12′ 5″	4′ 1½″	4′ 5″
40×10	100	740	115	865	40″	10′	47″	25	20	25-50	60-100	6660	7260	13′ 4″	14′ 5″	4′ 1½″	4′ 5″
44×10	120	900	144	1080	44″	10′	52″	25	20	25-50	60-100	7800	8500	13′ 4″	14′ 5″	4′ 5½″	4′10″
50×10	155	1160	186	1400	50″	10′	59″	25	20	25-50	60-100	9500	10300	13′ 4″	14′ 6″	4′11½″	5′ 5″
54×10	180	1350	203	1520	54″	10′	61″	25	20	25-50	60-100	10500	11400	13′ 4″	14′ 7″	5′ 5″	5′ 7″
54×12	215	1600	246	1850	54″	12′	61″	25	20	30-50	60-100	12300	13300	15′ 7″	16′ 9″	5′ 5″	5′ 7″
60×12	270	2000	298	2225	60″	12′	66″	20	15	30-50	60-100	14000	15100	15′ 7″	16′10″	5′11″	6′ 1″
66×12	325	2425	357	2670	66″	12′	73″	20	15	30-50	60-100	15000	16200	15′ 7″	16′11″	6′ 6½″	6′ 8″
72×12	385	2875	421	3154	72″	12′	80″	20	15	30-50	60-100	17000	18300	15′ 7″	16′11″	7′ 1″	7′ 3″
72×16	515	3850	562	4200	72″	16′	80″	20	15	30-50	60-100	26500	28000	19′ 9″	21′ 2″	7′ 1″	7′ 3″

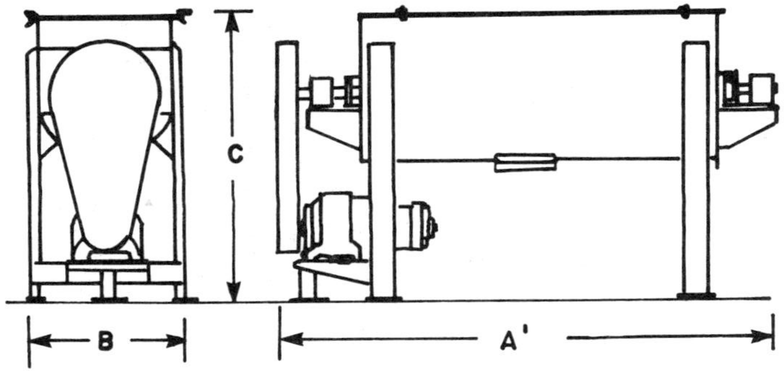

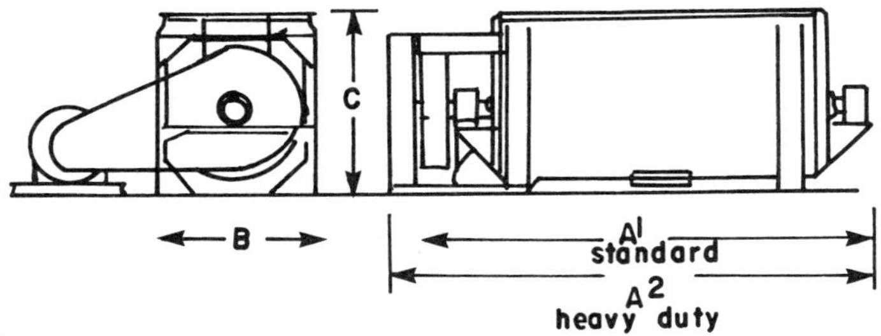

Courtesy of Day Mixing Co., Nautamix Europa, Cincinnati, OH.
[a]H.P. availabilities based upon product characteristics and densities.
[b]Dimensions are for nonjacketed tanks. For jacketed tanks add 6″.
[c]With standard legs (measured from bottom of discharge valve) of 30″ on smaller blenders and of 2″ on larger blenders.

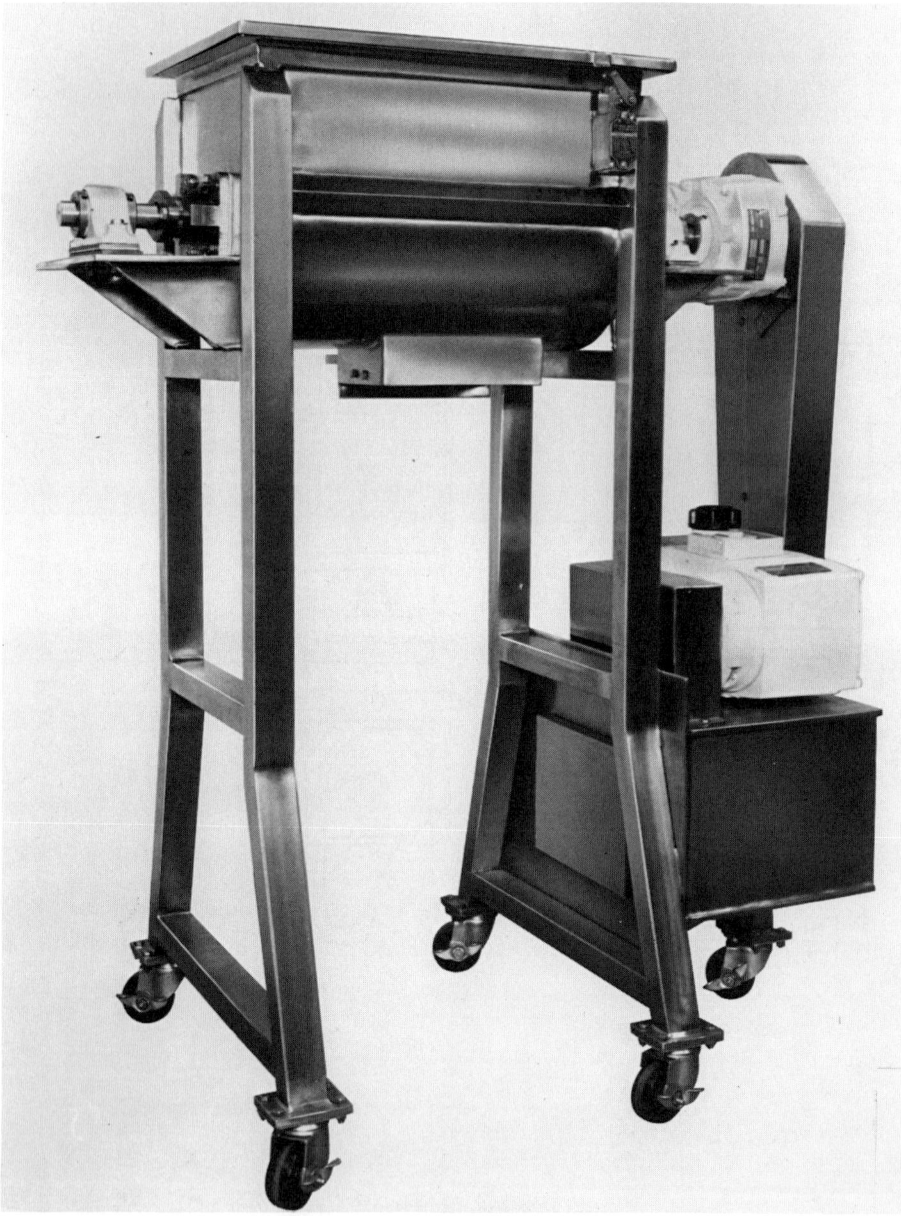

Figure 27 A 1 ft^3 capacity ribbon blender with variable-speed drive. (Courtesy of Charles Ross & Sons Co., Hauppauge, NY.)

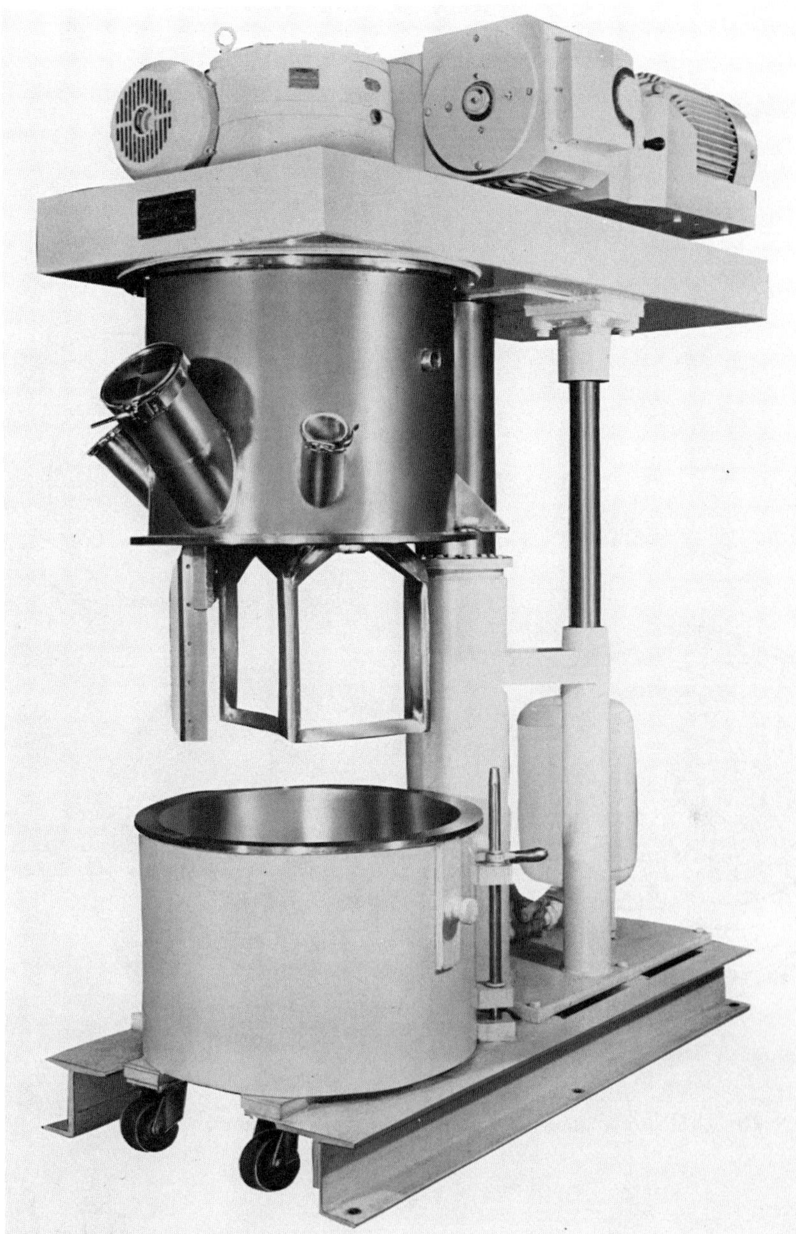

Figure 28 A planetary mixer with a vacuum hood. (Courtesy of Charles Ross & Sons Co., Hauppauge, NY.)

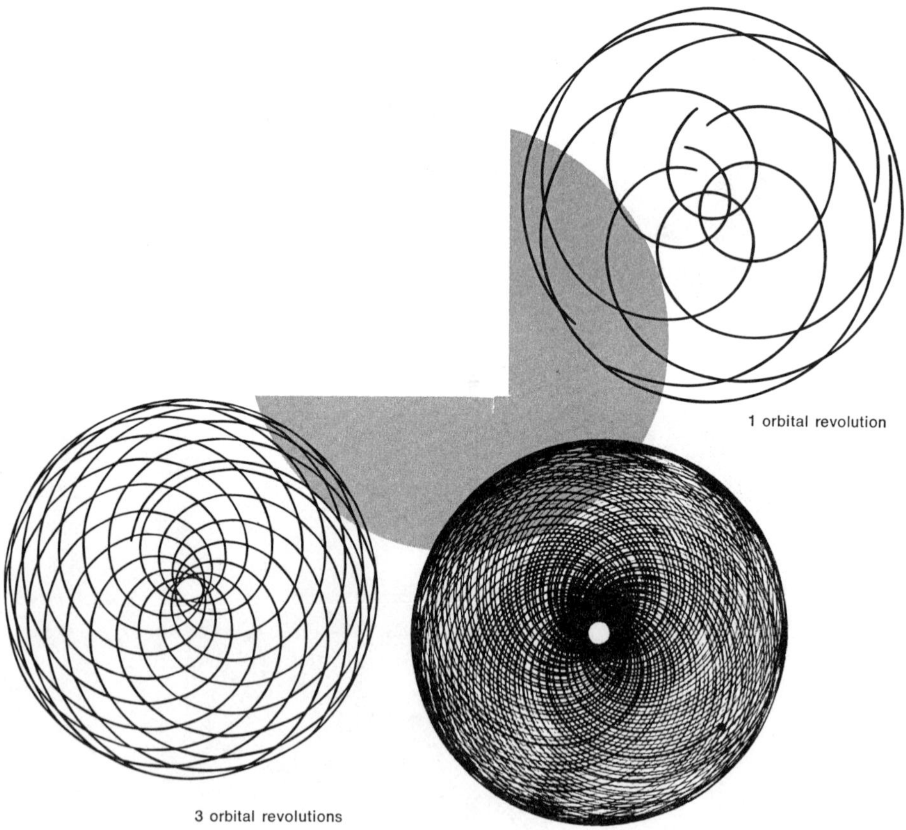

Figure 29 Orbital revolutions of a double planetary mixer stirrer.

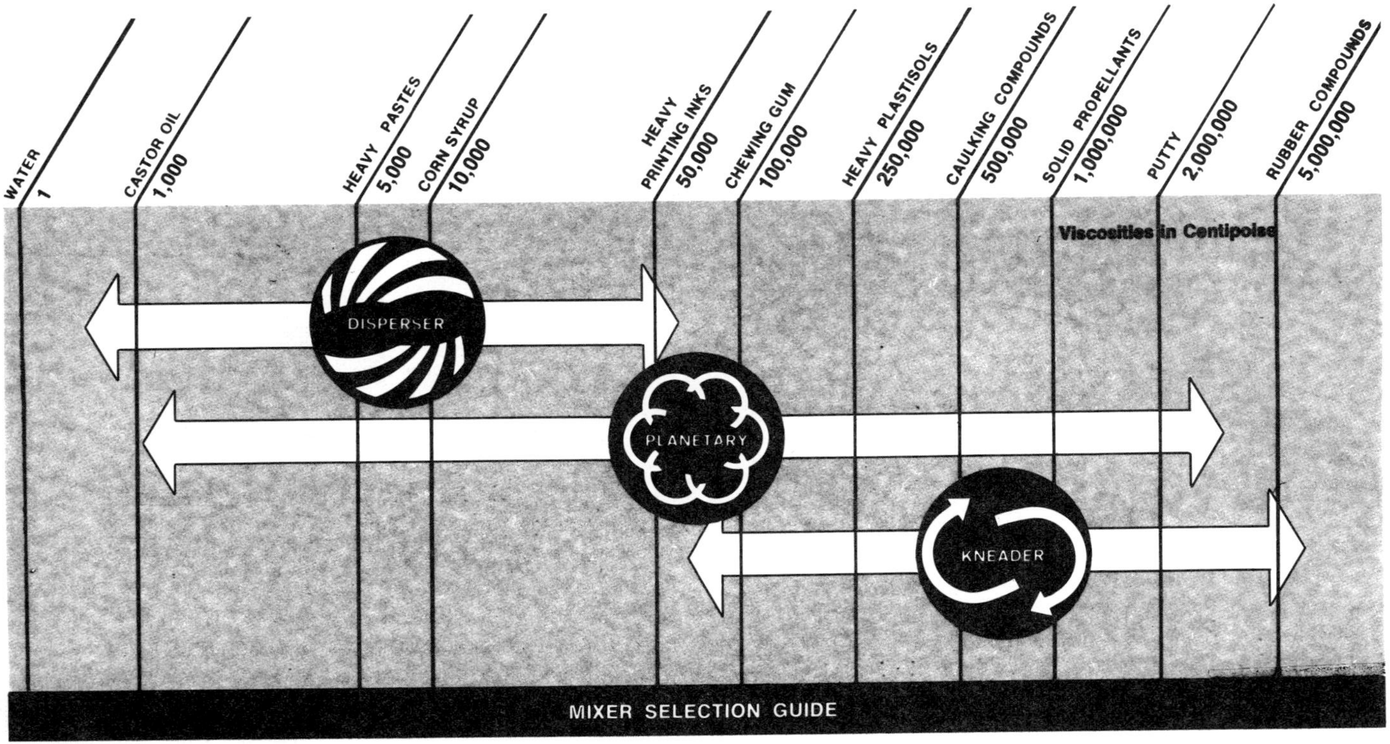

Figure 30 Guide to mixer selection based on viscosity.

viscous pastes and doughlike materials. Figure 30 can serve as a rough guide to mixer type selection.

EMULSIFERS AND BLENDERS

There are a number of mixing systems which have widespread applications to both liquid–solid and solid–solid blending operations. Although design principles for these mixers are not well treated in the literature, their description merits attention.

Mixer-emulsifiers are sometimes used as an alternative to slow-speed impeller mixing or high-pressure homogenization for a wide range of processing requirements. Typical applications include the preparation of adhesives (e.g., asphalts, carbon dispersions, clay dispersions, dyestuffs, paints, and inks, lacquers), cosmetics (e.g., creams, emulsions, hand lotion, perfumes, shampoos, deodorants), foods (e.g., chocolate coatings, mustard, soft drinks, sugar emulsions), pharmaceuticals (e.g., antibiotics, ointments, reducing animal tissues), plastics (e.g., cold cutting resins, polyester dispersions, resin solutions), and various miscellaneous mixtures such as floor polishes, gum dispersions, lubricants, petroleum emulsions, and so on. These mixers are normally used in dished or conical-bottom vessels.

An example of a mixer-emulsifier is shown in Figure 31. The mixing process can be thought of as being performed in three stages. In the first stage the high-speed rotor operating at close clearance to the stator draws material in from the bottom of the mixing vessel and subjects it to intense mixing. In stage 2 the rotor accelerates the product toward the blade periphery. There it is expelled through openings in the stator into the body of the mix while undergoing an intensive mechanical and hydraulic shearing action. At the same time new material is drawn into the center of the rotor. In the third stage the expelled mixture is deflected by the tank wall, completing the circulation. To increase circulation or to create a vortex for the incorporation of light solids, a downthrust propeller may be mounted on the rotor shaft.

The operating sequence for emulsifiers varies with the intended application. An operating guide for these mixers is given in Table 6. Standard batch units are available through 100 hp. For most applications the rotor head should be mounted 2 to 3 head diameters above the vessel floor.

Maximum mixing efficiency is generally achieved on mixtures under 10,000 cP. The principal criterion for achieving the best efficiency is to maintain a maximum circulation of all materials through the rotor/stator at all times during the process cycle. As viscosity increases, flow through the head decreases, thus reducing the work required for a given volume of material while it is being circulated within the vessel. In-line emulsifiers are less susceptible to this problem as they may be fed by means of positive displacement pumps.

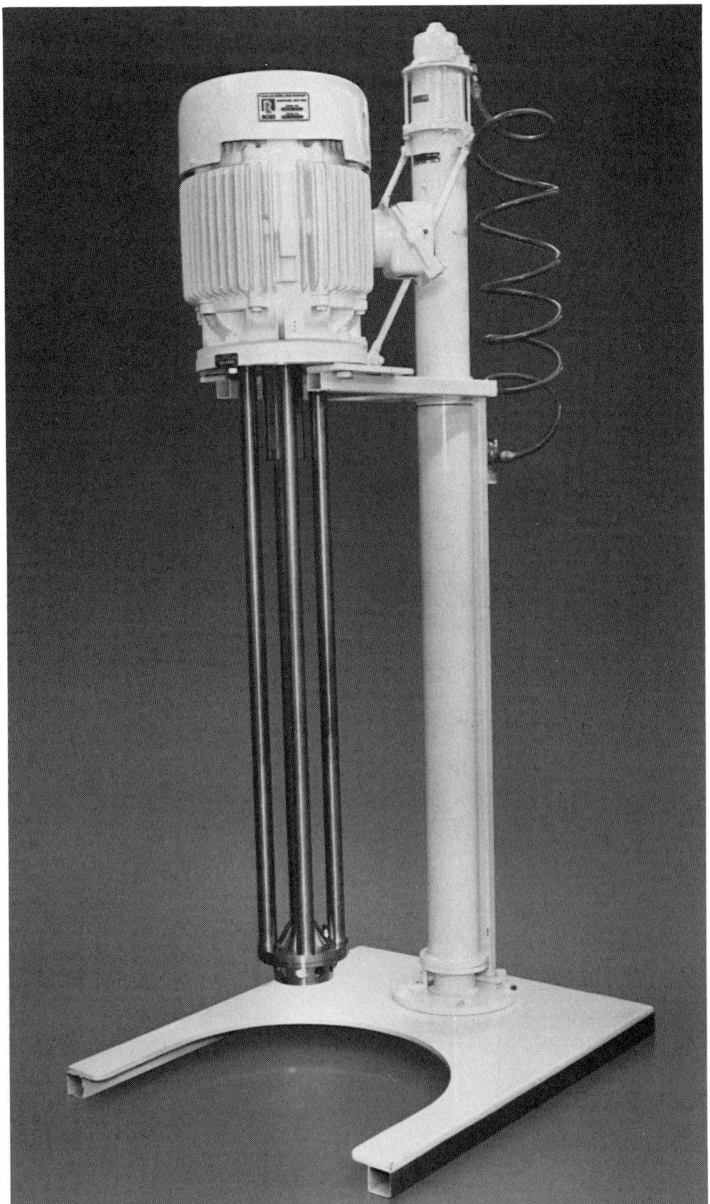

Figure 31 Mixer-emulsifier mounted on an adjustable lift. (Courtesy of Charles Ross & Sons Co., Hauppauge, NY.)

Table 6 Operating Guidelines for Emulsifiers

Application	Procedure	Benefits
Dissolution	1. Charge all liquids into vessel.	1. Short turn-around time permits the use of smaller equipment to accomplish the required end result.
	2. Add any minor ingredients which require long dispersion time.	2. Complete wetting and subsequent dissolution of all solids.
	3. Add solids to be dissolved. If the solids are of low density, down thrust propellers may be advantageous to create a vortex and assist in initial wetting.	3. Low heat build-up as a result of the short cycle time.
	4. Run unit until all powders are completely wetted out and dissolved in the liquid components.	4. Significant energy savings.
Suspensions	1. Charge all liquid components of the formulation into the tank.	1. Uniform particle size reduction.
	2. Start mixer and charge solid materials until a uniform particle size has been achieved.	2. One-step operation to suspend solids and add dispersant.
	3. Add the dispersant as rapidly as possible.	3. Interchangeable stator heads may be used with solids of varying size.
		4. Short cycle time.
Emulsification	1. Charge continuous phase and heat.	1. Short cycle time.
	2. Add surfactant.	2. Intimate and immediate contact of dispersed phases.
	3. Externally heat and slowly add the dispersed phase through an open pipe line terminating below the rotor/stator assembly.	3. Fine droplet size.

continued

Table 6 (Continued)

Application	Procedure	Benefits
	4. Mix until emulsion completed.	4. Potential surfactant savings through efficient use of high shear agitation.
Disintegration	1. Charge all liquids into vessel.	1. Single step process with potential materials handling and time savings.
	2. Add to liquids any solids that are easily liquified and begin mixing.	2. Drastically reduced cycle time.
	3. Add slowly the solids that are to be reduced in size.	3. Substantial savings in electrical costs resulting from shorter cycle times.
	4. Continue operation until solids are reduced to a desired uniform particle size.	4. Floorspace savings. Less capital equipment required.

Courtesy of Charles Ross & Son Company, Hauppauge, NY.

These designs are capable of handling a wide range of materials where the density and rheological properties vary greatly. The pumping capabilities of a unit on materials that have the same apparent viscosity may not be the same in actual practice. For example, many polymers have a low apparent viscosity and appear ideally suited to a particular unit's design/application criteria. However, upon testing, it may be observed that the flow is poor and that the end result is only marginal. Often, modified tank configurations, auxiliary agitation, different rotor speeds, or head diameters will improve efficiency. Pilot testing is almost always recommended for accurate scale-up.

Paddle mixers are devices consisting of two or more blades mounted on a vertical or inclined shaft. The basic paddle impeller configurations are shown in Figure 32, and a description of each is given in Table 7. The main advantages of paddle mixers are simplicity and low cost. A disadvantage is their small pumping capacity (a slow axial flow), which does not provide a thorough mixing of the tank volume. Perfect mixing is attained only in a relatively thin stratum of liquid in the immediate vicinity of the blades. The turbulence spreads outward very slowly and imperfectly into the entire contents of the tanks; hence, circulation of the liquid is slow. Therefore, paddle mixers are used for liquids with viscosities only up to about 1000 cP. Because of a concentration gradient often created in

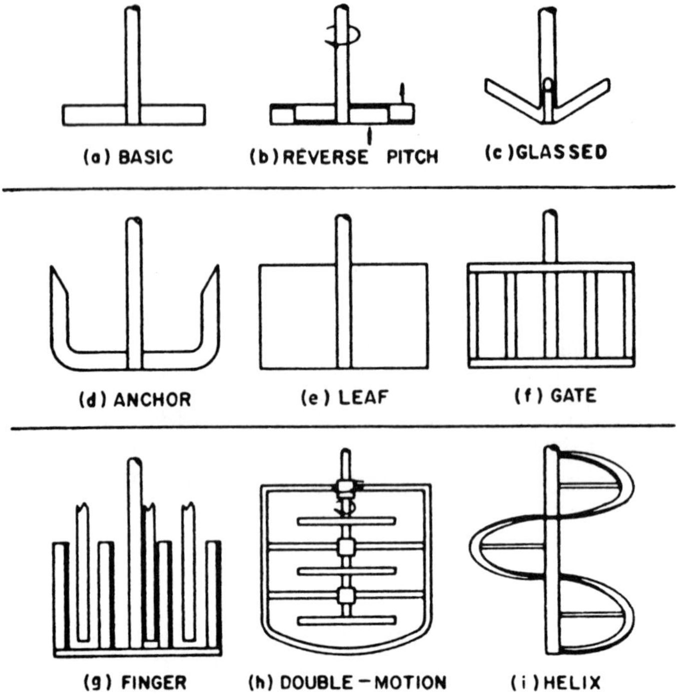

Figure 32 Different paddle impeller configurations.

the liquid when these mixers are used, they are unsuitable for continuous operation. This can be remedied by tilting the paddle blades by 30 to 45° to the axis of the shaft, resulting in an increase in axial flow and consequently a decrease in concentration gradients. Such a mixer is capable of maintaining particles suspended, provided settling velocities are not high. Mixers with tilted blades are used for processing slow chemical reactions which are not limited by diffusion.

To increase the turbulence of the medium in tanks with a large height-to-diameter ratio, a configuration consisting of several paddles mounted one above the other on a single shaft is employed. The separation between individual paddles lies in the range of 0.3 to 0.8d (where d is the diameter of the paddle) and is selected according to viscosity of the mixture. For mixing liquids with viscosities up to 1000 cP, as well as for heated tanks in cases where sedimentation can occur, anchor or gate paddle mixers are employed (Figure 32D and F). Paddle diameters are almost as large as the inside diameter of the tank in such applications, so that the outer and bottom edges of the paddle scrape (or clean) the walls and bottom.

Table 7 Description of Paddle Impeller Configurations Shown in Figure 32

Basic paddle	The simplest design is a single horizontal flat beam. The ratio of impeller diameter to tank diameter is usually 0.5–0.9. The peripheral speed range is generally 250–450 ft/min. Paddles used in the United States have generally had a ratio of width to diameter (w/D) of one-sixth to one-twelfth; European practice is generally one-fourth to one-sixth.
Reverse pitch	To improve the top-to-bottom turnover characteristic of a simple paddle, the reverse pitch design in Figure 11.8B is used. The 45° blade angle is reversed at a diameter of approximately three-fourths of the impeller diameter and rotation is generally set to produce up-flow in the outer section.
Glassed steel	This impeller is a three-blade style common in glass-coated vessel applications. Blade form either has a pitch of about 30° from the vertical or is curved. Usually, D/T is 0.55–0.65. For low-viscosity fluids its operation is definitely in the turbine category, but the difficulty of achieving a fully baffled condition in a glass-lined vessel causes it to perform like a paddle in many cases.
Anchor	Contouring a simple paddle to the shape of a tank bottom gives the anchor or "horseshoe" style. Extent of the blade may be limited to the lower vessel tangent line or the blades may continue upward along the straight side. Clearance between blade and vessel shell may be from $\frac{1}{8}$ to 3 in., depending on tank diameter and heat transfer requirements.
Leaf	This has an extreme w/D ratio for a paddle, a geometry used frequently in European practice.
Gate	A multiple-arm paddle with connecting vertical members, this design is often adopted for structural reasons in large tanks.
Finger	Also known as a paste mixer, this combination of vertical blades meshing with stationary baffles has been in use for many years but no data on power or performance have been published. The application is restricted to small batch sizes (less than 1000 gal) because of structural design difficulties.
Double-motion	This design combines a gate and anchor and a multiple-pitched paddle. Rotation of the two assemblies is countercurrent. Since a special drive with concentric counterrotating output shafts is required, use is restricted to a few applications requiring intensive mixing of very viscous or non-Newtonian fluids.
Helix	This configuration least resembles the basic paddle. It does, however, operate in the laminar range at normally high D/T ratios and is an important member of the paddle group. One traditional use of a helix or screw is in a vertical draft tube.

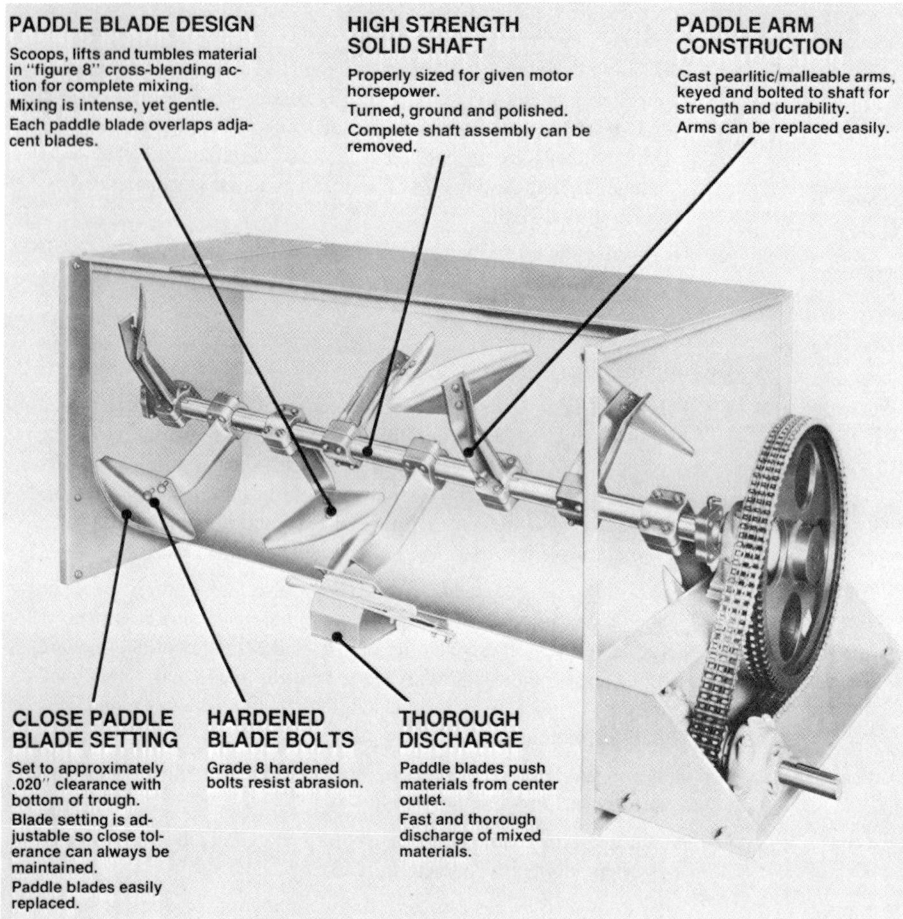

Figure 33 Cutaway view of a paddle-mixer. (Courtesy of Marion Mixers, Rapids Machinery Co., Marion, IO.)

Leaf-shaped (broad-blade) paddle mixers provide a predominantly tangential flow of liquid, but there is also turbulence at the upper and lower edges of the blade. Leaf-type blades are used for mixing low-viscosity liquids, intensifying heat transfer processes, to promote chemical reactions in a reactor vessel, and for dissolution. For dissolving applications, leaf blades are generally perforated. During the mixer rotation, jets are formed at the exits from the holes which promote the dissolution of solid materials.

Figure 34 Semicylindrical paddle-mixer. The unit is of stainless steel construction, 12 in. diameter by 2 ft long, jacketed, and equipped with variable-speed drive. (Courtesy of Marion Mixers, Rapids Machinery Co., Marion, IO.)

The rotational velocity of paddle mixers is in the range of 15–45 rpm. Under these conditions pumping action is small and there is no danger of vortex formation. Therefore, paddle mixers are mostly used without baffles. However, for broad-blade paddles, which operate at speeds up to 120 rpm, baffles are incorporated into the design to minimize vortex formation. Azbel and Cheremisinoff (1983) provide scale-up guidelines for paddle mixers.

Figure 33 shows a cutaway view of one type of paddle mixer and summarizes the principal design features. Figures 34 and 35 show photographs of semicylindrical and cylindrical units, respectively. Both designs are applicable to vacuum and cryogenic services.

Figure 35 Cylindrical paddle-mixer of stainless steel construction, 20 in. diameter by 4 ft long. The unit is designed for full vacuum and cryogenic service. It has power-operated charge, is equipped with discharge and sampling valves, and has variable-speed drive. (Courtesy of Marion Mixers, Rapids Machinery Co., Marion, IO.)

NOTATION

A	area
$A'B'$	constants
a	exponent or variable
a'	interfacial area
c	velocity
C	concentration
c_1, C_2	coefficients
D, d	diameter
Eu	Euler number
e	elasticity coefficient
F	area
F'	force
Fr	Froude number
G	modulus of rigidity, also gravitational force
g	gravitational acceleration
H	depth or head
h, h_e	heat transfer coefficients
J	heat transfer coefficient
k', k	constants
k_n	power number
L	length
L_p	turbulent mixing length
M	torque moments
N_p	mixer power
N_t	mixing number
N_L	modified power number
N	rotational speed
n, n_0	rotational speed
P	pressure, also centrifugal force
Pr	Prandtl number
p	power input
Q	volumetric rate or pumping capacity
q	mass or volumetric throughput
Re	Reynolds number
R, r	radius
s	pitch
T	absolute temperature
t	temperature or time
U	overall heat transfer coefficient
u	velocity
V	velocity or volume
W	dimension
We	Weber number

Greek Symbols

α, β	angle or coefficients
$\dot{\gamma}$	shear rate
γ	specific weight or concentration
δ	clearance
ϵ	deformation
λ	friction
λ_1	time constant
λ_v	heat of vaporization
μ	Newtonian viscosity
η_p	plastic viscosity
η_c, η_d, η'	effective viscosities of continuous and disperse phases, and emulsion, respectively
η_e	drive efficiency
θ	characteristic time
τ_i, τ'	mixing time
τ	shear stress
τ_y	yield stress
ρ	density
ν	velocity
ϕ	parameter or angle
ψ	angle, stream function
ψ'	void fraction
ψ_0	angle

REFERENCES

Azbel, D. S., and N. P. Cheremisinoff, *Fluid Mechanics and Unit Operation,* Ann Arbor Science Publishers, Ann Arbor, MI, 1983.

Azbel, D. S., *Two-Phase Flows in Chemical Engineering,* Cambridge University Press, New York, 1981.

Baker, C. K., and G. H. Walter, *Heat Transfer Eng.* 1, 2, 28 (1979).

Bekin, N. G., and V. A. Nemytkov, *Kauchiek i Resina* (USSR), *10,* 31, 1966.

Bernhardt, E., *Pererabotka Termoplasticheskikh Materialov,* M. Goskhinisdat, USSR, 1962.

Bingham, E. C., *Fluidity and Plasticity,* McGraw-Hill, New York, 1922.

Bott, T. R., and S. Azoory, *Chem. Proc. Eng., 50*(1), 85–90 (1969).

Calderbank, P. H., and M. B. Moo-Young, *Trans. Inst. Ch.E.* (London), *39,* 337–347 (1961).

Calderbank, P. H., *Mixing,* Vol. 2, V. H. Uhl and J. B. Gray, eds., Academic Press, New York, 1967.

Carreau, P. J., I. Patterson, and Yap, *Can. J. Chem. Eng., 54,* 135 (1976).

Chavan, V. V., and R. A. Mashelkar, in *Advances in Transport Processes,* Vol. 1, A. S. Mujudar, ed. Wiley, New York, 1980, pp. 210–252.

Chavan, V. V., and J. Ulbrecht, *Chem. Eng. J., 6,* 213–223 (1973).

Chavan, V. V., M. Arumugam, and J. Ulbrecht, *AICHE J., 21,* 613 (1975).

Coyled, C. K., H. E. Hirschland, B. J. Michal, and J. Y. Oldshue, *AICHE J., 16,* 903 (1970).

Ford, D. E., and J. Ulbrecht, *AICHE J., 21,* 1230 (1975).

Frohlich, H., and R. Sack, *Proc. Roy. Soc. Lond.*, A185, 415 (1946).

Guber, F. B., *Kauchuk i Resina* (USSR), *9,* 28 (1966).

Hagedorn, D., and J. J. Salamone, *Ind. Eng. Chem. Proc. Des. Dev., 6,* 469 (1967).

Harriott, P., *AICHE J., 8,* 93 (1962).

Harriott, P., *Chem. Eng. Prog. Symp. Ser., 55*(29), 137 (1959).

Houlton, H. G., *Ind. Eng. Chem., 36,* 522 (1944).

Huggins, R. E., *Ind. Eng. Chem., 23,* 749 (1931).

Kapitonov, E. N., *Kauchuk i Resina* (USSR), *4,* 20 (1968).

Karpachev P. S., M. M. Maisel, and N. A. Plevako, *Mashiny i Apparaty Proisvodstv iskustvenoy kozhy i Plenochnykh Materialov* M. Legkay Industria, USSR, 1964.

Kelkar, J. V., R. A. Mahelkar, and J. Ulbrecht, *J. Trans. Inst. Chem. Eng.*, London, *50,* 343 (1972).

Kharkhuta, N. Ya., M. I. Kapustin, and V. P. Semenov, *Dorozhnye Mashiny i Oborudovanie,* M. Mashgiz, USSR, 1968.

Lukach, Yu. E., D. D. Ryabinin, and B. N. Metlov, *Valkovye Mashiny dlia Pererabotki Plastmass i Resinovykh Smesey,* M. Mashinostroenie, USSR, 1967.

Magnusson, K., *IVA, 23*(2), 86–99 (1952).

Metzner, A. B., and J. S. Taylor, *AICHE J., 6,* 109 (1960).

Metzner, A. B., and R. E. Otto, *AICHE J., 3,* 3 (1957).

Metzner, A. B., *Non-Newtonian Technology, Fluid Mechanics, Mixing and Heat Transfer,* Academic Press, New York, 1956, pp. 77–153.

Metzner, A. B., R. D. Vaughn, and G. L. Houghton, *AICHE J., 3,* 92 (1957).

Mizushina, J., et al., *Kagaku Kogaku, 31*(12), 1208 (1967).
Mohr, W. D., R. L. Saxton, and C. H. Japson, *Ind. Eng. Chem., 49*(11), 1857 (1957).
Nagata, S., *Mixing Principles and Application,* Wiley, New York, 1975.
Oldroyd, J. G., *Proc. Roy. Soc. Lond., A128,* 122 (1953).
Orr, C., and J. M. Dalla Valle, *Chem. Eng. Progr. Symp.,* Service No. 9, 50, 29 (1954).
Ostwald, W., *Kollaidzchr, 38,* 26 (1926).
Pikovsky, Ya. M., S. M. Polosin-Nikitin, and N. N. Votshchinia, *Dorozhnye Mashiny i Oborudovanie,* M. Mashgiz, USSR, 1960.
Pryce-Jones, J., *Coll. Zeits., 96,* 129 (1952).
Ramdas, V., V. W. Uhl, M. W. Osborne, and J. R. Ortt, *Heat Transfer Eng., 1*(4), 38–46 (1980).
Reiner, M., *Deformation and Flow,* Lewis Pub., London, 1949.
Rushton, J. H., E. W. Costich, and H. J. Everett, *Chem. Eng. Prog., 46,* 395, 467 (1950).
Schofield, R. K., and G. W. Scott-Bair, *Proc. Roy. Soc. Lond., A138,* 707 (1932).
Skelland, A. H. P., and G. R. Dimick, *Ind. Eng. Chem. Proc. Des. Dev., 8,* 267 (1969).
Skelland, A. H. P., and R. Seksaria, *Ind. Eng. Chem. Proc. Des. Dev., 17,* 56 (1978).
Skelland, A. H. P., "Mixing and Agitation for Non-Newtonian Fluids," in Handbook of Fluids in Motion, N. P. Cheremisinoff and R. Gupta, eds., Ann Arbor Science Publishers, Ann Arbor, MI, 1983, pp. 179–211.
Skelland, A. H. P., D. R. Oliver, and E. Tooke, *Brit. Chem. Eng., 7,* 346 (1962).
Skelland, A. H. P., *Ind. Eng. Sci., 7,* 166 (1958).
Soroka, B. S., and B. A., *Khimichestioe Mashinostroenie,* Kiev, Tekhnika, USSR, 2, 15–19 (1965).
Starov, I. M., et al., *Kauchuk i Resina,* USSR, *6,* 19 (1961).
Toms, B. A., and D. J. Strawbridge, *Trans. Faraday Soc., 49,* 1225 (1953).
Treybal, R. E., *Liquid Extraction,* 2nd ed., McGraw-Hill, New York, 1963.
Ulbrecht, J. *Chem. Eng.,* London, 347–353, *367,* (1974).
Wilkinson, W. L., *Non-Newtonian Flow,* Pergamon Press, New York, 1960.
Winding, C. C., F. W. Dittmann, and W. L. Kranich, Cornell University Report, Ithaca, NY (1944).
Zakharkin, O. A., et al., *Kauchuk i Resina,* USSR, *7,* 8–11 (1966).

4
Application of Screw Mixers

OPERATING PRINCIPLES AND GENERAL CHARACTERISTICS

Screw-type mixers, or extruders, of which there are numerous design configurations, are widely employed for the extrusion of molten or plasticized polymers to semifinal- and final-product forms. These machines are also used for preparing additives such as resins for plastification, pigments, and others. Components to be mixed are normally doped with additives in a feed tunnel to the machine, thus providing a relatively homogeneous mass which is then extruded in the main working chamber (screw channel). Mixing and material extrusion can be accomplished in single or double rotating screw machines.

In single-screw machines, mixing quality is established by the total shear of deformation for a given material volume. Shear deformation is a measure of the relative particle displacement and, as such, is equal to the product of the velocity gradient and the material's residence time. The total shear deformation may be increased by decreasing the height of the screw channel, by varying the angle of the pitch helix, and/or by increasing the countercurrent or sinks through an increase in the head resistance.

The shearing strain of any material element (e.g., a polymer melt) depends on its initial location in the screw channel. The streamlines of material displacement, and consequently the residence time and shear rate, depend on the elements location in the channel's cross section. Material elements at the center of the channel undergo less mixing than elements near the screw's periphery or the

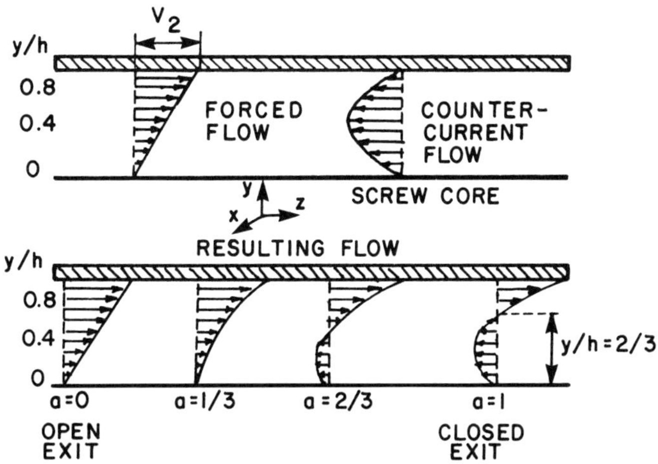

Figure 1 Velocity profiles along channel axis.

channel walls. The reason for this is a smaller residence time for material located at the center of the channel. This can be illustrated by examining the velocity distribution in the working chamber.

There are two types of flows in the channel of a single-screw mixer: longitudinal (along the helical axis of the channel) and transverse (circulatory). Figures 1 and 2 show typical material velocity profiles in the longitudinal and transverse mixing directions, respectively. In the "melt zone," there are four components of motion:

1. A forward motion of melt along the screw axis due to the screw's motion relative to the cylinder (working chamber).
2. A second component which characterizes the braking of the forward motion due to head resistance against the primary flow direction. This component, referred to as the velocity of countercurrent motion, is essentially an ideal-

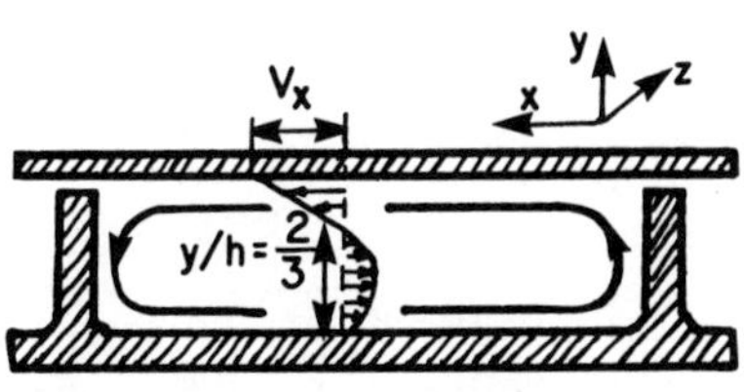

Figure 2 Velocity profiles for circulation flow.

ized phenomenon that will be shown later to simplify analysis of the mixer's productivity and flow patterns.

3. A component which characterizes the flow of sink taking place in the gap between the body and the external surface of the screw.
4. The last component, which characterizes material circulation in the normal plane of the channel, that is, in the plane parallel to horizontal axis. The circulation flow is in fact a transverse flow. The velocity distribution of the transverse flow does not depend on pressure in the head but rather on the channel geometry and rotational velocity of the screw.

To realistically represent the velocity distribution in three dimensions, it is necessary to vectorially sum the velocity components at each point. Figure 3 shows the velocity distribution in the flow region of a screw in a rectangular cross section. Given are values of the forward motion for the melt at the closed exit, open exit, and an intermediate section. The plots reveal an insignificant displacement along the axial direction. Bernhardt (1962) and Lukach et al. (1967) recommend Moore's equation, among others, for determining the shearing strains in the planes parallel or perpendicular to the screw axis. These equations can be used to evaluate the shearing strain as a function of material element location along the height of the channel (i.e., along the y-axis):

$$D_z = \frac{L}{h}\left(A - \frac{B}{1 - a}\right) \tag{1}$$

$$D_x = \frac{L}{h}\left(\frac{A}{\tan \psi} + \frac{B \tan \psi}{1 - a}\right) \tag{2}$$

where

D_z = shearing strain in the plane parallel to the screw axis
D_x = shearing strain in the plane perpendicular to the screw axis
L = screw length
h = height of the screw channel
ψ = helix angle
a = ratio of countercurrent rate to that of forced flow
A and B = dimensionless functions of the ratio to the thread base and height of the channel y/h.

By integrating these equations over the flow's cross-sectional area, an average value of the shearing strain is obtained. Differential configuration and size mix-

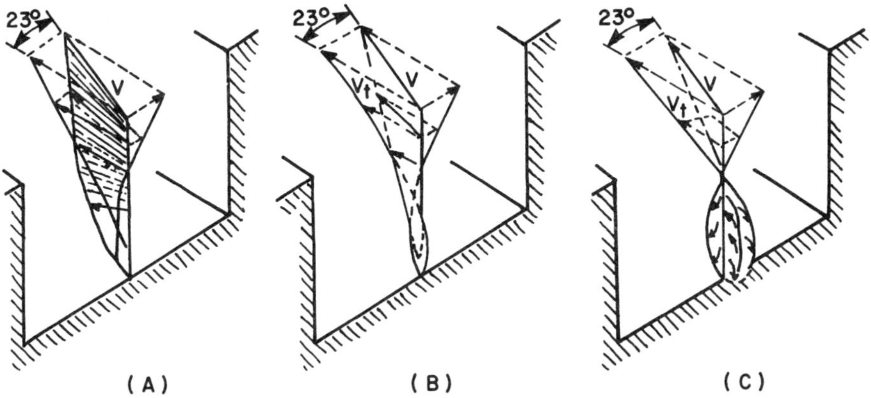

Figure 3 Three-dimensional representation of velocity vector and its two components at melt flow of polymer in the screw channels. (A) free exit; (B) intermediate regime; (C) closed exit (v, velocity vector of particle; V_t, vector velocity component, tangent to the axis helix channel; V_b, vector velocity component normal to helix channel axis).

ing nozzles are provided at the exit of screw mixers in order to increase the shearing strain on the extruded material.

Twin-screw mixers are more effective because they process material by the action of intermeshing screws. The material streamlines are interrupted when they pass through a zone of low-velocity gradients with low mixing effects. However, overall, the bulk of the material volume enclosed between the wall of the casing and screw surfaces undergoes a highly efficient form of mixing. A successful modification of the screw mixer consists of the working member comprising one of the screws with an interrupted helix (the teeth and gaps are located on the helix). These teeth on the rotating screw pass between the stationary teeth on the casing. Material between the meshing teeth is thus subjected to longitudinal, axial, and radial shears.

Another variation is the *KO-kneader,* which can handle light and medium stiff pastes. In this design a worm runs along a horizontal casing fitted with teeth. The worm is interrupted at regular intervals by gaps. The shaft not only turns, but reciprocates in the axial direction so that the teeth periodically clean the gaps on the casing. This periodic reciprocation minimizes the material flow in one direction only. Note, however, that the intensive shearing strain in this machine is periodic and the gaps tend to reduce the effective shear stress. The large surface areas of material exposed are a result of shear stresses generated by great torques. This limits the viscosity or rigidity of material to be mixed.

PARAMETRIC CALCULATIONS

The first parameter of interest in design calculations for screw-type mixers is the rated capacity. In the case of polymers this requires knowledge of the material's non-Newtonian behavior in the dosing zone, that is, in its molten state. In some applications separate estimates of the capacity of the feeding zone are needed. This is important in the prevention of pulsating flows through the machine, whereby the designer must establish a feed capacity to exceed that in the dosing zone. Standard design calculations are outlined below.

The first case of interest concerns establishing the *capacity of the feeding zone of a single-screw mixer for polymer processing*. In this zone of the machine the polymer is normally introduced in solid granular or crumb form. The entrainment and transportation of solid granular material in this region depends not only on the equipment's geometry (pitch, rpm, helix angle), but also on the friction between the process material and working metal parts. The friction coefficient of plastics that are in contact with hot metal surfaces increases with increasing temperature up to the plastic's melting point and then decreases. Figure 4 defines the principal dimensions of interest in the transporting zone. The capacity in this zone may be computed from the following formula:

$$Q = \frac{\pi^2 D^2 \tan^3 \psi}{4} (F - \tan \rho' T)n, \quad \text{cm}^3/\text{min} \tag{3}$$

where

D = screw outside diameter, cm
ρ' = friction angle of material against metal
n = rotational velocity, rpm

The coefficients F and T are defined as follows:

$$F = 4.6 \log \frac{\sin \psi_1}{\sin \psi} + (\cos^2 \psi - \cos^2 \psi_1) \tag{4}$$

$$T = \frac{\sin(\psi_1 - \psi)}{\sin \psi \sin \psi_1} + \cos(\psi_1 + \psi) + \sin(\psi_1 - \psi)\frac{\pi}{90°} \tag{5}$$

where ψ is the helix angle based on the outer diameter of the coil;

$$\tan \psi = \frac{t}{\pi D} \tag{6}$$

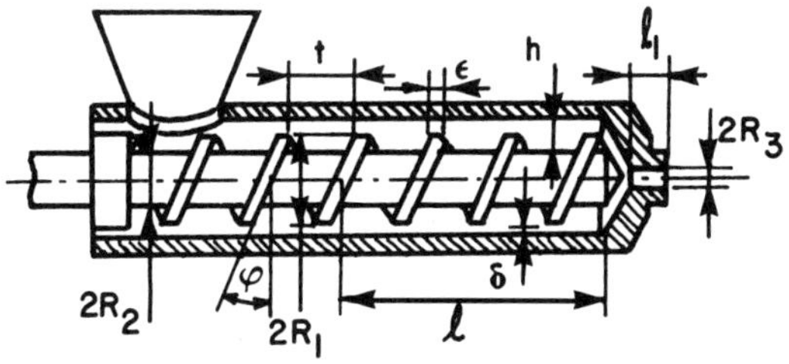

Figure 4 Important geometric relations for single-screw mixer extruders.

ψ_1 is the helix angle based on the core diameter;

$$\tan \psi_1 = \frac{t}{\pi(D - 2h_0)} \tag{7}$$

$$\tan \rho = f \tag{8}$$

where

t = helix pitch, cm
f = friction coefficient between material and metal

The volumetric efficiency of the transport zone can be calculated from the following formula:

$$\eta_v = \frac{\pi D^3 \tan^3 \psi}{(D^2 - d^2)t} (F - T \tan \rho) \tag{9}$$

Literature data for establishing the optimum screw size and specific dimensions are sparce. Ryabinin and Lukach (1965) report some specific criteria for this purpose along with the following formula for determining the screw's critical rotational velocity in the feeding zone:

$$n_{cr} = \frac{42.4 \cos \rho}{\sqrt{D} \sin \psi \sin (\psi + \rho)} \tag{10}$$

The capacity of the dosing zone characterizes the rated capacity of the machine. However, this zone depends not only on the screw's geometry and rotational velocity, but also on the arrangement of the head. Capacity calculations for the dozing zone are recommended as follows: first, determine the geometric characteristics of the screw, then design the capacity of passage through the head.

Capacity of Dosing Zone for Constant-Size Screws

For steady-state laminar flow of an isotropic noncompressible liquid, the equation of motion in the z-direction may be written as follows:

$$\frac{\partial p}{\partial z} = \mu_a \left(\frac{\partial^2 v_z}{\partial x^2} + \frac{\partial^2 v_z}{\partial y^2} \right) + \frac{\partial v_z}{\partial x} \cdot \frac{\partial \mu_a}{\partial x} + \frac{\partial v_z}{\partial y} \frac{\partial \mu_a}{\partial y} \tag{11}$$

From Eq. (11), the material's viscosity is a function of the location of each fluid element in the channel and consequently depends on both the temperature and velocity gradients.

If we assume that the fluid's viscosity varies insignificantly over the channel's cross section, which is reasonable for many cases, then $(\partial \mu_a / \partial x) = 0$ and Eq. (11) simplifies to

$$\frac{\partial p}{\partial z} = \mu_a \left(\frac{\partial^2 v_z}{\partial x^2} + \frac{\partial^2 v_z}{\partial y^2} \right) + \frac{\partial v_z}{\partial y} \times \frac{\partial \mu_a}{\partial y} \tag{12}$$

Equation (12) describes nonisothermal flow of a non-Newtonian fluid through the channel of the screw. Unfortunately, no analytical solution to Eq. (12) has been reported in the literature.

If we further assume the viscosity to be constant along the channel height, the last term in Eq. (12) is eliminated, that is, $(\partial \mu_a / \partial y) = 0$. The equation then becomes

$$\frac{\partial p}{\partial z} = \mu_a \left(\frac{\partial^2 v_z}{\partial x^2} + \frac{\partial^2 v_z}{\partial y^2} \right) \tag{13}$$

Further simplifying assumptions are needed in order to develop a working formula. Consider a fixed point on the surface of a screw. As the screw rotates, the position of this point changes relative to the cylinder, and the material is displaced along the helix line, which may be considered as the screw's mirror image. From the standpoint of material transport, this situation is easier to visualize by considering the screw to be stationary and the cylinder as rotating in the direction opposite to the actual rotation of the screw. The cylinder's velocity, U, may be expanded in two components: T, perpendicular to the direction of the screw thread, and V, parallel to it. All three velocity vectors can be related to each other through the angle of the helix slope of the screw:

$$T = U \sin \psi \tag{14}$$

$$V = U \cos \psi \tag{15}$$

If the material's density does not vary under pressure, then capacity may be defined as:

$$Q = Q_0 - Q_c - Q_l \tag{16}$$

where

Q_0 = volumetric capacity of longitudinal flow
Q_c = volumetric capacity of countercurrent flow
Q_l = volumetric capacity of sinks

The transverse velocity component [Eq. (14)] does not contribute to the system's overall capacity. However, where heat transfer is important, it does promote forced convection between flat layers of materials. In layers of high viscosity, natural convection is insignificant. The transverse velocity component is also important in applications involving the addition and mixing of fillers in polymer melts.

When the material is subjected to pressure, an additional flow pattern due to "leakage" arises. In this case the material flows through a very narrow ring between the screw and the internal surface of the mixer case. Considering this ring to be an infinitely long narrow gap, Eq. (13) may be restated to describe one-dimensional liquid flow under pressure:

$$\frac{\partial^2 v_z}{\partial y^2} = \frac{1}{\mu_a} \times \frac{dp}{dz} \tag{17}$$

Equation (17) is a second-order homogeneous differential equation.

The constants of integration, C_1 and C_2, can be obtained from the following boundary conditions:

At $y = 0$, $v_z = 0$; at $y = \delta$, $v_z = 0$.

From the first boundary condition, we find that $C_2 = 0$.

Using the second condition, we obtain

$$C_1 = \frac{1}{2\mu_a} \frac{dp}{dz} \delta \tag{18}$$

where δ is the height of the gap.

Substituting the expression for C_1 into the integrated form of Eq. (17), the velocity distribution is obtained:

$$V = \frac{1}{2\mu_a} \times \frac{dp}{dz} y^2 + \frac{1}{2\mu_a} \times \frac{dp}{dz} \delta y \tag{19}$$

or

$$V = \frac{1}{2\mu_a} \times \frac{dp}{dz} (\delta y - y^2) \tag{20}$$

To determine the volumetric rate, or the amount of leakage for a single-start thread, Eq. (20) is integrated along the gap height. The obtained equation is then multiplied by the gap width $D/\cos\psi$ to give

$$Q_y = \frac{\pi D}{\cos\psi} \frac{1}{2\mu_a} \frac{dp}{dz} \frac{\delta^3}{6} \tag{21}$$

or

$$Q_y \, dz = \frac{\pi D \delta^3}{12\mu_a \cos\psi} dp \tag{22}$$

Integrating Eq. (22) over the following limits: For the left-hand side from 0 to $l \cos\psi$ (i.e., the gap length); and for the right-hand side from p_1 to p_2 where p_1 and p_2 are the pressures on the sides of screw ridge:

$$Q_y l \cos\psi = \frac{\pi D \delta^3}{12\mu_a \cos\psi} (p_2 - p_1) \tag{23}$$

or

$$Q_y = \frac{\pi D \delta^3 \epsilon}{12\mu_a \cos^2\psi} \Delta p \tag{24}$$

where ϵ = eccentricity; $\epsilon = 1$ for ideal centering of screw in the casing. ϵ may vary up to a value of 2.5 depending on the friction of the screw against the casing.

For practical calculations ϵ may be assumed to have a value of 1.2. The pressure drop across a single-start thread can be estimated from the total pressure drop across the unit (Δp_0):

$$\Delta p = \frac{\pi D \tan \psi}{L} p_0 \cos^2 \psi \tag{25}$$

Substituting Eq. (25) into Eq. (24) for $\epsilon = 1.2$, we obtain

$$Q_y = \frac{\pi^2 D^2 \delta^3 \tan \psi \sin \psi}{10 \mu_a e l_0} \Delta p_0 = \gamma \frac{\Delta p_0}{\mu_a} \tag{26}$$

The gap will typically account for about 10% of the countercurrent flow rate.

Depending on the size of the gap, the mean bulk temperature of the melt can be affected. As a rough guideline, the gap height should be approximately 0.2 to 0.3 mm in order to minimize thermal gradients. For radial gaps larger than 0.2 mm, the leaking layer of polymer will experience a temperature rise usually no greater than 10°C. For very narrow gaps, temperature gradients are more pronounced. For examples, at a gap size of 0.1 mm, the temperature increase can be as high as 40°C, and at 0.05 mm, it may be 100°C. The designer should always check the gap size in screw mixers, particularly for materials susceptible to thermal destruction.

The radial gap has no influence on the total horsepower requirements; however, if leakage rates exceed 10% of the countercurrent flow, mixer capacity suffers. It is the forced flow along the axis of the screw which basically establishes the unit's capacity. This flow is strongly dependent on the conditional countercurrent flow. Both flows are described by the following differential equation:

$$\frac{d^2 v}{dy^2} + \frac{d^2 v}{dx^2} = \frac{1}{\mu} \frac{dp}{dz} \tag{27}$$

As shown by Eq. (27), the material's velocity at any point along the screw axis depends on the local depth (i.e., the flowing fluid's free surface profile). The two-dimensional velocity distribution given by this equation is difficult to solve even for the simplest of screw geometries because of the complexity of flow patterns. To simplify the expression, we may assume that the channel walls do not influence the velocity distribution. This is a limiting-case assumption, analogous to flow between infinite parallel plates. Equation (27) then reduces to a one-dimensional expression for the velocity profile. For relatively shallow channels, this is a good approximation. For aspect ratios (i.e., ratio of channel width to mean depth) greater than or equal to 10, the absolute error between predicted and measured profiles is no more than 10%.

To derive the one-dimensional velocity distribution, the second derivative along the x-axis in Eq. (27) is taken to be zero:

$$\frac{d^2v}{dy^2} = \frac{1}{\mu_a}\frac{dp}{dz} \tag{28}$$

whence the equation is solved:

$$v = \frac{1}{\mu_a}\frac{dp}{dz}\left(\frac{y^2}{2}\right) + C_1 y + C_2 \tag{29}$$

The boundary conditions for C_1 and C_2 are

At $y = 0$, $v = 0$; at $y = h$, $v = v_0$.

Thus,

$$C_1 = \frac{v_0}{h} - \frac{1}{\mu_a}\frac{dp}{dz}\left(\frac{h}{2}\right) \tag{30}$$

and

$$v = \frac{v_0 y}{h} + \frac{(y^2 - hy)}{2\mu_a}\frac{dp}{dz} \tag{31}$$

The first term in Eq. (31) is the longitudinal or forced-flow velocity. The second term describes the countercurrent flow.

The forced flow follows a linear law, whereas the countercurrent flow is defined by a parabolic relationship. The volumetric velocity in this case is

$$Q = \int_0^h bv\,dy \tag{32}$$

where b is channel width;

$$b = (t - e)\cos\psi \tag{33}$$

where

t = helix pitch
ψ = helix angle

or

$$Q = \frac{bh}{2}v_0 - \frac{bh^3}{12\mu_a}\frac{dp}{dz} \tag{34}$$

The first term on the right-hand side of Eq. (34) is the volumetric rate of longitudinal (forced) flow, and the second term is the volumetric countercurrent flow.

The linear velocity at rotation is $U = \pi Dn/60$, and its component is

$$v_0 = \frac{\pi Dn}{60} \cos \psi, \quad \text{cm/s} \tag{35}$$

where

D = outside diameter of the screw, cm
n = rotational velocity, rpm

Taking into account the relations for v_0 and b, we obtain

$$Q = \frac{\pi Dnh(t - e) \cos^2 \psi}{120} - \frac{h^3(t - e) \cos \psi}{12\mu_a} \frac{dp}{dz} \tag{36}$$

where z is the axis along the helical channel of the screw; it is convenient to change it by the longitudinal axis of the screw, x. Then

$$dz = \frac{dx}{\sin \psi} \tag{37}$$

The volumetric capacity thus becomes

$$Q = \frac{\pi Dn(t - e) \cos^2 \psi}{120} - \frac{h^3(t - e) \sin^2 \psi}{24\mu_a} \frac{dp}{dz} \tag{38}$$

The volumetric capacity of a multithreaded screw having constant size is expressed by the following equation:

$$Q = \frac{i\pi Dnh(t/i - e) \cos^2 \psi}{120} - \frac{ih^3(t/i - e) \sin^2 \psi}{24\mu_a l_0} p_0 \tag{39}$$

or

$$Q = \alpha n - \beta \frac{\Delta p_0}{\mu_a} \tag{40}$$

Taking into account leakage:

$$Q = \alpha n - \beta \frac{\Delta p_0}{\mu_a} - \gamma \frac{\Delta p_0}{\mu_a} \tag{41}$$

where

i = number of threads
D = outside diameter of the screw
h = thread depth
n = rotational velocity of the screw
t = helix pitch
e = thread width
ψ = helix angle
l_0 = zone length
α = coefficient of the injection flow
β = coefficient of countercurrent flow
γ = coefficient of leaking

Equation (41) characterizes the flow in a screw channel of constant dimension and geometry.

Dosing Zone Capacity of Screws of Variable Pitch

The efficiency of a screw mixer varies with the volume of the pitch in the helical channel of the screw. The volume reduces in the direction from the inlet hopper to the extruder head. Capacity can be computed from the following formula:

$$Q = \alpha_1 n - \frac{\beta_1 + \gamma_1}{\mu_a} \Delta p_0, \quad \text{cm}^3/\text{min} \tag{42}$$

where α_1 is the coefficient of flow injection

$$\alpha_1 = \frac{\pi^3 D(D - h)^2 h J_1}{2\pi^2 (D - h) J_2 - J_3} \tag{43}$$

β_1 is the coefficient of countercurrent flow

$$\beta = \frac{h^3 \pi (D - h)\, \Delta t}{3[4\pi^2 (D - h)^2 J_2 - J_3]} \tag{43}$$

and γ_1 is the coefficient of leaking

$$\gamma_1 = \frac{\pi D^2 \delta^3\, \Delta t}{5e(D - h)[\pi(D - h) J_4 + t J_5]} \tag{44}$$

where

n = rotational velocity
μ_a = apparent viscosity
Δp_0 = pressure drop over the screw
D = screw outside diameter
h = depth of screw thread
Δt = pitch increment along screw length, defined as: $\Delta t = (t_{in} - t_{out})/(m - 1)$
δ = gap between screw threads and case
J_1, J_2, J_3, J_4, J_5 = coefficients characterizing a screw design having variable pitch,

$$J_1 = t_{in} - t_{out} + \Delta t\ 2.3 \log(t_{in}/t_{out}) \tag{45}$$

$$J_3 = \frac{\Delta t + 2e}{2e}\ 2.3 \log\left(\frac{t_{in} - e}{t_{out} - e}\right) - \frac{\Delta t}{2e}\ 2.3 \log\left(\frac{t_{in}}{t_{out}}\right) \tag{46}$$

$$J_3 = (t_{in} - t_{out})(t_{in} - t_{out} + 2\Delta t + 6e) + 4.6e(\Delta t + 2e) \log\left(\frac{t_{in} - e}{t_{out} - e}\right) \tag{47}$$

$$J_4 = 2.3 \log\left(\frac{\tan \psi_{in}/2}{\tan \psi_{out}/2}\right) + 2\frac{\sin\dfrac{\psi_{in} + \psi_{out}}{2}\sin\dfrac{\psi_{in} - \psi_{out}}{2}}{\cos \psi_{in} \cos \psi_{out}} \tag{48}$$

$$J_5 = 2\frac{\cos\dfrac{\psi_{in} + \psi_{out}}{2}\cos\dfrac{\psi_{in} - \psi_{out}}{2}}{\sin \psi_{in} \sin \psi_{out}} - 2.3 \log\left\{\frac{\tan[(90 - \psi_{in})/2]}{\tan[(90 - \psi_{out})/2]}\right\} \tag{49}$$

where

t_{in} = initial screw pitch
t_{out} = outlet screw pitch
ψ_{in} = initial helix angle
ψ_{out} = outlet helix angle
m = number of pitches, $m = 2e/(t_{in} + t_{out})$
e = width of screw thread
l = total length of the screw

Dosing Zone Capacity of Screws with Variable Depth

The volume of material in the mixer chamber decreases along the length of the helical pitch because of a decrease in depth. A decreasing material profile is observed in mixer designs where the conical core diameter of the screw is a maximum at the head of the mixer. The channel depth along the axis of the mixer can be approximated by the following linear relation:

$$h_x = h_1 - (h_1 - h_2)\frac{x}{l} \tag{50}$$

where h_1 is the depth of the helix channel at $x_2 = 0$, that is, at the entrance to the zone where the material starts melting, and h_2 is the depth of the helix channel at $x_1 = l$, that is, at the end of the channel, near the head.

Defining the rate of compression as

$$i' = \frac{V_{\text{in}}}{V_{\text{out}}} \tag{51}$$

and

$$h_1 = 0.155D \tag{52}$$

The available chamber volume per screw pitch at the loading funnel is then

$$V_{\text{in}} = \frac{\pi(D^2 - d_1^2)}{4}(t - e) \tag{53}$$

where

D = outside diameter of the screw
d_1 = core diameter

The available chamber volume per screw pitch at the head is

$$V_{\text{out}} = \frac{\pi(D^2 - d_2^2)}{4}(t - e) \tag{54}$$

where d_2 is the core diameter of the head, $d_2 = D - 2H_2$.

The rate of compression is given as

$$i' = \frac{h_1(D - h_1)}{h_2(D - h_2)} \tag{55}$$

The total capacity of the screw with variable core depth is

$$Q = \alpha_2 n - \frac{(\beta_2 + \beta_2)}{\mu_a} \Delta p, \quad \text{cm}^3/\text{min} \tag{56}$$

where

n = screw's rotational velocity
μ_a = apparent viscosity
Δp = pressure drop over the limits of the screw
α_2 = coefficient of injecting flow

$$\alpha_2 = \frac{\pi^3 (t - \lambda e)\sigma}{a + t^2 b}, \quad \text{cm}^3 \tag{57}$$

σ = constant based on screw dimensions:

$$\sigma = 1 - \frac{6.9D}{2(h_{\text{out}} - h_z)} \log \frac{h_{\text{out}}}{h_z} + \frac{D^2}{2h_{\text{out}} h_z} \tag{58}$$

β_2 = coefficient of countercurrent flow

$$\beta_2 = \frac{\pi t (t - \lambda e)}{12\, L_{\text{in}} (a + t^2 b)}, \quad \text{cm}^3 \tag{59}$$

γ_2 = coefficient of leaking

$$\gamma_2 = \frac{\pi D \delta t^2}{10\, e L_{\text{in}} \sqrt{\pi D^2 + t^2}}, \quad \text{cm}^3 \tag{60}$$

a, b are coefficients characterizing a mixer with variable core dimensions:

$$a = \frac{\pi^2}{h_{\text{in}} h_z} \left(\frac{D(h_{\text{in}} + h_z)}{2h_{\text{in}} h_z} - 1 \right), \quad \text{cm}^{-2} \tag{61}$$

$$b = \frac{2.3}{(h_{\text{in}} - h_z)} \log \frac{h_{\text{in}} (D + d_2)}{h_z (D + d_1)} + \frac{2h_{\text{in}} h_z + (h_{\text{in}} + h_z)D}{2D^2 h_{\text{in}}^2 h_z^2}, \quad \text{cm}^{-4} \tag{62}$$

$$h_{\text{in}} = h_1 - \frac{h_1 - h_2}{L} L_0 = h_1 (0.6 \sim 0.4)(h_1 - h_2) \tag{63}$$

ψ = helix angle based on the screw's outer diameter.

Dosing Zone Capacity and Head Resistance

A screw mixer's operation depends not only on the geometry of working parts and their boundaries, but also on the geometry of the head and profiling elements.

The volumetric flow capacity through the head or some other device at the mixer's inlet is directly proportional to pressure drop and inversely proportional to the material's dynamic viscosity:

$$Q = K \frac{\Delta p}{\mu} \tag{64}$$

Combining Eq. (64) with Eqs. (41), (42), or (56), we obtain

$$Q = \frac{\alpha K}{K + \beta + \gamma} n \tag{65}$$

$$Q = \frac{\alpha_1 K}{K + \beta_1 + \gamma_1} n \tag{66}$$

$$Q = \frac{\alpha_2 K}{K + \beta_2 + \gamma_2} n \tag{67}$$

where K is the coefficient depending on head geometry, cm^3.

From information on K for different screw mixer heads the capacity of a screw mixer can be readily determined. Figure 5 shows the relationships for volumetric capacity in terms of different process parameters. Figure 5A shows that the capacity of the head and screw increases as a function of pressure induced by the screw. The working point A is the intersection of the head characteristics (line 1) with the screw characteristics (line 2). Figure 5B shows that the capacity of the head and screw is process dependent. Curves 1 and 4 characterize an adiabatic process, and curves 2 and 3 show the change in capacity for an isothermal process. The region limited by curves 1, 2, 3, and 4 represents a polytropic mixer operation.

Figure 5C shows the location of the working points of the mixer which depend on the head resistance (curves K_1 and K_2) and rotational velocity (curves n_1 and n_2). Figure 5D shows the influence of material temperature on mixer capacity. In this plot the characteristics of the screw and head are represented at temperatures T_1 and T_2 (where $T_2 > T_1$). If the melt temperatures in the screw and head are the same and equal to T_2, the regime of the mixer is determined by point 1. By simultaneously decreasing the melt temperature in the screw and the head to value T_1, the pressure in the head is increased up to point C. The capacity in this case remains unchanged. However, if the melt temperature in the screw is

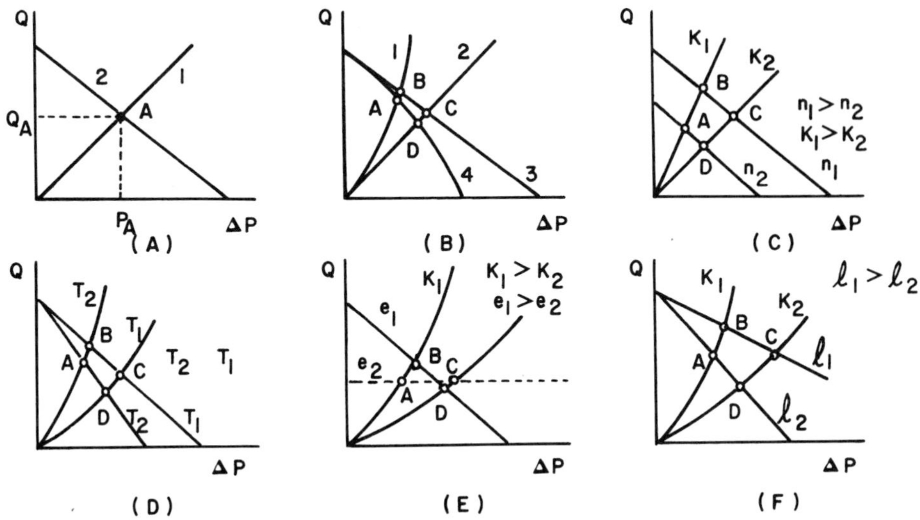

Figure 5 Relationship between capacity and pressure drop.

T_2, but the temperature in the head is decreased to T_1, capacity will decrease to some value as denoted by point D. On the other hand, if the melt temperature in the screw is T_1 and the head is heated to the higher temperature T_2, the mixer's capacity is increased slighty and is given by point B.

Figure 5E illustrates capacity in the screw and the head as a function of the depth of the helical channel (e_1, deep channel; e_2, shallow channel) and head resistance. For the head having the larger cross-sectional area (i.e., smaller resistance coefficient), point A denotes the operating regime for a shallow channel and point B denotes that of a deep channel. In this case the screw having a deeper channel provides higher capacity. For a head with a higher resistance (i.e., smaller cross-sectional area) point C determines the operating regime for a shallow channel, and point D that of a deep one. In the example shown, the screw having a shallow channel and higher pressure provides greater capacity.

For each head there corresponds an optimum screw channel geometry which provides maximum capacity. Figure 5F shows the influence of length on the discharge zone. An increase in screw length results in a decrease in capacity.

Design of Heads

The forming elements of heads are comprised of orifices of specified geometries through which polymer can be extruded, thus producing a finished part. The design method for these forming elements consists of specifying a flow area to meet maximum capacity at a given pressure. That is, the design should be based on a predetermined resistance coefficient for the head or an inverse value for K, referred to as the shape factor. From this information, the geometry and size of the forming elements required to deliver the desired product can be established. Forming elements can be classified into two groups according to the flow patterns of the melt:

1. Head-forming elements where the melt is in one-dimensional flow, that is, the flow velocity varies only in one direction (simple section)
2. Head-forming elements where the flow is two- or three-dimensional, that is, the velocity varies in two or more directions (sections of complex profiles)

In forming elements having two-dimensional flow the acting forces distort the shape of the extruded profile in comparison to the profile of the forming elements because there are different extrusion velocities in sections of different sizes. The shape factor is defined as

$$K = \frac{Q\mu_a}{\Delta p} \tag{68}$$

where

Q = volumetric capacity, cm^3/s
Δp = pressure drop across the head, kg/cm^2
μ_a = apparent viscosity, $kg \times s/cm^2$

Neglecting time, local temperature variations due to adiabatic effects, and changes in viscosity, the shape factor K (cm^3) can be considered a constant. In addition, several approximating assumptions can be made in order to establish a design basis. First, the melt may be considered to be in laminar flow.

Second, we assume that the melt "wets" the inside walls of the head casing. Thus, the outer film of the flow sticks to the walls and the flow velocity on them is zero. The equilibrium equation between the pressure of extrusion and shear stress is

$$\int p\, dq = \int \tau\, dF \tag{69}$$

where

p = exit melt pressure, kg/cm^2
q = channel cross-sectional area, cm^2
τ = shear stress, kg/cm^2
F = boundary area between the two layers of laminar flow, cm^2

The flow expression is

$$\frac{dv}{dl} = -\frac{1}{\mu_a}\left(\frac{\tau}{\tau_0}\right) \tag{70}$$

where

v = velocity at a specified point, cm/s
l = linear size, cm
μ_a = melt viscosity, kg-s/cm^2
τ = shear stress between two adjacent layers, kg/cm^2
τ_0 = shear stress, kg/cm^2

For many non-Newtonian thermoplastics, a modified form of Newton's law may be applied:

$$\frac{dv}{d\rho} = -\frac{1}{\mu_a}\left(\frac{\tau}{\tau_0}\right)^n \tag{71}$$

where

n = exponent ($1 < n < 4$)
ρ = flow coordinate, cm

The extrusion rate is obtained by integrating this expression:

$$v = \int \frac{dv}{d\rho}\, d\rho \tag{72}$$

A second integration yields an expression for the extrusion capacity:

$$Q = \int v\, dq \tag{73}$$

where

q = cross section of the flow, cm^2
v = velocity at a specified point along the channel, cm/s

Typical channel shapes and design formulas for their coefficients are given in Table 1.

Table 1 Screw Mixer Channel Shape Factors K (cm³) and Shear Rates S (1/s)

Channel shape	Design scheme	Design formulas
Round cylindrical	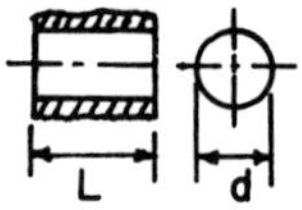	$K_c = \dfrac{\pi d^4}{128 L_c}$; $Sc = \dfrac{32Q}{\pi d^3}$
Round conical	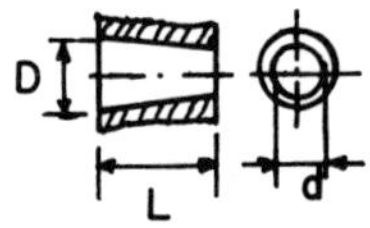	$K_k = \dfrac{3\pi D^3 d^3}{128L(D^2 + Dd + d^2)}$ $S = \dfrac{256Q}{\pi(D - d)}$
Round circular	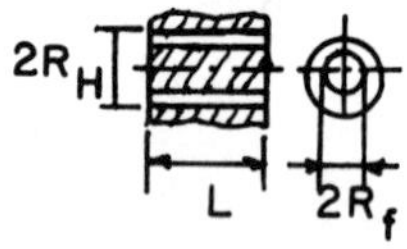	$K = \dfrac{\pi}{8L}\left[R_{out}^4 - R_{in}^4 - \dfrac{(R_{out}^2 - R_{in}^2)^2}{\ln(R_{out}/R_{in})}\right]$ $S = \dfrac{5.58Q}{\pi(R_{out} + R_{in})(R_{out} - R_{in})^2}$
Conical circular with a conical gap	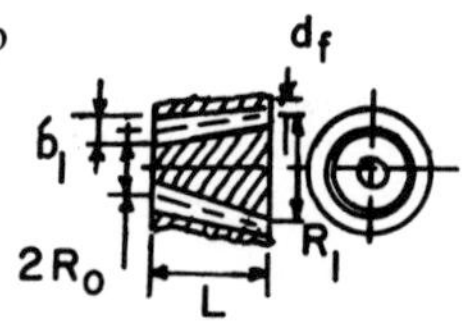	$K_{b'} = \dfrac{\pi(R_0 b'_2 - R_1 b'_1)}{6Lm}$ $m = \dfrac{(R_0 - R_1)^2}{(R_2 b'_2 - R_1 b'_1)^2}\ln(R_0 b'_2/R_1 b'_1) - \dfrac{(R_0 - R_1)(b'_1 - b'_2)}{(R_0 b'_1 - R_1 b'_1)b'_1 b'_1} - \dfrac{b_1'^2 - b_2'^2}{2b_1'^2 b_2'^2}$ $S = \dfrac{22.32Q}{\pi(R_0 + R_1)(b_1 + b_2)^2}$
Flat gap	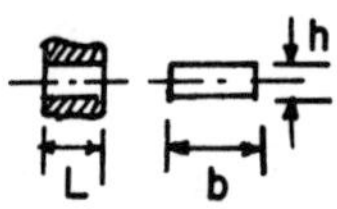	$K_{fg} = \dfrac{bh^3}{12L}$, $S = \dfrac{5.58Q}{bh^2}$
Wedge gap	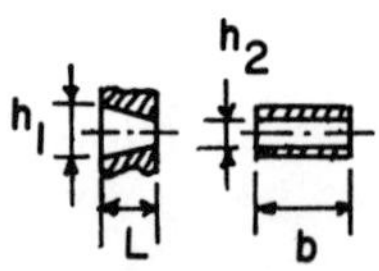	$K_{wg} = \dfrac{bh_1^2 h_2^2}{6L(h_1 + h_2)}$ $S_{wg} = \dfrac{11.16Q}{6(h_1 + h_2)^2}$
Any cross section	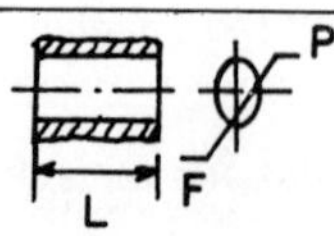	$K = \dfrac{F}{2LP^2}$, $S = \dfrac{2QP}{F^2}$

The head profile may be considered to consist of channels of simple shapes. Then the total shape factor may be calculated as:

$$K = \frac{1}{1/K_1 + 1/K_2 + \cdots + 1/K_i} \tag{74}$$

This simplified method of calculating profile elements of heads is satisfactory for specifying head profiles for mixing and can be used for granulation or semi-product manufacturing. Other design methods for more complex shapes are outlined in the literature [see Lukach et al. (1967), Ryabinin et al. (1965)]. In addition to the shape factor, the head must be specified. The following formula can be used:

$$\Delta p = \frac{Q\mu_a}{K} \tag{75}$$

where

Q = capacity, cm^3/s
μ_a = viscosity, $kg\text{-}s/cm^2$
K = shape factor of the head, cm^3

Capacity of Twin-Screw Mixers

The flow of molten material in this type of machine is more complex than in single-screw mixers. Design specifications are primarily based on experimental data. The mixer's capacity is determined by both forced and countercurrent flows, as defined by the following equation:

$$Q = n(2\pi - d')(\pi D \tan \psi - b')(D - h)h - \frac{(2\pi - \alpha')D\delta^3 \tan \psi \, \Delta p}{12\mu_a b'} \tag{76}$$

where

n = screw rotational velocity
α' = an angle determined by the following formula:

$$\cos \alpha' = 1 - \frac{2h}{D - h} + \frac{h^2}{2(D - h)^2}$$

h = depth of helix channel

D = screw diameter
ψ = helix angle
b' = thickness of the thread
δ = radial gap between the coils and core of the screw
Δp = mass pressure difference on both sides of the coil
μ_a = apparent viscosity of the mass

The effective capacity as computed from Eq. (76) does not account for the geometric compatibility of the screws. Consequently, the formula given below is recommended for characterizing the capacity of double-screw mixers with the screw rotation in one direction:

$$Q = 2Kn\left[\pi h(t - by)(Df + DF_d)\frac{\cos^2\psi}{2} - DF_c f - hf - F_3 t\right] \tag{77}$$

where

K = coefficient considering the influence of the mixing cams, $K \simeq 0.6$
n = rotational velocity, rpm
h = depth of helix channel
t = helix pitch
b_y = conditional width of coil ridge
D = outside diameter, cm
ψ = helix angle
F_d = coefficient considering the influence of the side walls, $F_d \simeq 0.64$
F_c = coefficient considering the influence of the channel curvature, $F_c \simeq 1.05$
f = correction coefficient, $f \simeq 0.28$
F_3 = area of segment gearing, cm^2

Screw Mixer Capacity for Rubber Processing

The capacity of screw mixers employed in processing rubber mixtures can be computed from the following correlation:

$$Q = 60\nu_c n i \rho \beta, \quad \text{kg/hr} \tag{78}$$

where

ν_c = volume between two coils of one-screw mixer, m^3
n = screw rotational velocity, rpm

i = number of screw threads
ρ = mixture density, kg/m^3
β = space factor between coils, $\beta = 0.05 \sim 0.3$ (for a single-screw mixer with constant pitch and for double-screw mixers with variable pitch).

In this case the force acting on the screw is

$$P_z = \frac{91700N}{nr_0 \tan(\alpha + \rho)}, \quad \text{kg} \tag{79}$$

where

P_z = axial force in the helix channel
N = horsepower transmitted by the screw, kW
n = rotational velocity, rpm
r_0 = average coil radius, cm
α = helix angle
ρ = friction angle of rubber mixture, $\rho = 26 \sim 38°$

Further details on this design case are given by Ryabinin and Lukach (1965).

Horsepower Drive

The horsepower consumed by screw mixers is expended on material displacement along the channel and for material shear in the gap between the screw ridge and the internal wall of the case:

$$N = \frac{N_1 + N_2}{\eta} \tag{80}$$

where

N_1 = horsepower spent for forced displacement along the screw channel
N_2 = horsepower spent on material shear in the gap between coil ridge and wall of the case
η = efficiency, $\eta = 0.4 \sim 0.6$.

$$N_1 = \frac{\pi^3(t - e)LJ\mu_a \times 10^{-4}}{36t} n^2 + \frac{\alpha_2 \Delta pn \times 10^{-3}}{6}, \quad \text{kg-m/s} \tag{81}$$

where J is a coefficient obtained from the following formula:

$$J = \frac{\pi^2 D^2 - 4t^2}{\pi^2} + \frac{(D - d)^3 - (D + d_1)^3}{3(d_2 - d_1)} + \frac{2.3\pi^2 D^5}{[t^2 + \pi^2 D^2 (h_{in} - h_{out})]} \log\left(\frac{h_{in}}{h_{out}}\right) \tag{82}$$

where

n = rotational velocity, rpm
p = pressure drop across the head, kg/cm^2
α_2 = constant for a specified flow, cm^3
d_1 = core diameter at the inlet of the dosing zone, cm
d_2 = core diameter at the outlet, cm
h_{in} = thread depth at inlet of the dosing zone, cm
h_{out} = thread depth at the screw outlet, cm

The following formula can be used to estimate the shear rate in the screw helix, which in turn can be used to evaluate the material's dynamic viscosity:

$$S = \frac{\pi^2 (D - h_{th})(D - 2h_{th})^n}{60 h_{th} \sqrt{\pi^2 (D - 2h_{th})^2 - t^2}}, \quad 1/\text{s} \tag{83}$$

where

h_{th} = average depth of the thread,

$$h_{th} = \frac{h_1 + h_2}{2} \tag{84}$$

h_1 = thread depth at the inlet of the dosing zone, cm
h_2 = thread depth at the end of the screw, cm

$$N_2 = \frac{10^{-4} \pi^3 D^3 e L \mu_a n^2}{36 \delta t}, \quad \text{kg-m/s} \tag{85}$$

where

D = screw outside diameter, cm
e = width of a ridge, cm
L = length of dosing zone, cm
δ = radial gap, cm

t = helix pitch of the channel, cm
n = screw rotational velocity, rpm
μ_a = apparent viscosity, kg-s/cm^2

The apparent dynamic viscosity can be determined from the relationship between shear rate and viscosity. The rate of shear can be computed from the following formula:

$$S = \frac{\pi^2 D^2 n}{60\delta\sqrt{\pi^2 D^2 + t^2}}, \quad s^{-1} \tag{86}$$

The temperature rise due to the transformation of mechanical energy into thermal energy may be approximated from the following formula:

$$\Delta t = \frac{860[N - (Q\ \Delta p/36 \times 10^4)]}{QC_p\rho} \tag{87}$$

where

N = horsepower consumed by a screw, kwt
Q = capacity, m^3/hr
Δp = pressure drop in the head, kg/m^2
C_p = heat capacity, kcal/kg-°C
ρ = density, kg/m^3

The value of density used in this calculation should be based on the material's melt temperature since density decreases with increasing temperature.

Horsepower Calculations for Twin-Screw Mixers in Plastics Processing

The principal design equation for this example is

$$N = P\nu = \frac{\mu\nu^2 F}{a} \tag{88}$$

where

P = shear force during the motion of the working surface and acting on the mass, g
F = area of working surfaces (shear surface), cm^2
ν = relative velocity between adjacent working surfaces, cm/s
a = distance between adjacent working surfaces, cm

The ratio of P/F is a shear stress, and v/a is a gradient or shear rate. Equation (88) may be rewritten in an alternate form:

$$N = \frac{\mu}{a} \int v^2 \, dF \tag{89}$$

The power consumption in wedged sections for screws rotating in one direction is

$$N_{\text{zone}} = Z \frac{\mu}{a}\left(\frac{\alpha}{2}\right)(2\pi h)^2(D - h + \delta)\frac{2\alpha'}{2}(D - h)h \tag{90}$$

where Z = total amount of wedge-shaped sections depending on the number of coils in gearing; $\delta = \delta_1 - \delta_2$ – radial gap between coils; a = side gap between coils.

The drive horsepower spent on the work in the radial gap may be determined considering this gap to be parallel between two coils, that is, the same as in the rolls. In this case there is a great inaccuracy because in contrast to rolls in the given gap before and after there is the pressure of the mass. For counterrotating screws the following formula is applicable:

$$N_{\text{counter}} = \mu\pi^2 n^2 (D - h)^3 \left(\frac{1}{\delta} - \frac{1}{\omega}\right) L \tag{91}$$

where ω is the material thickness before entering the gap.

Horsepower Calculation of Screw Mixers for Processing Rubber Mixtures

The energy transmitted to the screw is spent for overcoming the following resistances:

1. Extrusion forces, T_1
2. Friction forces of material against the screw, T_2
3. Friction forces of material against the machine case, T_3
4. Friction forces in the bearings of the drive, T_4

Taking into account these forces the horsepower is

$$N = \frac{1}{60 \times 102} \sum_1^4 Tv = \frac{1}{6120} \sum_1^3 Tv \frac{1}{\eta}, \quad \text{kW} \tag{92}$$

where

η = drive and bearing efficiency which is accounted for by T_4.
ν = extrusion velocity, m/min

The extrusion force is

$$T_1 = pF_0 \times 10^{-2}, \quad \text{kg} \tag{93}$$

where

p = pressure of the mixture at extrusion, kg/cm^2.

The friction force of the mixture against the screw is

$$T_2 = \frac{\tan(\alpha + \rho) - \tan \alpha}{\tan \alpha} T_1, \quad \text{kg} \tag{94}$$

The friction force of the mixture against the casing walls is

$$T_3 = \frac{\pi DLfp\nu_1}{2\nu}, \quad \text{kg} \tag{95}$$

where

D = diameter of the cylindrical working part, cm
L = length of the cylinder working section, cm
f = friction factor of the mixture against the cylinder
ν = velocity of the helix motion; ν_1 = tni
ρ = friction angle, $\rho = \tan^{-1} f; f = 0.3$ (96)
n = rotational velocity of the screw
t = distance between coils
i = number of screw threads

The overall efficiency of a screw mixer is

$$\eta_0 = \frac{\eta T_1}{T_1 + T_2 + T_3} \tag{97}$$

or

$$\eta_0 = \frac{\eta}{[\tan(\alpha + \rho)]/\tan \alpha + \pi DLfv_1/2F_0 v \times 10^{-2}} \tag{98}$$

Substituting average values for variables in the denominator of Eq. (98), it is observed that the two terms differ by only 5%, and hence, we may write

$$\eta_0 = \frac{2F_0 v\eta \times 10^{-2}}{\pi DLfv_1} \tag{99}$$

Substituting this expression into the horsepower formula, we obtain

$$N = \frac{T_1 v}{6120\eta} = \frac{pDLfni}{3900\eta}, \quad \text{kW} \tag{100}$$

The horsepower of a small-screw mixer for processing rubber mixtures may be computed from the following formula:

$$N = \frac{T_1 v}{6120\eta_0} \tag{101}$$

where η_0 is efficiency. Substituting into Eq. (101),

$$T_1 v = F_0 pv \times 10^{-2} = \frac{pG}{60\gamma_0}$$

then

$$N = \frac{Gp}{3.68 \times 10^5 \gamma_0 \eta_0} \tag{102}$$

As shown by Eq. (102), horsepower is proportional to the pressure and extrusion capacities.

The energy of extrusion transformed into heat is calculated from the following formula:

$$Q_T = N\eta_T 860, \quad \text{kcal/hr} \tag{103}$$

or

$$Q_T = GC_p(t_2 - t_1), \quad \text{kcal/hr} \tag{104}$$

where the temperature of the mixture increases to

$$t_2 = t_1 + \frac{810\eta_T}{GC_p} \tag{105}$$

where

t_1 = temperature of the mixture before extrusion
C_p = heat capacity of the mixture, kcal/kg-°C
η_T = heat efficiency for heating the mixture, $\eta_T = 0.9$

The horsepower of an electric motor may be approximated from

$$N = 0.00353D^2n, \quad \text{kW} \tag{106}$$

where

D = outer screw diameter, cm
n = screw velocity, rpm

Determination of Forces Acting on the Screw

Two calculation procedures are available for estimating the specific pressure at the end of the screw (i.e., near the head). The first consists of computing forces transmitted by the screw, and the second is based on the determination of the maximum specific pressure generated at the closed opening in the head.

From the first method we have

$$P_z = \frac{97{,}000N}{nr_s \tan(\alpha + \rho)}, \quad \text{kg} \tag{107}$$

where

P_z = axial force developed in the screw thread, kg
N = horsepower transmitted by the screw, kW
n = screw velocity, rpm
r_s = average thread radius, cm
α = helix angle at r_s
ρ = friction angle, = 26–38°

and specific pressure

$$P = \frac{P_z}{F_z} = \frac{P_z}{\pi(R^2 - r^2)} \tag{108}$$

where

F_z = cross-sectional area of screw thread, cm²
R = outside thread radius, cm
r = inside thread radius, cm

The axial force acting from the forming head side is

$$A = F_s p \tag{109}$$

where

A = axial force, kg
F_s = screw cross-sectional area, cm²
p = specific pressure at the screw end

The total axial force acting on the thrust bearing is the sum of the axial forces and the tangential component of material friction against the screw surface:

$$K = A + T \tag{110}$$

where

K = total axial force, kg
A = axial force, kg
T = tangential component of friction

$$T = F 2\pi r_s \tan \rho \tag{111}$$

where F is the specific force,

$$F = \tfrac{1}{3} p \frac{l}{2} = \frac{pl}{6}$$

l = length of the filled screw thread (assumed half of threaded part)
p = specific pressure at the screw end, kg/cm²

r_s = average radius of the screw thread, cm
$\tan \rho = f$ − friction factor (for rubber against steel $\rho = 26 \sim 30°$).

Heat Balances

The heat balance about a screw mixer may be written as follows:

$$G_m C_m t_{in} + Q_N + Q_K + Q_H + G_w C_w t_{w_1} = G_m C_m t_{out} + G_w C_w t_{w_2} + Q_l + Q_3 \quad (112)$$

where

G_m = amount of material entering the machine, kg/hr
C_m = heat capacity of the material, kcal/kg-°C
t_{in} = initial temperature of material, °C
t_{out} = final temperature of material, °C
Q_N = amount of heat due to energy transformation, kcal/hr
Q_K = amount of heat introduced to the machine (by steam, heated water, or electrical heaters), kcal/hr
Q_H = amount of heat introduced to the head (by steam, heated liquid, or electrical heaters), kcal/hr
G_w = amount of water for cooling the screw, kg/hr
C_w = heat capacity of water, kcal/kg-°C
t_{w_1}, t_{w_2} = initial and final temperature of cooling water, °C
Q_l = heat losses by the screw mixer, kcal/hr
Q_3 = amount of heat introduced in feeding zone, kcal/hr

Solving for the temperature of the material leaving the mixer, we obtain

$$t_{out} = \frac{Q_N + Q_K + Q_H - G_w C_w (t_{w_2} - t_{w_1}) - Q_l - Q_3}{G_m C_m} + t_{in} \quad (113)$$

The components of the heat balance are determined as follows:

$$Q_N = 860N, \quad \text{kcal/hr} \quad (114)$$

where N is determined from Eq. (80) for the entire machine.

The amount of heat Q_K introduced to the mixer from outside depends on the system of heating which may be electric, liquid, or steam.

Since the material discharge temperature is known, Q_K may be determined from Eq. (113):

$$Q_K = G_m C_m(t_{out} - t_{in}) + G_w C_w(t_{w_2} - t_{w_1}) + Q_l + Q_3 - Q_N - Q_H \tag{115}$$

The machine's capacity is determined from parametric calculations. Selecting an average heat capacity $\overline{C}$ we determine the amount of heat absorbed by the cooling water:

$$Q_x = G_w C_w(t_{w_2} - t_{w_1}) \tag{116}$$

The proper size of the screw channel can be determined from design considerations, and the cooling water pipe diameter is established by the water channel cross section.

Assuming a water velocity of $0.1 \sim 0.8$ m/s, we obtain

$$G_w = f \times \nu \times 3600\rho, \quad \text{kg/hr} \tag{117}$$

where

f = cross-sectional area, mm^2
ν = water velocity, m/s
ρ = water density, kg/m^3

The temperature differential $(t_{w_2} - t_{w_1})$ may be assumed to be around $5 \sim 10$°C for preliminary calculations.

Heat losses from the mixer are

$$Q_l = F\alpha(t_b - t_w), \quad \text{kcal/hr} \tag{118}$$

where

F = outside surface of the mixer, m^2
α = heat transfer coefficient at free convection
t_b = temperature of outside surface of the mixer (insulated), t_b = 50–80°C.

The amount of heat absorbed in the feeding zone, Q_3, is calculated similarly to that for screw cooling, provided that the cooling medium is water.

The amount of heat introduced to the head is

$$Q_H = G_m C_m(t_{out} - t_h) \tag{119}$$

where

t_{out} = temperature of the material after the head, °C
t_h = temperature of the material before the head °C

The amount of heat for heating the body and head of the mixer for liquid heating agents is

$$Q = G_T C_T (t_{\text{in}} - t_{\text{out}}), \quad \text{kcal/hr} \tag{120}$$

where

G_T = amount of heating agent, kg/hr

$$G_T = 3600 F \nu \rho, \quad \text{kg/hr} \tag{121}$$

F = cross-section area for introducing the heating agent, m^2
C = heat capacity of the agent, kcal/kg-°C
ν = velocity of the agent in the channel, m/s
ρ = density of heating agent, kg/m^3

Knowing Q, C_T, F, and ρ and selecting the velocity of the heating agent we may determine the temperature difference necessary for heat transfer Q.

For saturated steam as the heating agent:

$$Q = G(i_s - i_c), \quad \text{kcal/hr} \tag{122}$$

where

G = amount of steam entering the mixer, kg/hr
i_s = enthalpy of steam, kcal/kg
i_c = enthalpy of condensate, kcal/kg

The steam pressure is determined by the temperature needed for processing the material. The use of steam is limited by the available pressure drop rather than temperatures. For example, a steam temperature of 250°C is possible at a pressure of about 40 kg/cm^2, which makes the design of the mixer heavier and more complicated. The method considered above is related to determining the thermal characteristics of screw mixers, associated with a heat balance. However, provisions should also be provided for the transfer of quantities of heat. During the final stages of design it is necessary to check whether the obtained heat transfer surface provides the necessary amount of thermal energy. From the basic equation of heat transfer:

$$Q = KL\, \Delta t_{\text{av}} \tag{123}$$

it is possible to determine the desired heat transfer zone length (surface), or, from a known length, to determine the amount of heat transferred, Q, from information on the heat transfer coefficient K and average temperature difference at Δt_{av}.

If electrical energy is used for heating, the heat transfer coefficient for a cylindric wall is determined from the following formula:

$$K = \frac{1}{\dfrac{1}{\pi d_{in}\alpha_1} + \dfrac{1}{2\pi\lambda}\, 2.3 \log \dfrac{d_{out}}{d_{in}}}, \quad \text{kcal/m-hr-°C} \tag{124}$$

If steam or a liquid carrier is used, then

$$K = \frac{1}{\dfrac{1}{\pi d_{in}\alpha_1} + \dfrac{1}{2\pi\lambda}\, 2.3 \log \dfrac{d_{out}}{d_{in}} - \dfrac{1}{\pi d_{out}\alpha_2}} \tag{125}$$

where

d_{out} = outside diameter of cylindrical body where the heater is mounted or comes in contact with the heating agent, m

d_{in} = inside diameter in contact with the material, m

λ = conductivity of the mixer body

α_1 = heat transfer coefficient from case surface to the material, kcal/m^2-hr-°C

α_2 = heat transfer coefficient from the carrier to the case outside surface, kcal/m^2-hr-°C.

Note that in the feeding zone, heat transfer from the wall to the surrounding air, as well as partially to the granular material, takes place. Because the size of this zone is small and heat transfer is not intensive, its importance to the preceding calculations is small. Physical constants, used in the formulas, should be based on the material's average melt temperature and the wall temperature. Because the material moves through the screw channel, the determining size is equivalent to the diameter of the screw channel:

$$d_{eq} = \frac{2\left[(t - b)\cos\alpha\, \dfrac{d_{out} - d_{in}}{2}\right]}{\left[(t - b)\cos\alpha + \dfrac{d_{out} - d_{in}}{2}\right]}, \quad \text{m} \tag{126}$$

where

t = helix pitch, m
b = coil width, m
d_{out} = screw outside diameter
d_{in} = screw inside diameter
α = helix angle

The material's velocity is

$$\nu = \frac{Q_\nu}{F \times i}, \quad \text{m/s} \tag{127}$$

where

Q_ν = volumetric capacity of the mixer, m^3/s
i = number of threads
F = cross-sectional area of screw channel, m^2

$$F = \left[(t - b)\cos\alpha\,\frac{d_{out} - d_{in}}{2}\right] \tag{128}$$

For heating calculations for the head, the following formulas are used:

$$d_{eq} = \frac{4F}{P'} \tag{129}$$

$$\nu = \frac{Q_\nu}{F} \tag{130}$$

where

F = cross-sectional area of discharge opening in the head, m^2
P' = perimeter of discharge opening, m

The thermal regime of the mixer can be checked on the basis of the heat balance equation; that is, if Q_K is determined from a heat balance, then the amount of heat calculated from Eq. (123) should be equal to Q_K. If such an equality is not realized, the process of heat transfer should be intensified or the heat transfer area F (for the case $F = \pi d_{in}L$) increased, that is, the size of the mixer or its head should be enlarged. Equation (123) should be used for determining the screw's cooling process, where the heat transfer coefficient is determined from the following formula:

$$K = \frac{1}{\dfrac{1}{\pi d_{in}\alpha_1} + \dfrac{1}{2\pi\lambda} 2.3 \log \dfrac{d_{out}}{d_{in}} + \dfrac{1}{\pi d_{in}\alpha_2}} \tag{131}$$

where

d_{out} = outside diameter of screw cylinder, m

d_{in} = inside diameter of screw cylinder, corresponding to the surface reached by the water stream moving in the circular gap, m

α_1 = heat transfer coefficient from material to the screw surface, kcal/m^2-hr-°C

α_2 = heat transfer coefficient from internal surface of the screw to the cooling water moving in the circular channel between the water feeding pipe and inside surface of the screw, kcal/m^2-hr-°C.

Coefficient α_1 is calculated in a manner similar to that in Eq. (125), where d_{eq} is obtained from Eq. (126) and the material velocity is obtained from Eq. (127). Coefficient α_2 (for flowing cooling water) is calculated from a dimensionless expression which depends on the flow regime. The determining size for this calculation is

$$d_{eq} = d_{in} - d_{out} \tag{132}$$

where

d_{in} = internal diameter of screw cylinder, corresponding to the surface reached by water, m

If the machine case is insulated, then the thickness of the insulated layer is calculated by using the formulas of conductivity through a flat or cylindrical surface. In this case the heat flux through the insulation layer is determined from the following equation:

$$Q = \alpha F(t_{out} - t_a), \quad \text{kcal/hr} \tag{133}$$

where

α = heat transfer coefficient from an outer surface of insulation to the air (use formulas of free convection)

F = outside surface of insulation, m^2

t_{out} = temperature of outside surface of insulation which should be assumed

t_a = temperature of air

The required insulation layer thickness may be determined from the heat flux and conductivity. In performing this calculation the temperature difference $\Delta t = t_{in} - t_{out}$ (where t_{in} is the temperature of the internal surface of the insulation layer) can be obtained by assuming t_{in} to be equal to the temperature of the surface that is to be insulated.

SINGLE-SCREW MIXER MODELING

The principals of similarity and modeling are widely used in the design of different unit operations [see Azbel and Cheremisinoff (1983)]. The study of the flow of polymer materials may be based on the following dimensionless relation:

$$\mathrm{Eu} = C\mathrm{Re}^n\left(\frac{l}{d}\right)^m \tag{134}$$

where

Eu = $\Delta p/\rho v^2$, the Euler number

Re = vd/ν, the Reynolds number

Δp = pressure drop in the channel

ρ = density of the polymer melt

v = average velocity of polymer

d = channel diameter

l = channel length

ν = kinematic viscosity

Silin (1964) applied similarity theory to developing design correlations for screw mixers. For a basis of design the process parameters of a well defined system must be known. A defined system is one where the mixer's dimensions, the screw's rotational velocity n_0, horsepower N_0, and so on, are known and the performance and limits of operation have been examined. From this basic information, a system of equations characterizing the process and design sizes of the prototype can be developed. The principal forms of these equations for dynamically similar machines are as follows:

$$\text{Capacity:} \qquad Q = Q_0\left(\frac{D}{D_0}\right)^{3-2\psi} \tag{135}$$

Screw rotational velocity: $$n = n_0\left(\frac{D}{D_0}\right)^{-\psi} \tag{136}$$

Depth of screw channel: $$h = h_0\left(\frac{D}{D_0}\right)^{1-\psi} \tag{137}$$

Pressure at the screw end: $$p = p_0\left(\frac{D}{D_0}\right)^{1-\psi} \tag{138}$$

Horsepower: $$N = N_0\left(\frac{D}{D_0}\right)^{3-\psi} \tag{139}$$

Resistance factor of the head: $$K = K_0\left(\frac{D}{D_0}\right)^{3-(1-\psi)} \tag{140}$$

Equations (135) through (140) were developed from successive model and prototype experimental units and define the principal design parameters in terms of the diameter ratio of prototype to model, and an additional parameter ψ defined as

$$\psi = \frac{1}{2}\frac{T_1 - T_0}{T_2 - T_0} \tag{141}$$

where T_0, T_1, and T_2 are the temperatures of the material in the three screw zones, respectively.

In certain applications it is necessary to maintain nearly identical design specifications between the prototype and model (e.g., constant screw length to diameter, and/or rate of compression). Parameter ψ is dependent on the thermal process in the mixer; that is, it is a function of the ratio between the energy introduced to the material via transformation of mechanical energy of the drive into thermal energy, and energy introduced by outside heaters through heat transfer. The extremes of operation are isothermal, where $\psi = 0.5$, and adiabatic, where $\psi = 0$.

TWIN-SCREW MIXER MODELING

Four scaling groups each consisting of different forces acting in a screw mixer were studied by Silin (1964). The first group relates to the scales of inertia forces, centrifugal forces, and head resistance. The second group concerns the scales of gravity forces. The third comprises the scales of internal friction, and the fourth refers to the scales of strength and elasticity of the polymer. The principal forces acting in plastics processing between two parallel screws are that of compression, rupture, shear, and elasticity (i.e., the fourth group of forces). The scale equality of these forces is expressed by the following equation:

$$C'_p = C_p^{IV} = \frac{\rho_0 C_e^4}{\rho_m D_e^2} = \frac{E_0 C_e^2}{E_m} \tag{142}$$

As follows from Eq. (142), for equal elastic moduluses $E_0 = E_m$ and densities $\rho_0 = \rho_m$ of processed materials (in the model and prototype) the geometric (linear) scale is equal to the time scale:

$$C_e = C_t \tag{143}$$

Thus, as a basis of similarity for double-screw mixers, the Cochy number $C_0 = \rho v^2/E$ can be used for corresponding scales of modeling. As follows from the scales of modeling, for two similar double-screw mixers, the product of rotational velocity and the height of the screw coils is constant:

$$C_n C_h = \frac{n_0 h_0}{n_m h_m} = \frac{C_e}{C_e} = 1 = C_e^a C_e^b \tag{144}$$

For the proper selection of modeling scales for the coil height and screw rotational velocity, it is necessary to maintain equality of the principal forces. In addition it is necessary to maintain conditions of viscosity at critical corresponding points in the model and prototype. The latter criterion may be achieved by maintaining the same temperature regimes and velocity gradients of the melt flow at corresponding points along the screw channels in both the prototype and model:

$$C_r = \frac{r_0}{r_m} = \frac{\pi D_0 n_0 h_0}{\pi D_m n_m h_m} = \frac{C_e C_n}{C_h} = 1$$

$$\mu_0 = \mu_m \tag{145}$$

Upon substitution of the modeling scale, velocity gradient and coil heights, we obtain:

$$C_r = \frac{C_e C_e^b}{C_e^a} = C_e^{1+b-a} = C_e^0 = 1 \tag{146}$$

Hence, from Eq. (146):

$$1 + b - a = 0$$

Simultaneously solving Eq. (144) and (146), we obtain

$$a = 0.5, \qquad b = -0.5$$

Thus the modeling of twin-screw mixers is achieved through the equality of modeling scales, height of coils, and rotational velocities between the model and prototype:

$$C_h = \frac{h_0}{h_m} = C_e^{0.5} \tag{147}$$

$$C_n = \frac{n_0}{n_m} = C_e^{-0.5}$$

The coil height of the screws ($h_0 = h_m\sqrt{C_e}$) at $n_0 = n_m/\sqrt{C_e}$ corresponds to the optimum modeling case.

The geometrical scale of coil and channel widths between coils is equal to the linear scale of diameter modeling:

$$C_e = \frac{b_0}{b_m} = \frac{a_0}{a_m} = \frac{L_0}{L_m} = \frac{D_0}{D_m} \tag{148}$$

It follows then that the proper scale of capacity is

$$C_Q = \frac{Q_0}{Q_m} = \frac{\pi D_0 h_0 H_0 n_0 K_0 \eta_0 \psi_0 \gamma_0}{\pi D_m h_m H_m h_m K_m \eta_m \psi_m \gamma_m} = C_e^2 \tag{149}$$

where K_0 and K_m are overlapping factors in the model and prototype.

The interaxial distances are determined from the conditions of equal velocity gradients and viscosities of material flow between the coil ridges and their cores in the model and prototype. Because in the gaps δ_0 and δ_m (refer to Figure 6), the material is subject to the largest strains and stresses, the following criteria must be maintained:

$$r_0 = r_m, \qquad \mu_0 = \mu_m$$

The scale for velocity gradients in the gaps between the coil ridges and screw cores can be computed from the following equation:

$$C_r = \frac{r_0}{r_m} = \frac{\pi D_0 n_0/\delta_0}{\pi D_m n_m/\delta_m} = C_e^{0.5}\frac{\delta_m}{\delta_0} = 1 \tag{150}$$

where $n_0/n_m = C_e^{-0.5}$; the scale of velocity modeling at the coil height is $h_0 = h_m C_e^{0.5}$.

Hence,

$$\delta_0 = \delta_m C_e^{0.5} \tag{151}$$

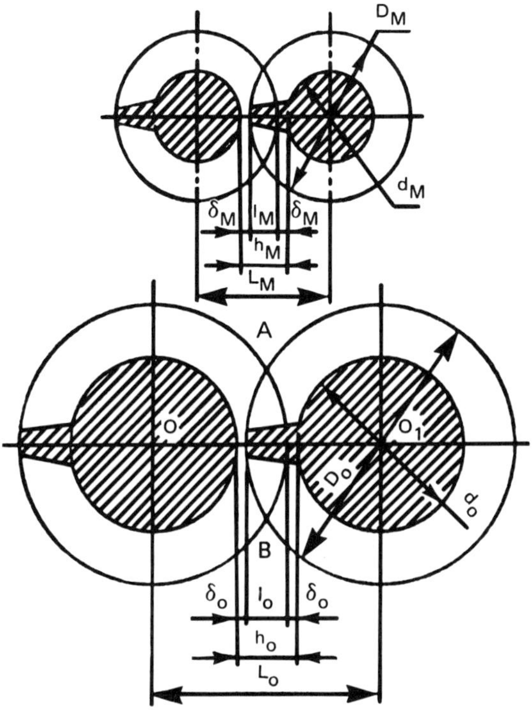

Figure 6 Cross sections of twin-screw mixers for scale-up development.

After establishing the coil height h_0 and the gap δ_0, the distance between the screw axes in the prototype can be computed as follows:

$$L_0 = D_0 - B_0 = D_0 - B_m C_e^{0.5} \tag{152}$$

where B_0 = width of coil overlapping, which is equal to

$$h_0 - \delta_0 = h_m C_e^{0.5} - \delta_m C_e^{0.5} = B_m C_e^{0.5}$$

MODELING OF UNIVERSAL OSCILLATING SCREW MIXERS

Universal screw mixers are used for processing a variety of polymer materials. These machines are characterized by a reciprocal motion of screws in the axial direction with simultaneous rotation. For intensive mixing, plastification, homogenation, and degassing of different plastics, the screw coils are interrupted in

Figure 7 Types of oscillating-screw mixers.

definite order forming clearances in which three rows of teeth, located on the internal surface of the mixer cylinder, enter these clearances. Because of variable axial and radial screw motion between the teeth of the mixer, the material is exposed to shear, compression, rupture, and mixing at different planes over small volumes between the coils.

A high degree of mixing is achieved in these designs as compared to other machines and at relatively low energy consumption (from 0.02 to 1.0 kW per 1 kg of mixture). Depending on the properties of the material, three design variations are possible. Figure 7 shows the three principal types. The *P*-type mixer is employed in processing very viscous materials (rigid and mild polyvinyl chloride, polystyrene, and other polymers). *K*-type mixers serve for processing different pastes and viscous materials (for manufacturing carbon electrodes, chocolate, dough). *M*-type mixers are used for processing different dry materials. The primary design differences between the three involve the shape and size of the teeth.

Mechanical similarity of oscillating screw mixers is achieved through geometric, kinematic, and dynamic similarities. Geometric similarity is reproduced by maintaining equality of ratios between the corresponding geometrical sizes of discontinuous screws and teeth of the cylinders of the model and prototype. Comparison of commercial and laboratory-size mixers reveals that geometric similarity is important. This implies the following constant geometric ratios:

1. Pitch of the left double-start thread (refer to Figure 8):

$$H_1 = 0.6D \tag{153}$$

where D is the outside screw diameter.

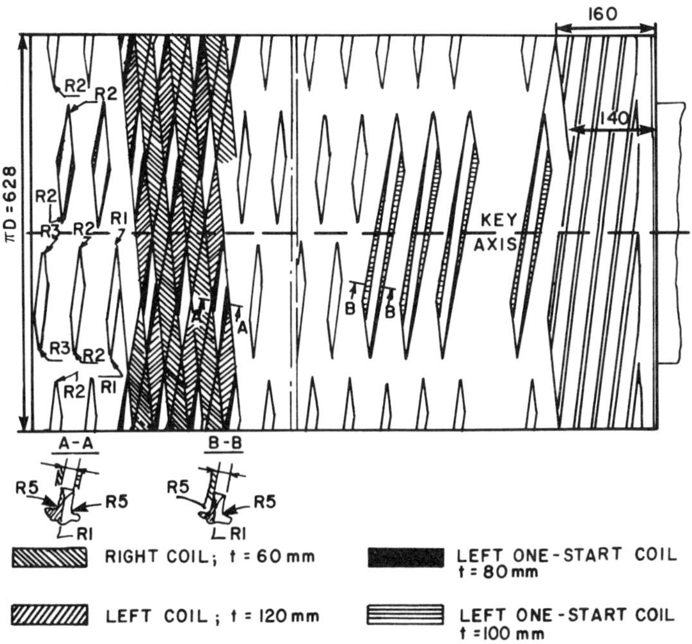

Figure 8 Involute of discontinuous coils of the screw of an oscillating mixer. Section A-A: One-tooth section along normal to plane is formed by the right coil with pitch $t = 60$ mm and the left two-start coil with pitch $t = 120$ mm; two-tooth section along normal to plane by left one-start coil with pitch $t = 80$ mm through four teeth with individual treatment of each tooth. Section B-B: One-tooth section along normal to plane formed by left two-start coil with pitch $t = 120$ mm; two-tooth section of tooth along normal to plane formed by left one-start coil with pitch $t = 100$ mm with individual treatment of each tooth.

2. Pitch of right single-start thread:

$$H_2 = \frac{H_1}{2} = 0.3D \tag{154}$$

3. Height of the thread:

$$h = 0.175D \tag{155}$$

4. The number of longitudinal grooves of the screw m is equal to the number of forming teeth K, located on the internal cylinder surface.

$$m = K = \frac{H_1 + H_2}{H_2} = 3 \tag{156}$$

5. Number of left-hand start threads:

$$Z = K - 1 = 2 \tag{157}$$

6. Length of screws and width of coils and grooves of thread are proportional to the diameter of the screws:

$$\begin{aligned} L_0 &= L_m C_e \\ b_0 &= b_m C_e \\ a_0 &= a_m C_e \end{aligned} \tag{158}$$

where

L_0, L_m = screw lengths of the prototype and model
C_e = geometric scale, D_0/D_m
b_0, b_m = coil width of the screw for prototype and model
a_0, a_m = groove width, for prototype and model

7. Teeth are located on the internal surface of cylinders between discontinuous screw coils in three rows; the angles between the teeth are $\alpha_0 = \alpha_m = 120°$.
8. The pitch between the teeth in each row on the cylinder surface is equal to the pitch of the right one-start screw thread:

$$H_{\text{tooth}} = H_2 = 0.3D \tag{159}$$

9. The forming teeth of mixers have different shapes depending on their location along the cylinder length: For the cylinder the tooth height is $h_{\text{tooth}} = 0.1D$, and its shape is $h_{\text{tooth}} = 0.15D$.
10. The axial path of the screw is equal to one-half of the pitch of a one-start thread of the screw:

$$l = \frac{H_2}{2} = 0.15D \tag{160}$$

11. For free passage of discontinuous coils of the screw between the forming teeth of the cylinder during reciprocal and rotational screw motion, the coils should be undercut with a milling cutter. The undercutting pitch in the

feeding zone is $H_3 = 0.5D$, and in the compression and plastification zones is $H_4 = 0.4D$.

The second condition of similarity of oscillating mixers is kinematic similarity, which is realized when in each compatible point of similar screws the directions of the velocities of moving particles of material to be mixed are parallel and proportional to each other. With this in mind, the kinematics of reciprocal motion together with the rotation of the screw is considered.

For simplifying calculations the teeth motion can be considered relative to a stationary screw. As we see from the involute of the surface of discontinuous coils of the screw and teeth (refer to Figure 9) at the first half turn and forward

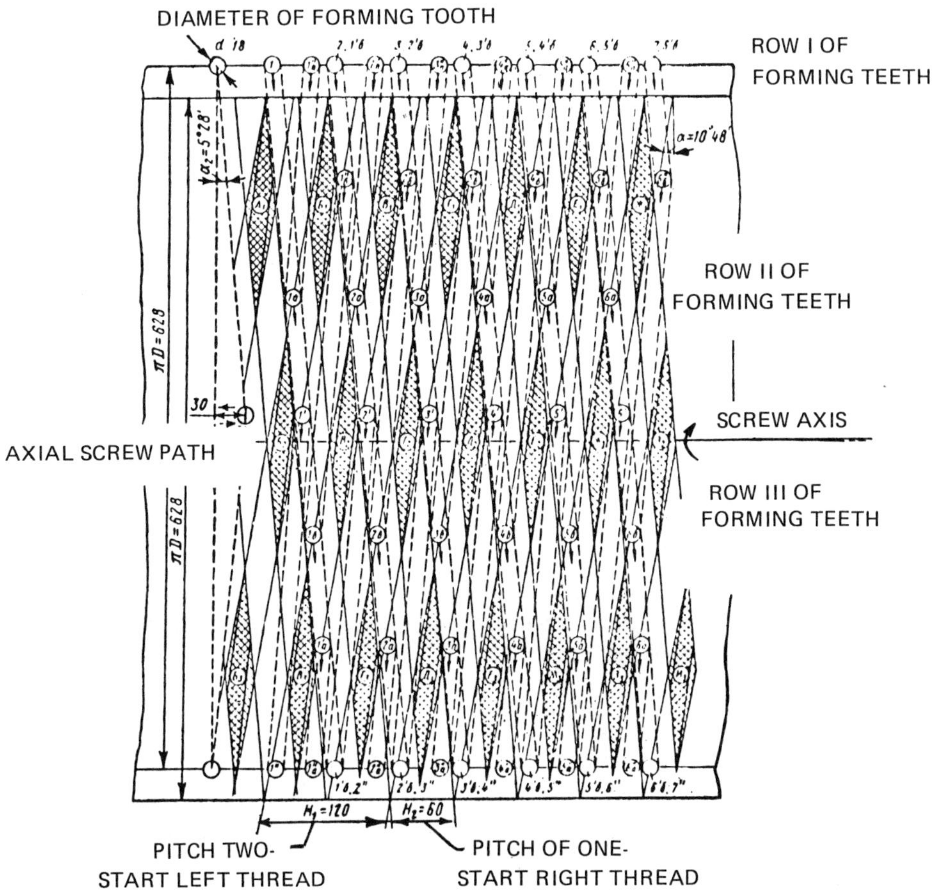

Figure 9 Geometric explanation of kinematics of discontinuous screw coils between forming teeth of an oscillating-screw mixer.

motion of the screw, the forming tooth 1 is moving along the helix line 1–1′. In this case tooth 1 crosses screw coil B in the slit between the coil (teeth) part B_1 and B_2. At the second half turn and reciprocal motion of the screw (backward) the forming tooth 1 is moving along helix line 1–1″ without crossing screw coils. The same motions relating to discontinuous screw coils B, F, D, . . . have forming teeth 2, 3, 4, . . . located in the same row with tooth 1 of the mixer cylinder.

For one turn of the screw they move along helix lines 2–2′–2″, 3–3′–3″, 4–4′–4″, . . . , crossing the screw coils B, F, D in the slits between the teeth (sections) of coils B_1 and B_2, F_1 and F_2, D_1 and D_2. The forming teeth of the second row of the mixer 1a, 2a, 3a, . . . , in the first half turn of the screw and its forward motion pass the path along their helix lines 1a–1′a, 2a–2′a, 3a–3′a, . . . , through coils B, F, and D in slits between the teeth (parts) of coils B_2 and B_3, Γ_2 and Γ_3, D_2 and D_3. In the second half turn the forming teeth 2a, 3a, 4a, . . . , pass the path (1′a–1″a) + (1″a–1a); (2′a–2″a) + (2″a–2a), . . . between coil teeth Γ_2 and Γ_3, D_2 and D_3. . . . Similar relative motions have the forming teeth of the third row of the mixer 1b, 2b, 3b, ... Thus sinusoidal reciprocal and rotating motion of the screw is due to the regular location of forming teeth in the coil slits of the screw and exactly limited axial displacement of the screw. The axial screw displacement (forward and backward) is equal to one-half of the pitch of a one-start right thread:

$$l = 0.5H_2 \tag{161}$$

In this case the helix angle relative to the motion of each forming tooth is equal to the helix angle of the right thread of the screw:

$$\psi = \tan^{-1}\frac{2l}{\pi D} = \psi_2 = \tan^{-1}\frac{H_2}{\pi D} \tag{162}$$

The number of longitudinal screw paths m which is equal to the number of rows of forming teeth K in the mixer cylinder is determined from the ratio of pitches of left and right threads of the screw:

$$m = K = \frac{H_1 + H_2}{H_2} = \frac{2H_2 + H_2}{H_2} = 3 \tag{163}$$

The number of starts of the left thread is

$$Z = K - 1 = 2 \tag{164}$$

The capacity of an oscillatory mixer is

$$Q = \frac{\pi(D^2 - d^2)HKA \cos \alpha n \eta \rho}{4\pi D} \tag{165}$$

where

D = screw outside diameter, cm
d = screw inside diameter, cm
H = pitch of left thread of the screw, cm
K = number of teeth in one coil of the screw which is equal to number of rows of forming teeth
A = length of one tooth of screw coil, cm
α = helix angle of left thread of screw, $\alpha = \tan^{-1}(H/\pi D)$
n = screw velocity, rpm
η = volumetric capacity efficiency, depending on head counterpressure, coefficient of charge, slit width between coils of screw, and properties of material to be processed; $\eta = 0.1 \sim 0.25$
ρ = density of material processed, kg/cm^3

The third condition of similarity of oscillating mixers is dynamic similarity. This is reproduced when the acting forces (pressure, gravity, inertia, elasticity, viscosity, friction, centrifugal, etc.) in the model and prototype are proportional and parallel to each other. Similarity is determined by the three scales: geometric C_e, time scale C_t, and force scale C_p. To achieve similarity between prototype and model based on all the acting forces corresponding to Froude, Reynolds and Cochy numbers is impossible. The qualities of scales are based on the forces of axial pressure acting on the oscillating screw, on centrifugal and inertia forces corresponding to the first group of forces, as well as on elastic forces and compressive strength, tensile strength and shear of material at mixing, plastification and strain between discontinuous coils and forming teeth corresponding to the fourth group of forces, and friction forces which are derivatives of the mentioned forces.

The equality of scales of these forces is expressed by the following equation:

$$C_p^{\mathrm{I}} = C_p^{\mathrm{IV}} = \frac{\rho_0 C_e^4}{\rho_m C_e^2} = \frac{E_0 C_e^2}{E_m} \tag{166}$$

which corresponds to a Cochy number defined as

$$C_0 = \frac{\rho v^2}{E} \tag{167}$$

Thus the modeling scales of oscillating mixers are the same as for screw mixers if geometric similarity is maintained. At equal rotational velocities in model and prototype units, the capacity and horsepower of the prototype are increased correspondingly by the values established by the modeling scales:

$$C'_Q = C_Q C_e = C_e^3$$
$$C'_N = C_N C_e = C_e^3 \tag{168}$$

DISPERSION-TYPE SCREW MIXERS

Dispersion-type screw mixers are in an entirely separate class from the extruder designs described above. They are applicable to processing dry materials or moderately viscous pastes, creams, or lotions. These machines normally consist of a conical vessel equipped with a vertical or inclined screw and they can be either single- or twin-screw models. Figures 10 and 11 show cutaway views of the design.

The screw flight agitator rotates on its own axis while orbiting the periphery of the conical tank. This simultaneously establishes three distinct intermixing currents in the batch, as illustrated in Figure 12. The screw agitator, turning on its own axis, produces a lifting action as it spirals the material in an upward flow in the vessel (Figure 12A). At the same time, orbiting the tank, the screw removes material from the wall and deflects it into the center of the tank, thus establishing a second, wider, spiral current throughout the batch (Figure 12B). Figure 12C shows the third flow pattern, where material lifted upward by the screw gravitates downward. This action causes downward-moving material to intermix with material being spiraled upward.

When dry materials are being mixed, the solids are charged into the unit through the top and ingredients may be added as desired without interrupting the mixing cycle. Commercial units are available for injecting liquid or paste additives into the mix through the top or bottom of the vessel. The screw's spiral action is in the upward direction, thus lifting ingredients from the bottom of the tank up through those at the top. The discharge nozzle is located at the vessel's bottom.

In mixing pastes, approximately half of the liquid ingredients are poured in from the top of the vessel. Then, with the screw agitator running downward, the dry ingredients are added. Dispersion of dry and liquid materials is accomplished by a highly efficient hydraulic shear action. After an initial dispersion is achieved, the remainder of the liquid is added until the proper consistency is achieved. These designs have proved to be very successful in mixing pastes, creams, doughs, and plastisols, and have the advantage of achieving fine dispersions with minimal agitation, low heat addition, and practically no aeration. In addition, designs can

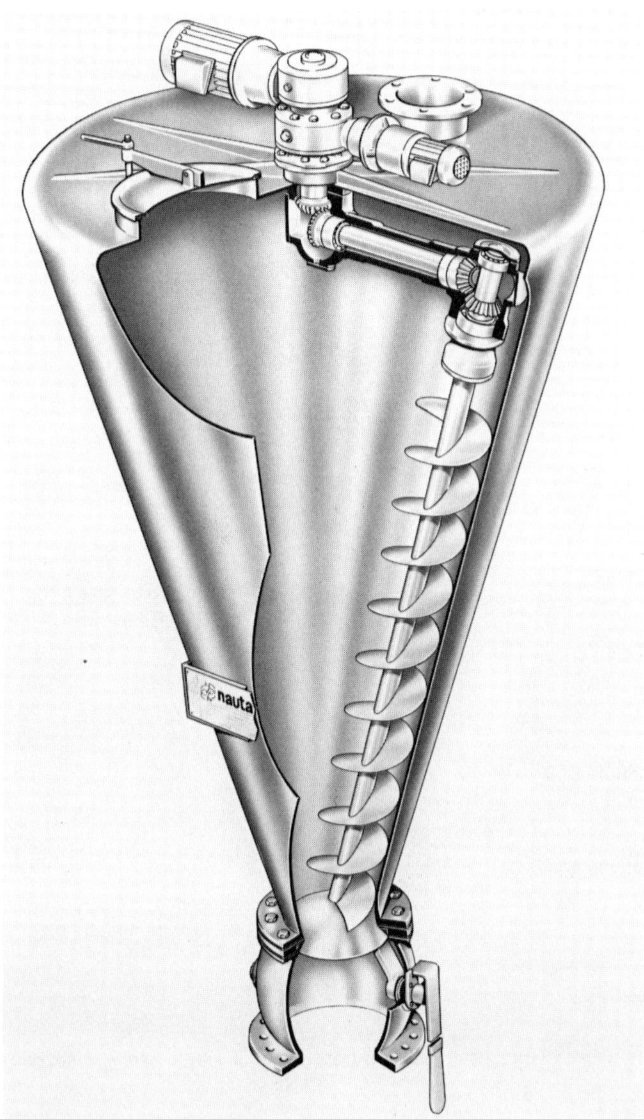

Figure 10 Cutaway view of a single-screw dispersion type mixer. (Courtesy of Day Mixing Co., Nautamix Europa, Cincinnati, OH.)

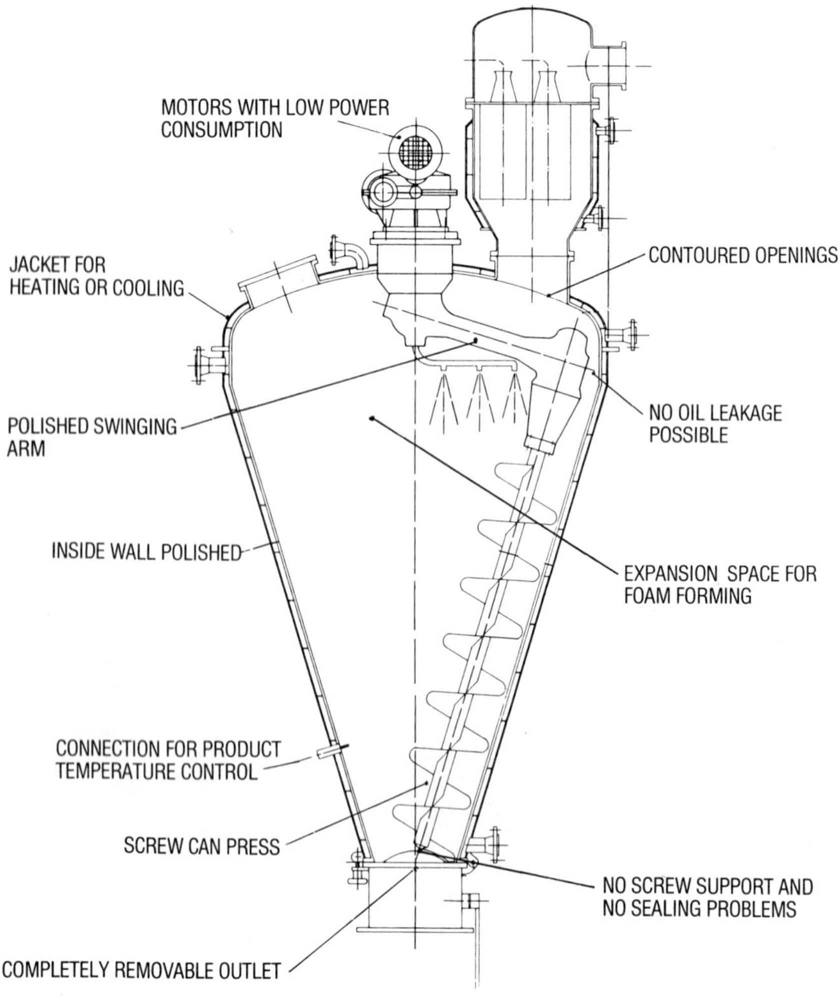

Figure 11 Cutaway view listing principal components of a single-screw dispersion mixer.

be equipped for vacuum applications, where deaeration is accomplished during mixing, thus eliminating the need for separate deaeration equipment.

These machines provide a high degree of blending action for moderately viscous materials and ingredients of widely differing densities and particle shapes and sizes. Commercial units can handle batch capacities up to 4000 pounds of some materials to complete homogeneity in less than 5 minutes.

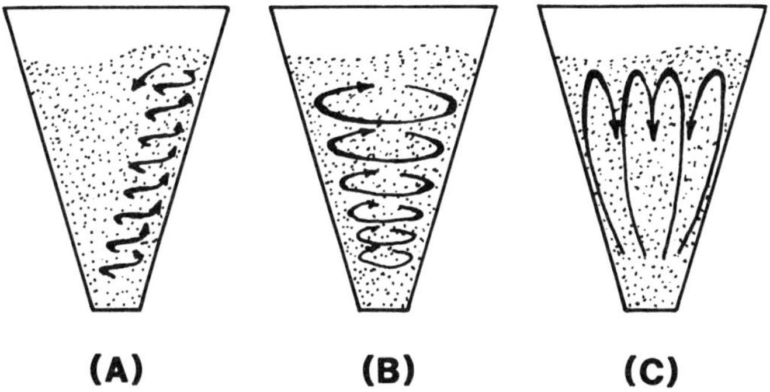

Figure 12 Three intermixing actions observed in a single-screw dispersion-type mixer.

These systems are not high shearing devices. On the contrary, they provide a gentle mixing action, and thus are applicable for handling materials sensitive to attrition or fiber disintegration (one example is in dry cell battery manufacturing, where acetylene black is lacelike in structure and very easily broken down).

The twin-screw design provides intermixing between the two screws in addition to the axial, orbital, and gravitational action provided by the single-screw design. Mixing times can be up to 90% faster than the single-screw system, depending on material properties.

Cantilevered arms are equipped with flexible couplings to the drive shaft so that the screws can be supported entirely from the top of the vessel. The critical portion of their design centers around the balancing forces of the two screws to assure long bearings life and driving assembly. The lower bearings (ring/pin lower bearings) carry no weight and serve only as guides.

These systems can be designed to operate in a continuous mode by employing two intersecting conical tanks with two orbiting spiral agitators. The axial and orbiting motions of the two agitators, along with the additional intermixing currents occurring through the juncture of the two tanks, provide a greater degree of blending action than the single-vessel configuration. Two motors and two gear boxes are required on small units. One set drives both screw flights axially and the other drives the two orbiting arms. On larger capacity units, separate motors are used for each screw flight and a single motor drives both orbiting arms. Heavy materials can be blended in larger quantities because the required horsepower is distributed over two balanced mechanisms. By employing proportional feeding, these designs can provide continuous mixing of a wide range of products.

Vacuum dispersion mixers are designed specifically for mixing heavy paste-like products requiring deaeration as part of the process. Deaeration is accom-

Figure 13 Two units for mixing and deaeration of pastes and creams. The units have a working capacity of 1000 liters. (Courtesy of Day Mixing Co., Nautamix Europa, Cincinnati, OH.)

plished as part of the mixing cycle, and thus there is no need for separate deaerating equipment. In most processes this results in lower overall equipment costs along with reduced cycle times and material handling costs. This type of mixer has found wide application in the preparation of plastisols. These machines offer the ability to produce partial as well as full batches of materials over a wide range of viscosities. These units are designed throughout for vacuum operation. Reinforcing rings are provided on the cone, and the top cover is ribbed for extra strength.

A typical installation of a screw-type dispersion mixer is shown in Figure 13. Because of the specific design features of this machine, no detailed studies of the flow patterns or modeling correlation have been reported in the journal and patent literatures.

DESIGN OF CENTRIFUGAL MIXERS

The design and scale-up of a centrifugal mixer by analytical means is difficult. One approach to the design of such mixers is to apply the calculation methods for propeller-type mixers operating in liquid mediums using appropriate correction factors established from experiments. Hence we rely again on the theory of similarity to develop scale-up criteria. Mechanical similarity, as determined by the three scales described earlier, should be applied. Recall that the scales are interdependent and, therefore, at geometric similarity between the model and prototype systems and at kinematic similarity of particle motion, the acting forces can be decreased in the prototype in accordance with a definite scale, C_p, which is uniquely related to the geometric scale, C_e, and the time scale C_t. The scales of derivative values of similar systems (e.g., velocity, acceleration, pressure, capacity, horsepower, etc.) can be readily derived from these three scales. Previously noted, an equality of scales is achieved when modeling similar systems through the largest acting forces in the process. Forces comprising groups I and II are the most important in centrifugal mixers. Forces comprising groups III and IV play a lesser role in the mixing process of these machines. Since mechanical similarity can therefore be achieved by equality of scales in a model and prototype for the first two groups of forces, we may write

$$C'_p = C''_p = \frac{\rho_0 C_e^4}{\rho_m C_t^2} = \frac{\rho_0 g_0 C_e^3}{\rho_m g_m} \tag{169}$$

If materials having the same densities are used in the two units (prototype and model mixer), then

$$\frac{g_0 C_e}{g_m} = \frac{C_e^2}{C_t^2} = C_v^2 \tag{170}$$

Substituting for the geometric and velocity scales (i.e., $C_e = l_0/l_m$ and $C_\nu = \nu_0/\nu_m$), we obtain

$$\frac{g_0 l_0}{g_m l_m} = \frac{\nu_0^2}{\nu_m^2} \tag{171a}$$

or

$$\frac{\nu_0^2}{g_0 l_0} = \frac{\nu_m^2}{g_m l_m} \tag{171b}$$

Equation (171) implies that at the transition from one similar system to another, ν^2/gl for all compatible points in model and prototype machines must be equivalent. That is, the Froude numbers

$$\mathrm{Fr} = \frac{\nu^2}{gl} \tag{172}$$

at corresponding locations are identical.

For equivalent densities ($\rho_0 = \rho_m$) and gravitation acceleration ($g_0 = g_m$), the time scale given in Eq. (169) simplifies to

$$C_t = C_e^{0.5} \tag{173}$$

Hence the velocity scale for the two mixers is

$$C_\nu = \frac{C_e}{C_t} = C_e^{-0.5} \tag{174}$$

Consequently, the scale of rotation speeds is

$$C_n = \frac{1}{C_t} = C_e^{-0.5} \tag{175}$$

We may now derive specific scale-up expressions from these scales.

An expression for the mixer body volume becomes

$$V_0 = V_m C_e^3, \quad \text{liters} \tag{176}$$

For the speed of the impeller (rpm):

$$n_0 = n_m C_e^{-0.5} \tag{177}$$

For the time required for one mixing cycle:

$$t_0 = t_m C_e^{0.5} \tag{178}$$

An expression for the required capacity at simultaneous operation of cold and hot mixes up to 120°C (in kg/hr):

$$Q_0 = Q_m C_e^{2.5} \tag{179}$$

The required horsepower of an electrical motor is

$$N_0 = N_m C_e^{3.5} \tag{180}$$

The forces acting on the material to be mixed are

$$P_0 = P_m C_e^3 \tag{181}$$

Finally, an expression for particle pressure:

$$P_0 = P_m C_e \tag{182}$$

DISK MIXERS

Disk mixers, or plasticizers, are used for mixing and plasticization of polymers. Despite the relatively low pressures (3–7 kg/cm^2) developed in these machines, they have proved to be advantageous over extruder-type screw mixers in many applications. In general, they provide good mixing and melt homogenity, short residence time (e.g., for a unit with 150-mm disk diameter and a capacity of 25 kg/hr, the average residence time is about 15 seconds); and they are two to three times smaller than screw-type extruders of similar capacities. The performance of these machines is strongly related to the deformation properties and behavior of the material. Polymer behavior during and after the application of shearing loads must therefore be described in detail before we proceed to parametric calculations of this type of mixer.

The total shear deformation imposed on a polymer is

$$\gamma_0 = \gamma_1 + \gamma_2 + \gamma_3 \tag{183}$$

where

γ_1 = elastic reversible deformation
γ_2 = high-elastic deformation
γ_3 = plastic deformation or viscous flow

Elastic deformation is characterized by a high modulus of elasticity, that is, by small strain at high stresses. The rate of these strains is independent of temperature; only the volume and internal energy of the system changes. These deformations follow Hooke's law and for polymers they generally are small (1~2%). High-elastic deformation is completely reversible. The restoration of the initial shape of the material takes place sometime after the elimination of the deformation load.

The shear modulus of a high-elasticity deformation is several orders of magnitude smaller than that of an elastic deformation. The phenomenon of high-elasticity deformation can be described as a straightening of long, flexible polymer chains under load, and restoration of their initial shape after the loading is removed. Thus a relaxation period exists where the material returns to its original state. If the polymer is raised to a temperature that exceeds its melting point, the material acquires flow characteristics, and the motion of smaller molecule chains retains sufficient inertia to impart motion to larger chains, with the help of a small external force. This displacement becomes irreversible. Under viscous-fluidity conditions, polymers acquire some of their properties from the linear structure of their molecules. For example, melt viscosity increases sharply with increasing molecular weight. The irreversible deformation of flow is accompanied by high-elasticity deformation.

Melt Elasticity and the Weissenberg Effect

The elastic component of deformation of a viscoelastic body undergoes a characteristic phenomenon known as the Weissenberg effect. The effect is most often illustrated by the example of a polymer climbing up a vertical rod inserted in a rotating cylindrical vessel (refer to Figure 14). The phenomenon is not viscosity dependent. Consider a stress applied to an elemental volume of rotating liquid of thickness dZ located at distance r from the vertical axis (refer to Figure 15). Assume a coordinate system in which the stresses σ_{xx}, σ_{yy}, σ_{zz} are directed

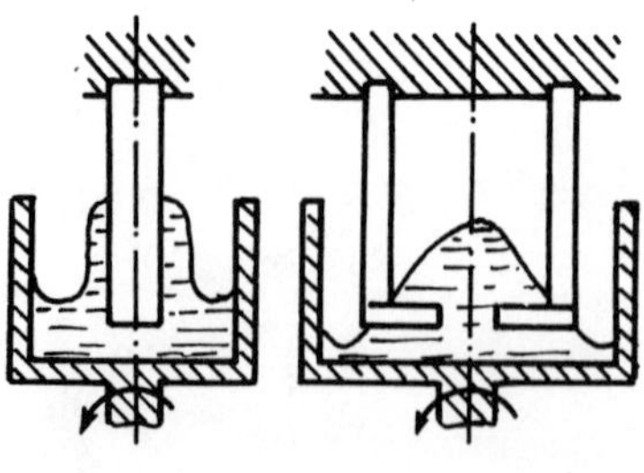

Figure 14 The Weissenberg effect.

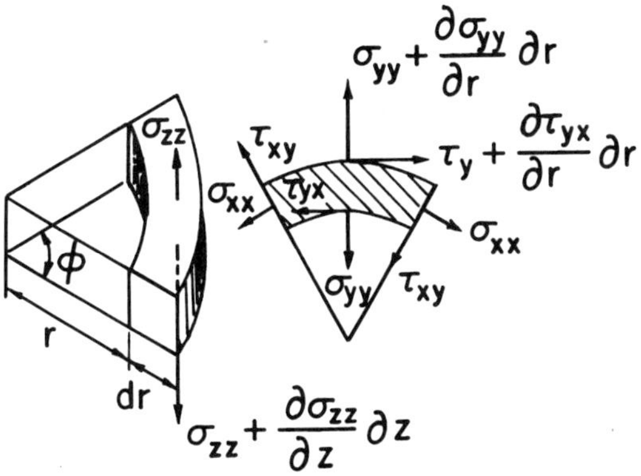

Figure 15 Forces acting on an elemental volume of fluid.

circumferentially, radially, and along the vertical axis, respectively. If the liquid is at a state of rest (i.e., not subject to shear), the stresses equal the hydrostatic pressure P_0. Under shear conditions the element may be considered as a bundle of fibers located along the circumference and submerged in the bulk of the viscous fluid. These fibers, then, become an elastic element. Tangential stresses, τ_{yx}, responsible for inducing laminar flow, will strive to elongate the fibers. If the material is incompressible, stretching of the longitudinal fiber will be accompanied by compression in the radial and axial directions, because σ_{yy} and σ_{zz} change at equal rates. This action creates a hydrostatic pressure which exceeds P_0 (at rest). Thus, for the element under consideration, the following expressions apply:

$$\sigma_{xx} - \sigma_{yy} > 0, \qquad \sigma_{zz} - \sigma_{yy} = 0 \tag{184}$$

When the forces are in equilibrium in the direction normal to the shear plane, and accounting for Eq. (184), we obtain

$$r \frac{d\sigma_{xx}}{dr} = \sigma_{xx} - \sigma_{yy} > 0 \tag{185}$$

Because σ_{zz} is negative, that is, it is a compressive stress, Eq. (185) may be restated as follows:

$$\frac{d(\sigma_{zz})}{dr} < 0 \tag{186}$$

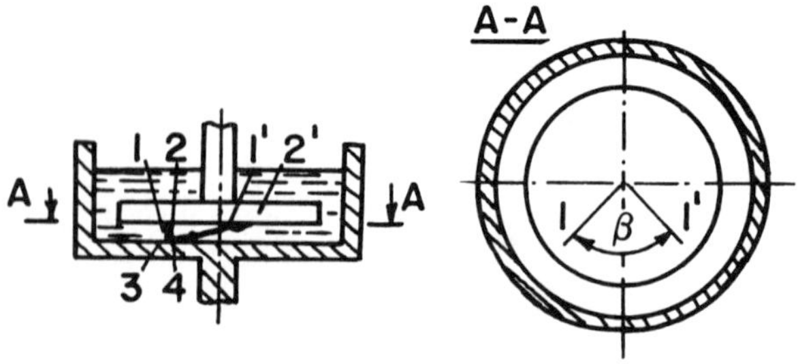

Figure 16 Scheme illustrating effect of normal stresses.

Equation (186) states simply that for a fluid under shear, hydrostatic pressure increases with decreasing radius. During shear action a dynamic equilibrium is established, whereby the pressure increase at the center is compensated by an increase in pressure drop (Δp) along the radius of a disk of material. From Eqs. (184) and (186), we have

$$-\frac{d\sigma_{zz}}{denr} = \sigma_{yy} - \sigma_{xx} \tag{187}$$

Equation (187) has been validated by experimental observations of the Weissenberg effect.

For a mechanical explanation of the Weissenberg effect the model shown in Figure 16 is of interest. In Figure 16 the rotating upper disk displaces material at point 1 by angle β, to position 1′. Because the melt has viscoelastic properties and is able to adhere to the surface of the rotating disk, elemental volume 1–2–3–4 is deformed and stretched along arch 1–1′. The stretched volume of material approaches elastic restoration along the chord and thus induces a force directed toward the center of rotation. The phenomenon is mathematically described as follows:

$$\frac{p_1 - p_2}{\lambda_1^2 - \lambda_2^2} = \frac{p_2 - p_3}{\lambda_2^2 - \lambda_3^2} = \frac{p_3 - p_1}{\lambda_3^2 - \lambda_1^2} \tag{188}$$

where

p = main stresses
λ = main rates of reversible elongation
G = shear modulus

Parametric Calculations

The basic design of a disk mixer is similar to the model concept illustrated in Figure 16. The capacity of a disk mixer can be computed from the following empirical formula developed by Sporyagin (1967):

$$Q = BRh - AR^3\mu_a nh^2, \quad \text{kg/hr} \tag{189}$$

where

μ_a = apparent viscosity, kg-s/cm^2
R = radius of the moving disk of the machine, cm
h = working gap between disks, cm
B = parameter (function of rotational disk velocity)
A = parameter (function of physicochemical properties of the material)

The relationships for the parameters A and B as functions of viscosity and disk rotational velocity are given in Figures 17 and 18. The optimum working gap at maximum capacity can be computed from the following formula:

$$h = (5.5 \sim 6.5)10^{-2}D \tag{190}$$

where D is the working diameter of the moving rotor.

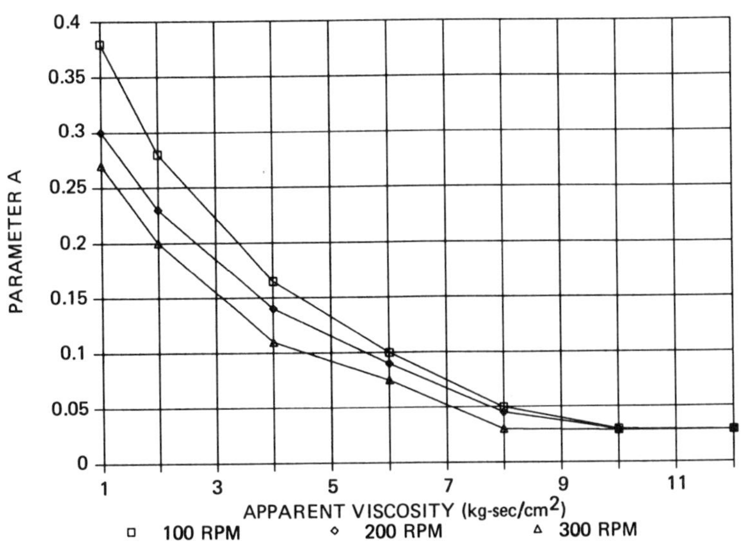

Figure 17 Parameter A as a function of viscosity and rotational velocity.

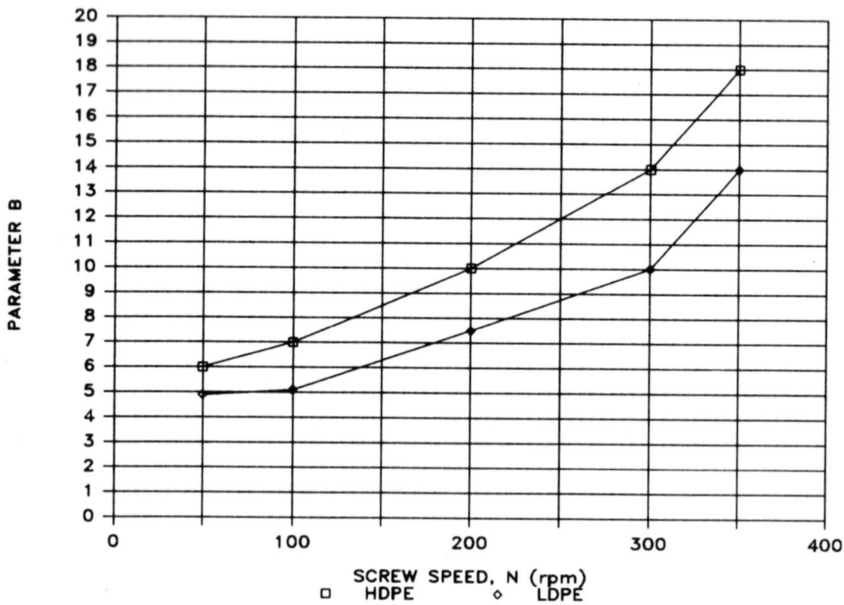

Figure 18 Parameter B as a function of rotational velocity.

The diameter of the mixer's discharge orifice is

$$d = (0.1 \sim 0.15)D \tag{191}$$

To estimate the pressure drop developed by the machine, the following formula is recommended (Dubovitzky, 1968):

$$\Delta p = \frac{2\pi(R + r)n\mu_a}{h} \ln \frac{R}{r} \tag{192}$$

where

R and r = outside disk radius and radius of discharge opening, respectively, cm
n = rotational velocity of the disk, rps
μ_a = apparent viscosity, kg-s/cm^2
h = gap between disks, cm

Formula (192) is valid when the normal stresses in the circular gap are equal to the tangential stresses. The average velocity gradient in the disk gap is used in the calculation. Rakhmanov (1969) gives a relationship for capacity as a function of disk shape factor K_d, disk diameter D, rotational velocity n, apparent viscosity μ_a, the specific weight of the material γ, and the shear rate ν:

$$Q = f(K_d, n, D, \mu_a, \gamma, \nu) \tag{193}$$

$$\nu = \frac{\pi D n}{3S_w} \tag{194}$$

and S_w is the working gap.

It is assumed in this case that the gap is a variable slit narrowing from the disk's periphery to its center at an angle of 4 ~ 9°. The shear rate is approximately constant over its cross section. In this calculation the value of the working gap S_R is used (i.e., the minimum distance between the disk end and the wall of the case at the transition from the case to the draw plate). This gap is usually located at $1/3R$ from the disk center. Thus, the diameter in Eq. (194) is $D/3$ in this case. The dimensional analysis relationship, Eq. (193), may be presented in the following form:

$$Q = K_d n D^{2.5}\sqrt{\mu_a \nu \gamma} = K_d n D^{2.5}\sqrt{\tau\gamma}, \quad \text{kg/s} \tag{195}$$

The disk's shape factor is determined experimentally at free flow through the head.

For a flat disk, $K_d = 5.1 \times 10^{-5}$
For a faceted disk, $K_d = 6 \times 10^{-5}$
For a helix-grooved disk, $K_d = 7.8 \times 10^{-5}$
The optimum gap is typically $(0.02 \sim 0.03)D$, and for flat disks it is $(0.05 \sim 0.06)D$.

The above relationships were developed for $D \leq 0.4m$ and $n \leq 400$ rpm.

Horsepower requirements for processing polyvinyl chloride, polyethylene, polystyrene, and similar materials can be estimated from the following empirical formula:

$$N = 0.96 \times 10^{-4} D^2 n, \quad \text{kW} \tag{196}$$

where

D = disk diameter, cm
n = rotational velocity, rpm

Formula (196) is valid at $D < 40$ cm and $n < 500$ rpm. Note that for each disk diameter there corresponds an optimum rotational velocity. The following rotational velocities can serve as rough guides for different disk diameters:

D (mm)	110	150	240
n (rpm)	180–220	250–300	340–400

From the theoretical analysis of Rakhmanov (1969), equations were derived for a disk mixer with a rotating-cone disk and a flat stationary disk applicable to describing the velocity of the melt.

$$v_\psi = v_r \frac{z}{h_r} + \frac{z^2 - h_r z}{2\mu_a r} \times \frac{\partial p}{\partial \psi} \tag{197}$$

where

v_r = peripheral velocity of a point of cone disk at distance r from axis of rotation
h_r = height of the gap corresponding to the peripheral velocity
p = discharge pressure

An equation for the capacity of the flow was also developed:

$$Q = \pi \frac{h_0 R_0^2}{3} n - \frac{h_0^3}{36\mu_a} \frac{\partial p}{\partial \psi} \tag{198}$$

where

R_0 = radius of stationary disk
h_0 = height of maximum gap between disks
n = disk rotational velocity
μ_a = viscosity of melt at the exit from mixer

For the case where both disks are flat and parallel the velocity equation transforms to

$$v_\psi = \frac{v_0 r}{H_0 R_0} Z + \frac{Z^2 - H_0 Z}{2\mu_a r} \times \frac{\partial p}{\partial \psi} \tag{199}$$

where H_0 is the gap height between disks.

Another design case occurs when the disk pair constitutes a combination cone–flat plane for the rotating cone and a stationary flat disk. This design has a central opening of radius R_1 for discharging the melt; consequently not all of the working surface of the cone disk takes part in the hydrodynamic process. The material to be processed is influenced only by the cone's surface limited by the outer radius R_0 and inside radius R_1. In this case, Eq. (198) is modified to

$$Q = \frac{\pi}{3}\frac{(h_0^3 - h_1^3)(R_0 - R_1)^2}{(h_0 - h_1)^2}n - \frac{h_0^3 - h_1^3}{36\mu_a} \times \frac{\partial p}{\partial \psi} \tag{200}$$

where h_1 is the height of the gap between disks at the point of maximum hole radius.

Scale-up is recommended on the basis of the largest acting forces generated during the processes of flow, friction, deformation, plastification, homogenation, and extrusion taking place between the narrow gaps of the rotor surface and the walls of the disk mixer. The principal forces acting in the machine are compression, shear, and elastic forces, which correspond to the fourth group of forces, and thrust forces, inertia, centrifugal forces, as well as friction forces, which are derivative forces of those previously mentioned.

The equality of scales of forces is

$$C_p^{\mathrm{I}} = C_p^{\mathrm{IV}} = \frac{\rho_0 C_e^4}{\rho_m C_e^2} = \frac{E_0 C_e^2}{E_m} \tag{201}$$

As follows from Eq. (201), at $E_0 = E_m$ and $\rho_0 = \rho_m$, in the model and prototype the geometric scale of modeling ($C_e = D_0/D_m$) is equal to the time scale ($C_t = t_0/t_m$) and the velocity scale:

$$C_v = \frac{v_0}{v_m} = \frac{C_e}{C_t} = 1 \tag{202}$$

Hence, the scale of rotational velocity is

$$C_n = \frac{1}{C_t} = \frac{1}{C_e} \tag{203}$$

The scale of capacity is

$$C_Q = \frac{Q_0}{Q_m} = \frac{C_e^3}{C_t} = C_e^2 \tag{204}$$

The scale of forces is

$$C_p = \frac{P_0}{P_m} = \frac{\rho_0 C_e^4}{\rho_m C_t} = C_e^2 \tag{205}$$

The scale of horsepower is

$$C_N = \frac{C_p C_v}{C_t} = C_e^2 \tag{206}$$

Thus, the Cochy number is used for the geometrical, kinematic, and dynamic similarity of disk mixers, $C_0 = P\nu^2/E$. To maintain the same rotor velocities in the model and prototype, the capacity of the prototype must be proportionally increased by the cube of the geometrical scale of modeling:

$$C'_Q = C_Q C_e = C_e^3 \tag{207}$$

Horsepower is increased by the same ratio:

$$C'_N = C_N C_e = C_e^3 \tag{208}$$

The scale of modeling of gaps between a rotor and the wall of mixer is assumed from the condition of equal gradients of shear rate and viscosity of materials at all compatible points in the model and prototype machines:

$$C_h = \frac{h_0}{h_m} = C_e^{0.5} \tag{209}$$

In this case the scale of modeling of the rotor velocity for maintaining similarity should be

$$C_n = \frac{n_0}{n_m} = C_e^{-0.5} \tag{210}$$

The effect of normal stresses on materials processed in disk mixers is greatly reduced by increases in disk diameters. Therefore, the modeling scales are valid over specified limits, which must be established from experiments.

Parameters Affecting Disk Mixer Operation

Both mixer capacity and pressure are influenced by the following factors:

1. Method of feeding
2. Disk diameter

3. Gap between disks
4. Shape and size of the feeding zone
5. Diameter and length of discharge
6. Hydraulic resistance of forming device
7. Viscoelastic properties of the melt
8. Shape of material particles

The viscous properties of the material characterize the apparent viscosity. The viscoelastic properties characterize the time of relaxation. The gap size is measured in the shear zone. These mixers are usually not designed with a transition zone between the feed and shearing sections. In this case the design radius of the disk is assumed to be considerably greater than the radius of the flat portion of the disk.

The diameters of the discharge orifices depend on the cross section of the extruded material. Because of low pressures in these machines the hydraulic resistance of the heads should be at a minimum. If the actual capacity is less than that of the maximum possible value, the residence time in the processing zone increases. This is undesirable for polymers of low thermal stability. The amount of heat dissipated in a unit volume increases, thus causing a lowering of the viscosity and elastic properties, which in turn decrease the unit's internal pressure. When this occurs there is a need to increase the intensity of heat transfer or to decrease the disk rotational velocity. The net result is a reduction in the velocity gradient in the gap, causing a lowering of elastic deformation and pressure of the melt. Figure 19 shows the effect of increasing shear strain on pressure in a disk mixer. With an almost insignificant increase in the maximum possible capacity over the actual, a large amount of dissipated energy is absorbed by the extruded material. This increase in the elastic reversible deformation results in the formation of high pressures. Experimental curves defining the relationships between capacity and gap sizes for different disks usually show a maximum. Such a plot is shown in Figure 20. The maximum on the curve may be explained by two simultaneous processes:

1. With a decrease in gap size, an increase in the shear gradient occurs. This increases the amount of heat evacuated from a thin layer of melt between the disks, which causes a reduction in the melt temperature, an increase in the melt viscosity, an increase in elastic deformation, and, consequently, an increase in the pressure due to Weissenberg effect.

2. A decrease in the gap between disks results in an increase in the hydraulic resistance of the melt flow from the periphery to the center of the disks. The largest hydraulic resistance to the melt flow is created at the transition from the shear zone to the central opening in the stationary disk. The relationship between these two processes from the gap between disks is not linear; therefore by their combined action the curve shows a maximum. An increase in the disk diameter is accompanied by an increase in the maximum possible capacity. Consequently, to

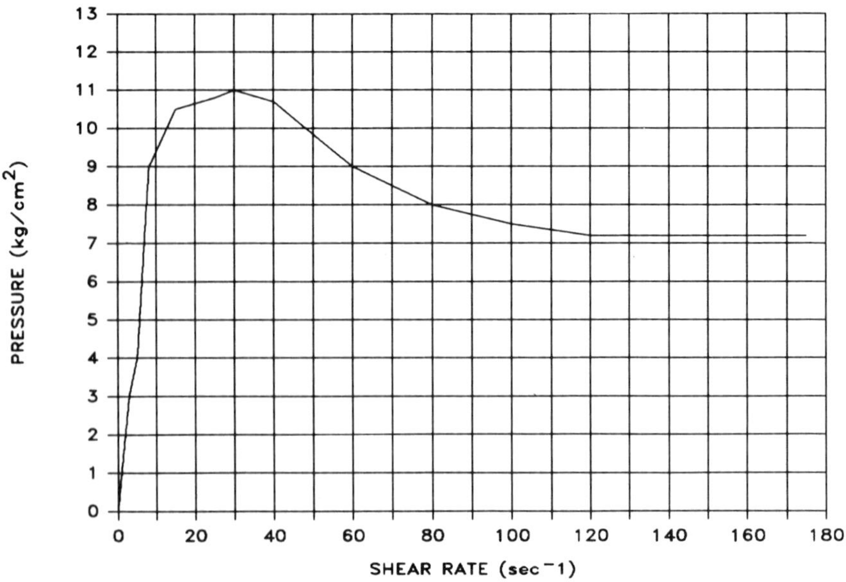

Figure 19 Plot of pressure versus shear rate.

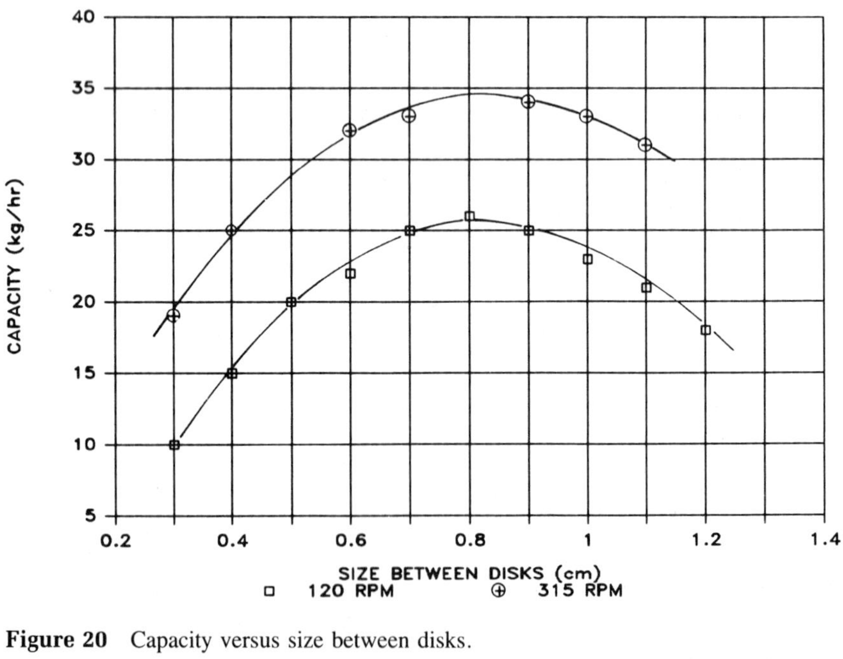

Figure 20 Capacity versus size between disks.

maintain conditions of maximum capacity, the distance between disks should be enlarged.

At higher melt temperatures, a reduction in elasticity and an increase in irreversible deformation is observed. The decrease of melt elasticity results in a reduction of the normal stresses in the melt, as well as pressure. This is illustrated by the plot shown in Figure 21. If the melt temperature is increased with simultaneous reduction of the elastic properties of the melt, the viscosity of the material lowers. Figure 22 shows an experimental relationship for capacity, whereby mixer capacity is directly proportional to temperature. It can be concluded from these data that viscosity decreases faster than the elastic properties.

The rate of decrease of hydraulic losses in the melt flow is greater than that of the normal stresses for any increase in temperature. This leads to high capacities accompanied by lower pressure. For larger gaps between disks, a greater torque and lower axial force at constant rotational velocity is observed. This can be explained as a result of the large increase in viscosity at lower velocity gradients when processing pseudoplastic materials having a low index of melt. The relationship between capacity and the opening in the forming head must be taken into consideration. For a large opening, capacity first increases because of a lower resistance. However, for larger openings, capacity tends to decrease because of the downstream effect in the shear zone.

Additional references given at the end of this chapter provide detailed design information on the various mixers discussed.

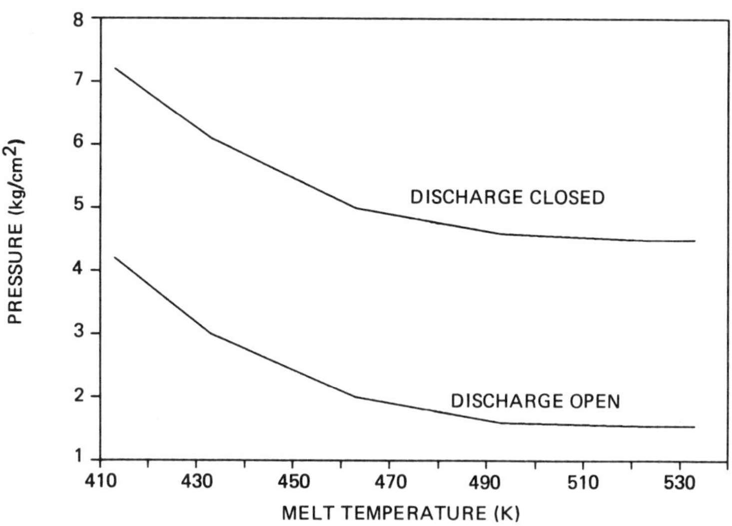

Figure 21 Pressure versus melt temperature (h = 1.6 mm, n = 385 rpm).

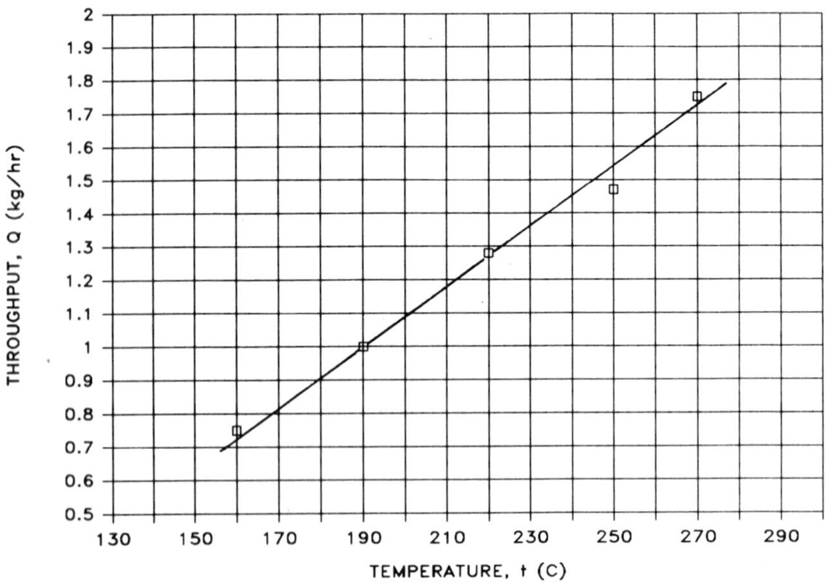

Figure 22 Capacity versus melt temperature for one speed.

NOTATION

A	axial force
A, B	dimensionless functions of ratio of thread base and channel height
a	ratio of countercurrent rate to forced-flow rate
b	channel width
b'	thickness of thread
C	force scale
C_e	geometric scale factor
C_w	heat capacity of water
D	diameter
$D_{x,z}$	shearing strain
d	diameter
Eu	Euler number
e	thread width
F	boundary area between laminar flow layers
$F_{c,d}$	friction and geometric coefficients
F_z	cross-sectional area of threads
Fr	Froude number
f	friction coefficient, or cross-sectional area
G	mass throughput
G_w	cooling-water mass rate
g	gravitational acceleration
H	pitch of left thread of screw
h	height of screw channel or thread depth
i	number of threads
$i_{s,c}$	enthalpy of steam and condensate, respectively
i'	compression ratio
J_i	coefficients characterizing screws of variable pitch
K	overall heat transfer coefficient or shape factor
L	screw length

l_0	zone length	q	channel cross-sectional area
m	number of pitches	Re	Reynolds number
m, n	exponents	R, r	radius
N	power	r_0	average coil radius
n	rotational velocity	S	shear rate in screw helix
n_{cr}	critical rotational velocity	T	parameter defined by Eq. (5)
P, p	pressure	t	helix pitch
Q	volumetric throughput	$t_{0,m}$	time
Q_H	heat flow to extruder head	U, V	velocity
Q_k	heat flow to extruder	$V_{0,m}$	volume
Q_N	heat flow due to energy transformation	z	axial distance

Greek Symbols

α	heat transfer coefficient, also coefficient of injection flow	λ	conductivity
β	coefficient of countercurrent flow	μ_a	apparent viscosity
γ	coefficient of leaking, also specific weight	ν	kinematic viscosity
$\gamma_{1,2,3}$	elastic deformation	ρ	density, also friction angle
δ	height	ρ'	friction angle (solid's angle of repose or slide)
η_0	efficiency	σ	stress
η_ν	volumetric efficiency	τ	shear stress
		ψ	helix angle

REFERENCES

Azbel, D. S., and N. P. Cheremisinoff, *Fluid Mechanics and Unit Operations,* Ann Arbor, MI, 1983.

Bernhardt, E., *Pererabotka Termoplasticheskikh Materialov,* M. Goskhinisdat, USSR, 1962.

Dolzhkov, A. D., *Khim i. Neft Mashinostroenie,* (USSR), *9,* 5–6, 21 (1973).

Dubovitzky, V. F., Sb., *Khimichestroe Mashinostroenie,* No. 7, 41, Tekhnika, MVSSO USSR (1968).

Ferns, A. W. D., *Plastics and Polymers, 42,* 160, 149 (1974).

Herrman, H., *Chemie, Ing. Tech., 38*(1), 25 (1966).

Jerchel, K. F., *Chemie Ing. Techn., 44*(8), 552 (1972).

Lukach, Yu E., D. D. Ryabinin, and B. N. Metlov, *Valkovye Mashiny dlia Pererabotki Plastmass i Resinovyk Smesey,* M. Mashinostroenie, USSR, 1967.

Mohr, W. D., R. L. Saxton, and C. H. Japson, *Ind. Eng. Chem., 49*(11), 1857 (1957).

Ostapchuk, Yu. G., and N. K. Zhuk, Sb., *Pererabotka Plastmas, 46,* Tekhnika, Kiev, USSR, 1969.

Pearson, J. R. A., *Chem. Eng., 317,* 19 (1977).

Prokepec, L., and J. Ulbrecht, *Chem. Ing. Techn., 42*(8), 530 (1970).

Rakhmanov, V. B., Sb. *khimicheskoe Mashinostroenie,* No. 9, p. 28, Tekhnika, MVSSO USSR, Kiev (1969).

Ryabinin, D. D. and Yu. E. Lukach, *Cherviachnye Mashiny dlia Pererabotki Resinovykh smesei i Plasticheskikh mass,* M. Mashinostroenie, USSR, 1965.

Silin, V. A., *Modelirovanie Chervyachnyk Mashin. sb. Trudov Ukrniplastmash, Oborudovanie dlia Pererabotki Polymerov,* Tekhnika, Kiev, USSR, 1964.

Sporyagin, E. A., and R. G. Mirsoev, *Mashiny i Tekhnologia Pereabotki Polimernykh Materialov, Trudy LTI im. Lensoveta,* USSR, 1967.

5

Rubber Compounding Practices

USES AND PROPERTIES OF CARBON BLACK

Applications

Carbon black belongs to an important class of industrial carbons employed as reinforcing agents in rubber and as black pigments in inks, coatings, and plastics. The large majority of carbon black manufactured for use in the rubber industry is used in tire production. The remainder is used in a variety of other rubber products such as inner tubes, conveyor belts, wire and cable covering, motor mounts, fan belts, hoses, gaskets, rubber heels, and a variety of specialty elastomeric products for the consumer market. Type and loading of carbon black vary with the application and service demands to which the product is to be subjected. In a tire tread, for example, the resistance to abrasive wear is of major concern; hence, highly reinforcing blacks are required. In contrast, in tire sidewalls and ply compounds a minimum of hysteresis and low heat generation is essential, and semireinforcing blacks are generally preferred. Extruded articles, such as automotive stripping and glass run channels, require blacks with structure which provides a minimum of shrinkage and close adherence to die dimensions during extrusion.

The preparation of carbon black–rubber "masterbatches" is common practice in the industry. Masterbatching consists of adding the appropriate loading of carbon black in the form of a slurry to the rubber prior to coagulation. Table 1 provides a summary of the various common grades of blacks used in the syn-

Table 1 Properties and Rubber Applications of Carbon Blacks

Type (designation)	Avg. particle size (Å)[a]	Typical applications
Channel blacks		
easy processing (EPC)	290	Tire treads, heels, soles of shoes, mechanical goods products
medium processing (MPC)	250	
Gas furnace blacks		
semireinforcing (SRF)	800	Tire carcass/bead insulation, footwear and soling, belts, hoses, packings, mechanical goods
high modulus (HMF)	600	Tire carcass/sidewalls, footwear, mechanical goods
fine furnace (FF)	395	Truck tire carcass, breaker, and cushion
Thermal blacks		
medium thermal (MT)	4700	Wire insulation/jackets, mechanical goods, belts, hose, packings, stripping mats
fine thermal (FT)	1800	Natural rubber inner tubes, footwear uppers, mechanical goods, inflations
Oil furnace blacks		
general purpose (GPF)	510	Tire tread base, sidewalls, sealing rings, cable jackets, hose, extruded stripping products
fast extruding (FEF)	400–450	Tire carcass, tread base, and sidewall, butyl inner tubes, hose, extruded stripping
high abrasion (HAF)	270	Camelback and tire treads, mechanical goods
high abrasion, high structure (HAF-HS)	250	Tire treads, products requiring high abrasion resistance
high abrasion, low structure (HAF-LS)	264	Channel black replacement in natural rubber applications
intermediate superabrasion (ISAF)	230	Tire treads and camelback, mechanical goods
intermediate superabrasion, high structure (ISAF-HS)	225	Tire treads, high abrasion service applications
intermediate superabrasion, low structure (ISAF-LS)	227	Tire treads, high abrasion service applications
superabrasion (SAF)	200	Tire treads, camelback, mechanical goods, heels/soles of shoes
conductive (CF)	230	Antistatic and conductive rubber products, belts, hose, flooring material

[a] $D_n = \Sigma\, nd/\Sigma\, n$, where n = number of particles of diameter d.

thetic rubber industry. Later in this chapter, discussions will largely be directed at carbon black incorporation into polymers such as elastomers and SBR (synthetic butyl rubber); however, the reader should be aware of applications in other industries.

In plastics applications, medium- and high-color channel blacks are employed for tinting and to provide maximum jetness at low loadings. Carbon black is also used as antiphoto- and antithermal-oxidation agents in polyolefins. As an example, clear polyethylene cable coating is susceptible to crazing or cracking, accompanied by rapid loss in physical properties and dielectric strength when exposed to sunlight for extended periods of time. The carbon black essentially serves as a blackbody which absorbs ultraviolet and infrared radiations. It also serves to terminate free radical chains and, hence, provides good protection against thermal degradation. Hawkins (1962) has shown that clear polyethylene cable jackets degrade to brittleness after a 2-year outdoor exposure, whereas the same compound containing 2% carbon black (medium-color channel) shows no change after the equivalent of 20 years exposure.

Carbon black is also employed in the paper industry in the manufacture of black papers, including album paper, leatherboard, wrapping and bag papers, opaque backing paper for photographic film, highly conducting and electrosensitive paper, as well as black tape for wrapping high-voltage transmission cables. This industry refers to channel black as Paris black. Both channel blacks and furnace blacks of appropriate jetness and wetting characteristics are widely used. The use of these blacks is particularly significant when conducting black papers are desired.

Black loadings vary with application, but a range of 2–8% by weight of pulp is typical. To obtain adequate dispersion the black is generally employed in the fluffy form, and it is commonly added directly to the beater in a dispersible bag to avoid dusting of carbon black to the atmosphere. In other applications it may be added in the form of an aqueous slurry.

Carbon blacks are commonly used as pigments in the preparation of paints, lacquers, enamels, and industrial finishes (classified as high, medium, and standard color). In these applications, jetness or masstone is of primary concern. Jetness depends upon particle size; therefore, the coating industry uses a wide range of channel and furnace carbon blacks varying from 100 to 4000 Å in diameter. High-color blacks provide exceptional jetness and gloss for automobile finishes and high-grade enamels. Medium-color blacks are used in industrial enamels, and the standard grades of black are used in general utility and industrial paints. The high- and medium-color blacks are prepared by the channel process, and the standard-color blacks may be either channel or furnace blacks. Lampblacks and large-particle-size furnace and thermal blacks are also used in paints and other coatings. These are deficient in jetness; however, they provide a desirable blue "tone" to the coating. Color blacks are available in pelleted or beaded form since the grinding techniques employed in the coating industry are adequate to ensure good dispersion of the pigment. Carbon black is also used in the lacquer industry in the form of masterbatch "chips." These chips consist of a high loading of carbon black dispersed in an appropriate resin.

Printing inks typically employ 10–12% black in petroleum oil. These inks are relatively fluid, and they dry by penetration of the oil into the fibers of the paper. Both channel blacks and HAF grades of oil furnace black are used for newspaper ink production. Inks for coated papers, such as those used in large-circulation publications, employ flash-drying inks, which consist of a volatile solvent, a resin, and 15–20% carbon black. The ink in this application is dried by passing it over a heat source such as steam or gas flames. In book printing, offset or lithographic inks are employed. These inks are characterized by maximum tinting strength and covering power. The oils used for dispersion are usually linseed varnishes which are relatively viscous and can employ black loadings as high as 25%. Flow in these ink dispersions is induced by increasing the combined oxygen or volatile matter on the black's surface to about 12%, which reduces the yield stress of the ink. Other examples of carbon black applications in this industry are pigments in carbon papers, typewriter ribbons, and various synthetic fibers.

There are a variety of other uses for carbon black which go beyond the subject of this volume, including uses in the production of dry cell batteries, in insulation materials for high-temperature applications, in ore reduction operations and carburization, in the production of antistatic phonographic records, as fuel and as an absorbent in liquid oxygen explosives, as a grinding agent in the cement industry, and as a coloring pigment in linoleum, leather coatings, plastic tile, and cement.

Properties

Carbon black is elemental carbon that is differentiated from commercial carbons such as coke and charcoals by the fact that it is in particulate form. Particles are spherical, quasi-graphitic in structure, and of colloidal dimensions. Figure 1 shows a photomicrograph of a typical commercial grade of carbon black used in rubber vulcanizates. Carbon blacks are manufactured either by a partial combustion process or by thermal decomposition of liquid or gaseous hydrocarbons. There are many commercial classifications of carbon black, but all blacks can be generally characterized by their method of production, as follows.

Lampblacks: produced via combustion of petroleum or coal tar residues in open shallow pans (typical sizes 600–4000 Å)

Channel blacks: produced by impingement of underventilated natural gas flames (100–300 Å)

Furnace combustion blacks: produced by partial combustion of natural gas or

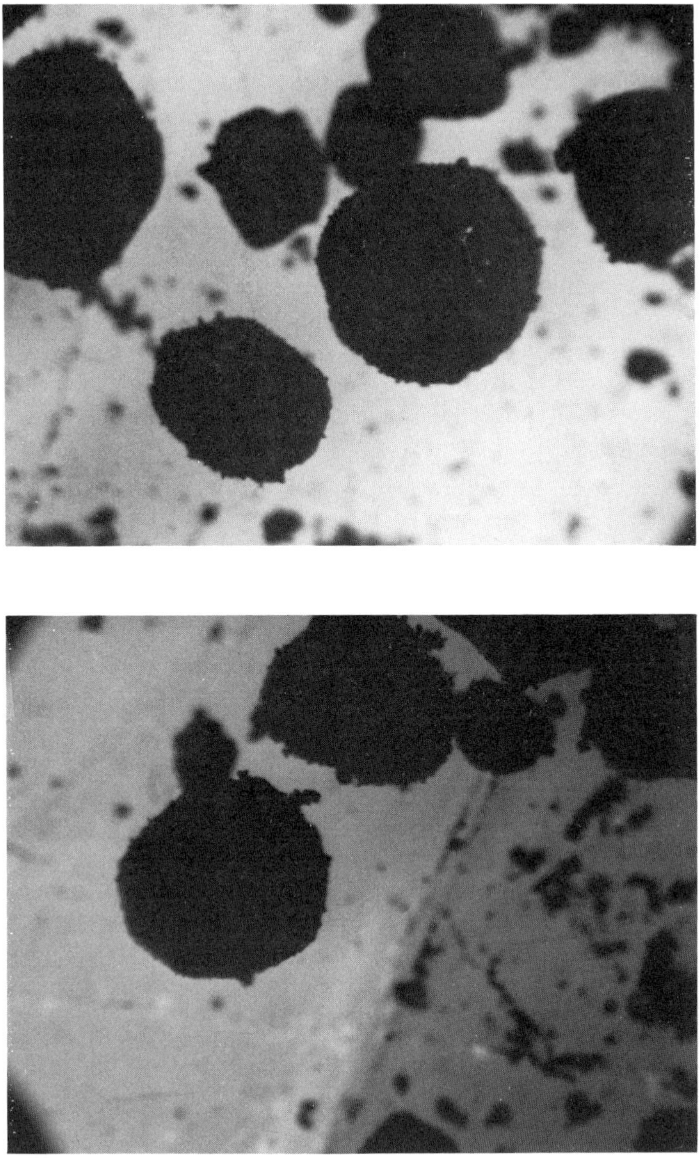

Figure 1 Photomicrograph of typical commercial grade of carbon black.

liquid hydrocarbons in retorts or furnaces (400–800 Å, coarsest variety 1400–4000 Å)

Thermal blacks: produced by the thermal decomposition of natural gas

Acetylene black: produced by the exothermic decomposition of acetylene.

The various commercial grades and the specific properties imparted to end-use products via their incorporation depend on particle size (i.e., surface area), chemical composition and relative activity of the particle surface, and degree of particle-to-particle association.

Carbon black essentially remains as a discrete phase when dispersed in mediums such as rubber. The extent of mixture reinforcement imparted to the rubber, or, for example, the intensity of color imparted to a coating formulation, mainly depends on the particle size or surface area.

The definitions used for characterizing average particle size vary. Arithmetic averages are defined as:

$$D_n = \frac{\Sigma\, nd}{\Sigma\, n} \tag{1}$$

where n is the number of particles of diameter d.

A more meaningful definition is the surface-average diameter

$$D_A = \frac{\Sigma\, nd^3}{\Sigma\, nd^2} \tag{2}$$

Particle size distributions of commercial grades can vary from Gaussian to skewed.

Specific surface areas are evaluated by absorption techniques and the arrangement of carbon atoms within a carbon black particle has been extensively studied by x-ray diffraction methods. Carbon black displays two-dimensional crystallinity, which is a structure defined as mesomorphic. Smith (1964) provides detailed discussions of the molecular and chemical properties of carbon blacks and graphite.

One of the most important properties of carbon black is its surface activity. Carbon itself has proved extremely useful as an adsorbent. All carbon blacks possess a distribution of energy sites, whereby those of highest energy are first occupied by adsorbate. As the adsorption process continues, less active sites become filled and the differential heat of adsorption decreases, eventually approaching the heat of liquefaction of the adsorbate at monolayer coverage. High-energy sites can be progressively reduced by heat treatment. At 3000°C all high-energy sites are reduced and the substance displays characteristics of a homogeneous surface of uniform activity. Carbon adsorption is treated in detail by Cheremisinoff and Ellerbusch (1978).

Chain structure is another property of importance, particularly in terms of the properties imparted to rubber. In particular, extrudability, elastic modulus, and electrical conductivity of rubber vulcanizates are sensitive. As general comments, lampblack and acetylene black have a high degree of structure, whereas channel and low blacks are typically low in chain structure.

Most commercial carbon black is available in bead or pellet form. For ink applications, black is supplied with bulk densities of 6-15 lb/ft^3, which is preferential for systems of low shear. The rubber industry typically employs carbon blacks having an average specific gravity of 1.8.

REINFORCEMENT OF ELASTOMERS

In the rubber industry, reinforcement by fillers, especially carbon black, is a major part of the technology. The relevant properties of carbon black include particle size–surface area, particle porosity, aggregate structure (i.e., bulkiness), the amount of carbon per aggregate, surface activity, and surface chemistry.

Examples of widely used elastomers are SBR, BR, acrylonitrile-butadiene copolymers, and EPDMs. These are examples of polymers in which the molecular chains are not highly oriented when their gum vulcanizates are stretched. Gessler et al. (1978) cites examples where upon stretching a gum vulcanizate, the tensile strengths are 0.4–0.6 MPa, whereas when the same elastomer is compounded with 50 phr (parts per hundred rubber) of a selected carbon black, a vulcanized tensile strength as high as 6.5 MPa is measured. Figure 2 gives another example of an EPDM, showing the dramatic improvement of tensile and modulus properties.

The preceding examples help to illustrate that the usefulness of elastomers greatly depends on the ability of fillers such as carbon black to impart reinforcement properties. The property of reinforcement is characterized in terms of the vulcanizates stiffness, modulus, tear strength, rupture energy, cracking resistance, fatigue stress, and abrasion resistance.

It is important to note that some fillers are nonreinforcing, but are instead used as extenders in rubber formulations to reduce raw material costs. These fillers are usually chemically inert and have large particle sizes. Reinforcing fillers are the opposite; that is, typically they are small in particle size and chemically active. Gessler et al. (1978) define fully reinforcing fillers as those solid additives having mean particle sizes below 50 nm. The two principal additives are carbon blacks and silica. Solid fillers having sizes greater than 50 nm are clays, silicates, calcium carbonates, SFR carbon blacks, and a range of semireinforcing blacks that are usually derived by thermal oxidation.

It is important to make a distinction between the carbon black–rubber inter-

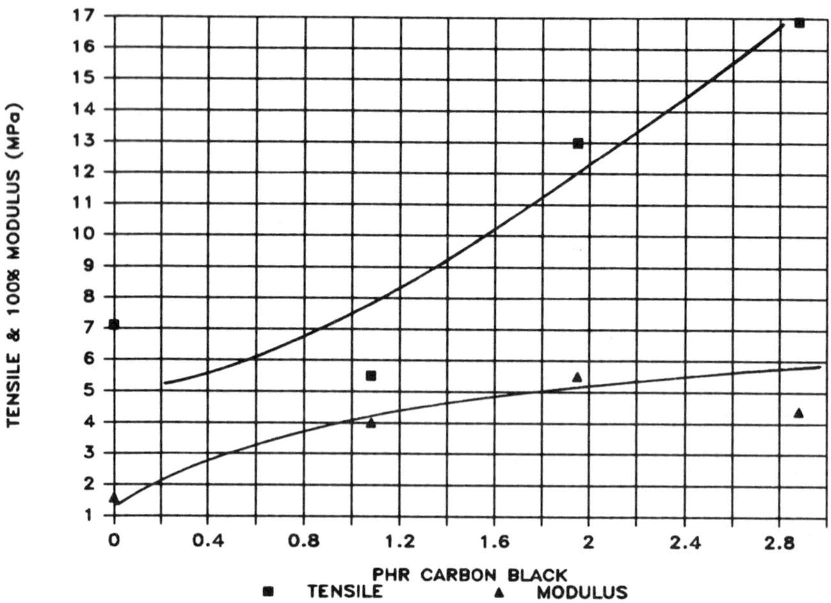

Figure 2 Tensile and modulus properties of sulfur-cured EPDM vulcanizate at different levels of carbon black.

actions of uncured and vulcanized materials. The effects of carbon black on such properties as resilience and stiffness will be different for these two states of the material. In the uncured state the incorporation of carbon black results in viscoelastic changes in the rubber system. In the vulcanized state, carbon black incorporation causes viscoelastic property changes, which in fact directly alter the rubber network. The specific properties affected in the vulcanizate are modulus, dynamic properties, and hysteresis.

Commercial carbon blacks used in rubber compounding applications are usually classified in terms of their morphology (i.e., particle size–surface area, vehicle absorptive capacity). In the ASTM nomenclature system for carbon black, which is strictly used in the United States, the first digit is based on the mean particle diameter as measured with an electron microscope. Other commonly used methods for classifying blacks by size are iodine number, nitrogen adsorption, and tinting strength. Important also are the shapes of individual particles and of aggregates. DBP absorption is the principal technique employed for measuring the irregularity of the primary aggregates.

The aggregates of carbon blacks can be classified into about nine categories (Gessler, 1970) as summarized in Figure 3. The dimensional aspects of aggregate size and shape are defined in terms of:

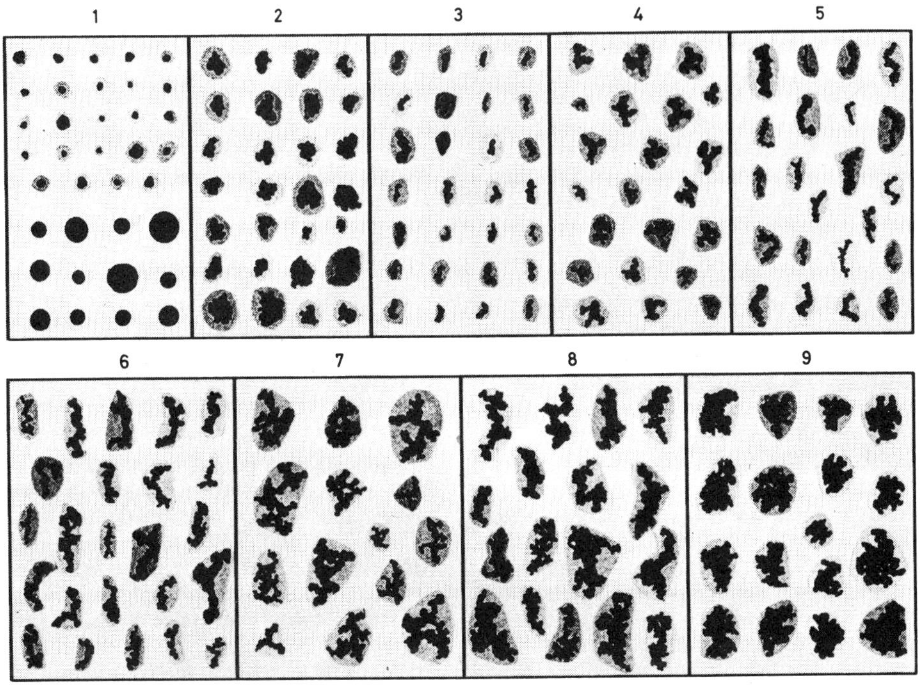

Figure 3 Aggregate shape classification defined by Gessler (1970).

Particle diameter, d
Average width, W
Longest length, L
Particle volume, V

Shape factors are usually defined in terms of dimensionless aspect ratios, for example,

Length-to-width ratio, F
Cube of aggregate perimeter/volume, P^3/V

The latter parameter is a measure of aggregate irregularity relative to that of a sphere. Gessler et al. (1978) recommend the following expression for estimated mean particle diameter from measurements of the aggregates:

$$\bar{d} \simeq c \times \left(\frac{W}{F^n}\right) \tag{3}$$

where F is the shape factor and c,n are constants. The number of particles in an aggregate is estimated from its measured volume, by microdensitometry, divided by the volume of a sphere having diameter d.

As a rough approximation of the mean particle size of carbon blacks, assume d is half the average aggregate width. An exception to this rule-of-thumb is thermal blacks, because of the high frequency of spheroidal particles. Table 2 lists several commercial blacks using the ASTM nomenclature system, along with estimated surface area values estimated by Gessler et al. (1978).

The term *structure* characterizes the bulkiness of individual aggregates. Often when carbon black is mixed into rubber, some of the aggregates fracture. In addition to fracturing, the separation of physically attracted or compacted aggregates (microagglomerates) can occur. Aggregates can be held together fairly rigidly because of the irregular nature of their structures. Microagglomerates can also be formed during mixing. That is, high shear forces can cause microcompaction effects wherein several aggregates can be pressed together. Excessive microagglomeration can lead to persistent black network structures which can affect extrusion and ultimate vulcanizate properties.

Baranival (1967) and Cotten (1975) have studied aggregate breakdown in ball milling and roll mill operations. Both investigators report methods of predicting breakdown.

Table 2 Various Rubber-Grade Carbon Blacks and Surface Area Estimates[a]

Type black	ASTM no.	Surface areas (m^2/g)[b]
SAF	N-110	144
ISAF	N-220	121
HAF-HS	N-347	94
HAF	N-330	85
HAF-LS	N-327	96
FEF	N-550	41
SRF-HS	N-765	32
SRF-HM	N-770	29
SRF-LM	N-761	24

[a]Reported by Gessler et al., 1978.
[b]Derived from electron microscopy analysis of aggregate size.

The relationship of carbon black morphology to the failure properties of rubber vulcanizates has been extensively studied. Strength properties are usually enhanced with increasing black surface area and loading (see Figure 4, for example). There are, however, limiting values. The upper limit of black loading for maximum tensile strength and tear resistance depends on carbon black fineness and structure, with the former usually having the greater effect. Coarser, lower-structure blacks generally show peak strength properties at higher loadings. In terms of the ultimate level of strength reinforcement for different blacks, structure is significant as a dispersing aid, which may be attributable to better bonding between black and polymer. High structure is also important at lower black loadings, particularly for tear resistance.

There is some evidence that the strength reinforcing properties of fillers are directly related to modulus development. These properties are derived from the large stresses held by the highly extended polymer chains attached to the immobile particles. Boonstra (1973) has related high tensile strength to energy dissipation by a slippage mechanism at the filler surface. This is supported by the fact

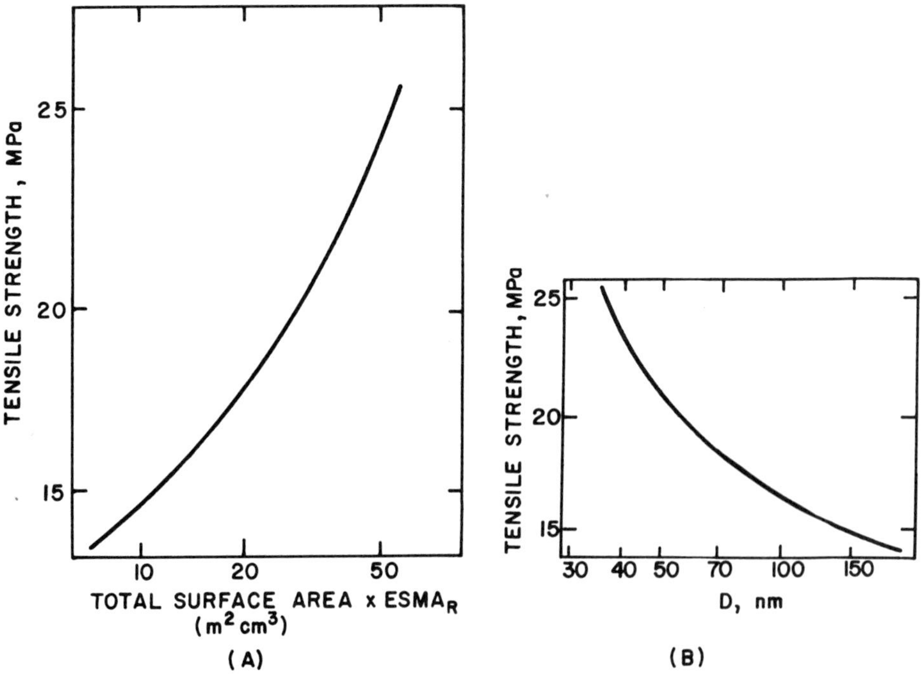

Figure 4 (A) Tensile strength as a function of total black surface area. (B) Tensile strength as a function of average interaggregate spacings.

that high-surface-area, inactive (partially graphitized) carbon blacks give very high tensile strength under standard testing conditions. Under more severe conditions, however, adhesion between black and polymer becomes important.

Andrews and Walsh (1958) have shown that the path of rupture through a filled vulcanizate passes from one filler aggregate to another, which are sites of high stress concentration. One approach to increasing strength is to lengthen the overall rupture path. The finer the filler and the higher the loading, the greater the effective increase in total cross section. There is a limiting point, however, when the packing of the filler aggregates becomes critical and they are no longer completely separated by the polymer. Preferential failure paths between aggregates may then be formed. These, combined with the fact that a smaller amount of polymer is being strained, can result in failure. The level of filler–polymer adhesion is important, especially as the upper limits of loading (aggregate packing) are reached.

Filler–polymer adhesion is important to the amount of elastic energy released by internal failure. Strongly adhering fillers enlarge the volume of rubber that must be highly strained during the process of rupture. It is worth noting that the relative importance of different carbon black parameters that influence vulcanizate properties greatly depends on the conditions of testing.

Carbon black fineness is the dominant factor governing vulcanizate strength properties. Tensile strength, tear resistance, and abrasion resistance all increase with decreasing black particle size. Cyclic processes such as cracking and fatigue are more complex because they may involve internal polymer degradation caused by heat buildup. Beatty (1964) points out that it is important to note whether or not the end-use application represents constant energy input or constant amplitude vibrations. The latter mode of service greatly reduces the fatigue life of rubber compounds containing fine or high-structure blacks. Beatty has reported no significant particle size effect on fatigue on the basis of equal energy input.

The relationship of carbon black fineness to failure properties is not straightforward since it is difficult to separate the effects of aggregate size and surface area. It is likely that both are important, since aggregate size relates to the manner in which the surface area is distributed. In general, decreasing aggregate size and increasing black loading reduce the average interaggregate spacing, thereby lowering the mean free rupture path. Increasing black fineness and loading can be viewed in terms of either increased interface between black and polymer, or increased total black cross section (lower aggregate spacing). Both show a strong effect on tensile strength up to the maximum tensile value that can be achieved. Gessler et al. (1978) measured the tensile strength properties of five blacks (SRF, HAF, HAF-LS, SAF, SAF-LS) in SBR-1712 over a loading range of 40–140 phr (optimum cure, room temperature). Figure 4A shows the positive correlation with total surface area and Figure 4B shows an inverse relationship with the average interaggregate spacing ($\overline{D}$). The values were derived from the volume fraction (ϕ) and the average aggregate volume ($V^{1/3}$) as measured in rubber. A simple cubic packing model was employed. Only the values up to the maximum tensile of each black are plotted; therefore the actual loadings vary for the different blacks. The tread blacks exhibited maximum tensile in the range of 70–80 phr while the SRF sample showed a peak at about 100 phr.

The ultimate value for tensile strength across all blacks appears to be determined by either aggregate size or specific surface area. Tensile strength can be improved to a limiting value by increasing the loading. It is not possible, however, to match the ultimate tensile of a fine black by increasing the loading of a coarse one, at least not without other modifications (e.g., improving the ultimate dispersion or the bonding between black and polymer).

Blends of two or more elastomers are used in a variety of rubber products. The compatibility of elastomers in terms of their relative miscibility and response to different fillers and curing systems is of great importance to the rubber compounder. From the standpoint of carbon black reinforcement, certain combinations of polymers can give less than optimum performance if the black is not proportioned properly between them. Note that equivalent volume proportionality of the black is not always desirable. This depends on the nature of the polymers and their relative filler requirements in terms of strength reinforcement.

The compatibility of elastomer blends has been studied by a variety of different techniques, including optical and electron microscopy, DTA, GPC, solubility, thermal and thermochemical analysis, and x-ray analysis. These and other methods have been discussed by Corish and Powell (1974).

Microscopic methods are especially useful in studying the phase separation (zone size) of different polymer combinations and in determining the relative amounts of filler in each blend component. Walters and Keyte (1962) studied elastomer blends by means of phase-contrast optical microscopy, where the contrast mechanism is based on differences in the refractive indexes of the polymers. Their work covered blends that included NR/SBR, NR/BR, and SBR/BR. Their results indicated that few, if any, elastomers can be blended on a molecular scale. They also provided direct evidence that fillers and curing agents do not necessarily distribute proportionately.

The term *compatibility* can be defined in terms of the glass transition temperature (T_g). Blends that have T_g values of the individual polymer components are considered incompatible, while those that result in a single, intermediate T_g can be viewed as compatible. Also, solubility parameters (δ) are considered a prerequisite to compatibility. In the category of compatible blend combinations Corish (1971) lists NR/SBR, NR/BR, and SBR/BR. Incompatible blends are NR/NBR, NBR/BR, and NR/CR.

The compatibility of elastomer blends can also be defined from the standpoint of interfacial bonding between the different polymer phases. The technique of differential swelling, using solvent systems and temperatures such that one elastomer is highly swollen and the other is below its θ temperature with the polymer chains remaining tightly coiled, can be used. Thus, the polymer below its θ temperature could be treated in the same manner as a filler with respect to the way it restricts the swell of the other polymer. A high degree of swelling restriction is indicative of interfacial bonding between phases, while the reverse is analogous to the dewetting that occurs with nonreinforcing mineral fillers in elastomer networks. Also, the type of curing system is important in achieving interfacial bonding.

The importance of the proper curing system in blends of low-functionality rubber with more unsaturated or polar elastomers must also be emphasized. Large amounts of polymer can be extracted from blends when unsatisfactory curing systems (e.g., sulfur-ZnO alone) are employed. This can be attributed to curative diffusion because it occurs whether or not the curatives are added to the blend or to the low-functionality rubber alone.

Good blending is favored by a similarity of solubility parameters and viscosity. For polymers varying in unsaturation or polarity, cure compatibility must be considered in terms of both properly cross-linking each polymer component and achieving satisfactory interfacial bonding. In contrast, the addition of fillers to elastomer blends can significantly alter the state of the polymer phases. Where large differences in unsaturation exist between the polymers in a blend, staining methods may be used to render the high-unsaturation polymer more opaque to provide contrast for either light- or electron-microscope analysis.

To the point, the contributing aspects of reinforcement are dispersion, aggregate breakdown and interactions, and distribution between the separate phases of a polymer blend. For blends of dissimilar elastomers (e.g., in unsaturation, polarity, crystallization, or viscosity), problems can arise in achieving optimum black distribution in the end product. When fillers are added to elastomer blends one very obvious change is a reduction in the size of the separate polymer zones. In other words, there is a preferential location of the black in one of the blend components. This phenomenon may be related to both molecular structure and relative viscosity. The great mobility and possibly more linear structure of one of the blend component molecules apparently enables better wetting of the black surface. This characteristic is often observed in SBR/BR blends. The lower initial viscosity of the BR would also tend to favor the acceptance of the black. It is, however, possible to alter the normal pattern of distribution in terms of how the filler is added, along with mixing conditions. Other polymer factors which favor preferential filler adsorption in elastomer preblends are higher unsaturation and polarity.

Figure 5A gives an example of hysteresis, indicating a minimum vulcanizate hysteresis with about 60% of the black in the BR phase of a 50/50 NR/BR. Failure properties are shown in Figure 5B. Note that although tear resistance paralleled the hysteresis results, both tensile and tread wear exhibited peak values with almost all of the black in the BR. This represented the BR master-

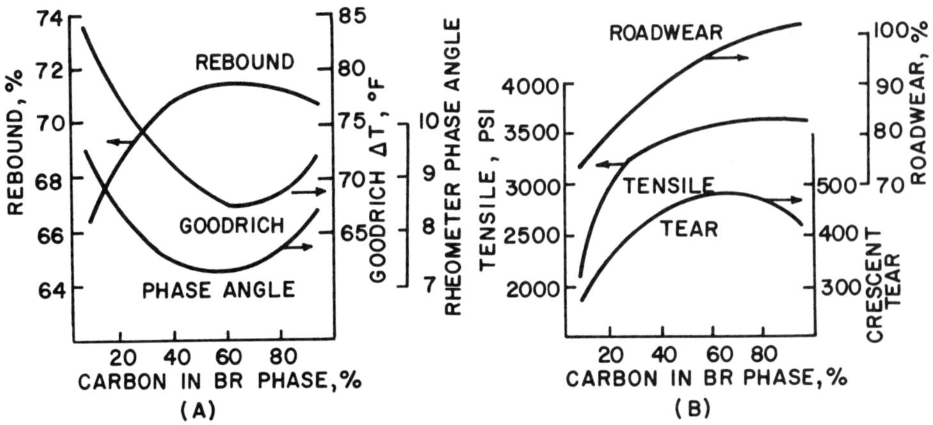

Figure 5 (A) Effect of carbon distribution on hysteresis. (B) Effect of carbon distribution on vulcanizate properties. (Data of Gessler et al., 1978.)

batch containing 80 phr of black, which was cut back with NR. In this example, high concentrations of black in the NR phase resulted in a sharp degradation of strength properties because of the specific need of BR for reinforcement, whereas unpigmented NR gives high strength through a crystallization mechanism.

Another phenomenon contributing to reinforcement is carbon black transfer. This is the ability of a carbon black to migrate from one polymer to another during mixing. Transfer is favored by low heat history and high extender-oil content. Both these parameters minimize interaction between black and polymer. Hence, solution or latex masterbatches tend to favor transfer, while hot mixing in a Banbury does not. Black transfer also occurs when a masterbatch of a low-unsaturation rubber is cut back with a high-unsaturation polymer. Polarity is also a factor.

CARBON BLACK MIXING

As noted above, carbon black is widely used as a reinforcing filler; however, the unique property of improving tear and abrasion resistance in rubber vulcanizates, particularly in synthetic rubber of the butadiene–styrene type, is not fully understood. It is likely that carbon black surface and rubber interact both chemically and by physical adsorption. The parameters important to this interaction are capacity, intensity, and geometry. The total interface between polymer and filler is expressed in units of square meter per cubic centimeter of vulcanizate or compound. The intensity of interaction is determined by the specific surface activity per unit area. It should be noted that adsorptive energies vary greatly in different locations on the black surface and it is likely that this distribution is

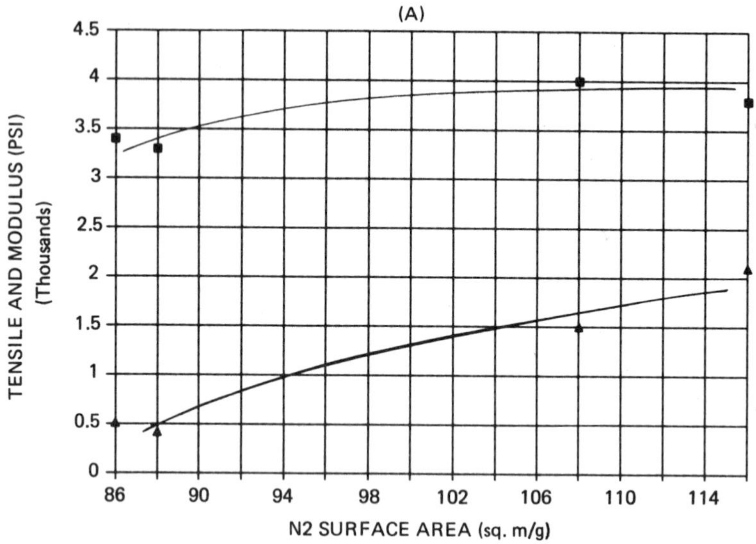

Figure 6 Influence of graphitization of black on vulcanizate properties. (Data for both high-structure and normal ISAF blacks.) (A) Influence on physical properties: ■ tensile; ▲ modulus. (B) Influence on vulcanizate elongation. (C) Influence on scorch.

responsible for the variability in properties among different types of carbon blacks. Geometrical properties are characterized by the structure, which is basically anisometric, and particle shape and porosity. Greater anisometry results in looser particle packing. The void volume is used as a measure of the packing density.

Carbon black particles form irregular structures which tend to break down during intensive mixing. It is therefore a combination of carbon black reactivity, rubber and black chemical, physical, and rheological properties, as well as the conditions of mixing which establish the final strength properties of the vulcanizate. As a rule of thumb, high-structure blacks (e.g., ISAF) impart a higher modulus (at 300% elongation) than a corresponding normal black. The high modulus is determined by both the anisometry and surface activity. The separate influence of these properties can be observed by heat-treating these blacks at 2000–3000°C. The crystallographic structure of carbon black is altered by heat treatment and properties approach that of graphite. Through recrystallization, highly active sites on the carbon black surface lose their high activity. In this situation the entire surface area becomes homogeneous and adsorption energies approach their lowest state. Figure 6 illustrates the effect of this reduction in surface activity on vulcanizate properties. Heat treatment typically results in a

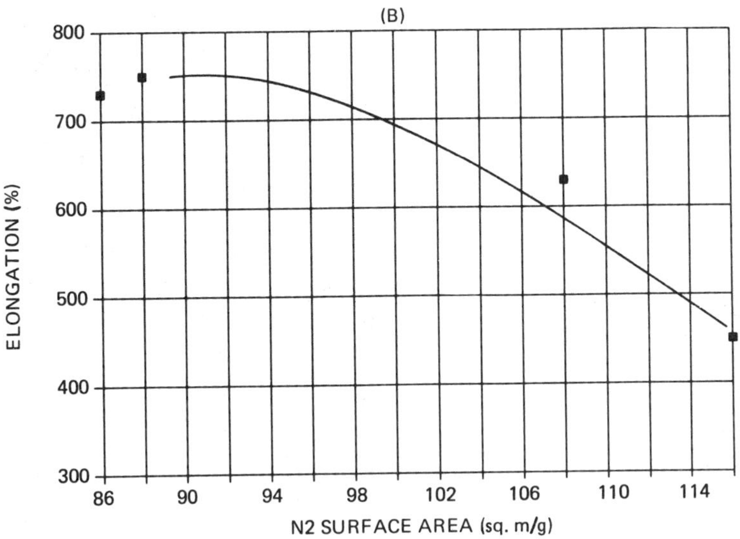
(B)
ELONGATION (%)
800
700
600
500
400
300
86
90
94
98
102
106
110
114
N2 SURFACE AREA (sq. m/g)

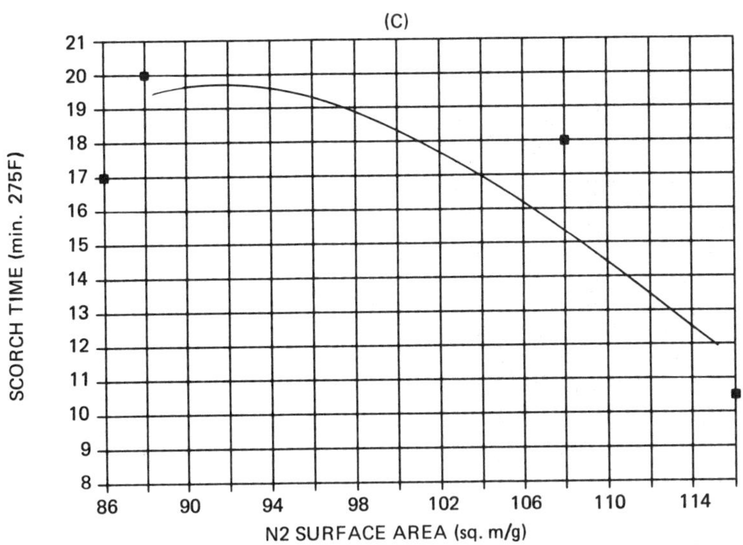
(C)
SCORCH TIME (min. 275F)
21
20
19
18
17
16
15
14
13
12
11
10
9
8
86
90
94
98
102
106
110
114
N2 SURFACE AREA (sq. m/g)

(20 RPM @ 5 min.)

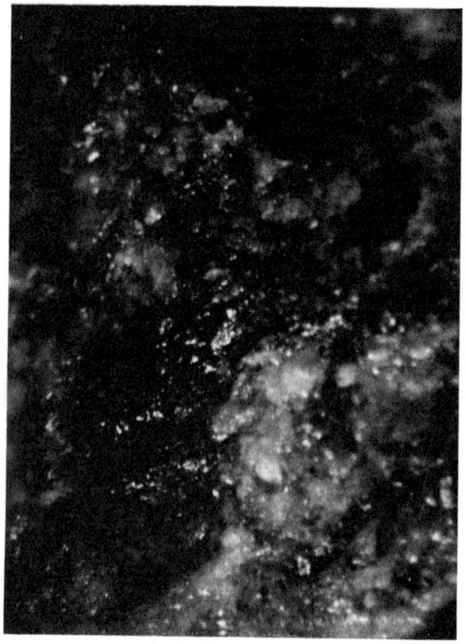

(30 RPM @ 7 min.)

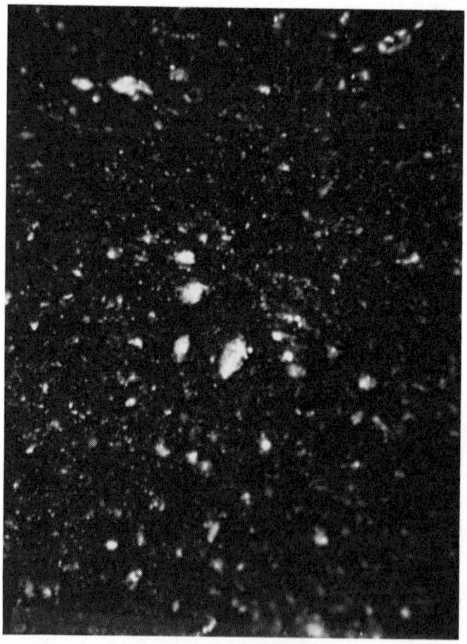

(40 RPM @ 9 min.)

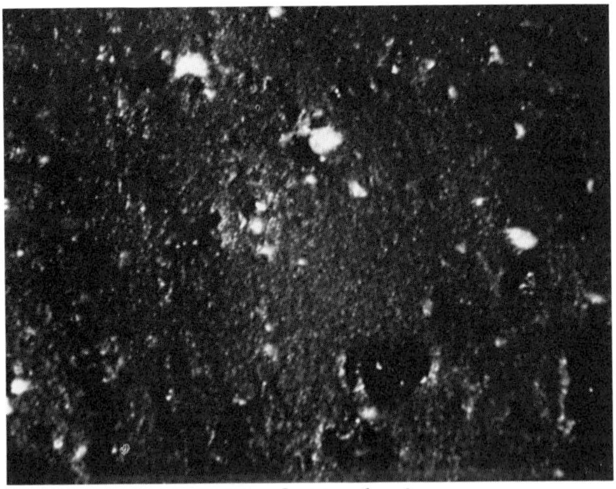

(50 RPM @ 11 min.)

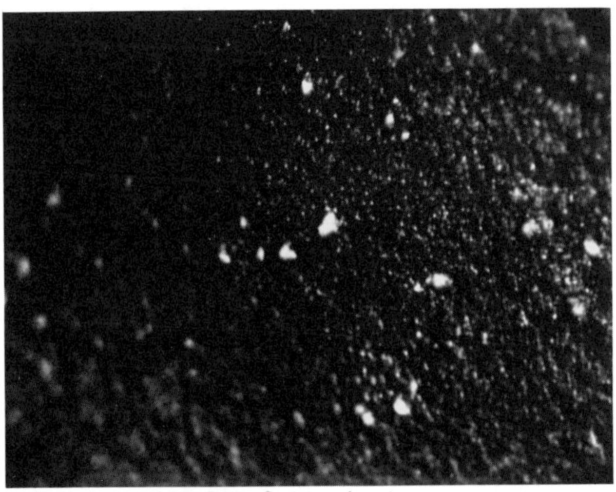

(75 RPM @ 13 min.)

Figure 7 Dispersion of mineral filler and carbon black improves with mix time and speed.

minor decrease in surface area and further reductions in tear resistance. Modulus and elongation can be decreased by a factor of 3 or 4. The decrease in properties can be attributed to the removal of highly active sites from the surface. Boonstra (1967) shows that this accompanied by a decrease in water adsorption and propane adsorption.

Microscopic examinations of thin sections of vulcanizate in which carbon black or other fillers and polymer were mixed for a short time only reveal that this additive exists as coarse agglomerates in an almost pure rubber matrix, without almost any colloidal dispersion of the additive. Only after mixing has been continued at greater intensity and/or for longer time periods can one observe how these agglomerates gradually disappear, making room for an increasing amount of additive. In all cases the initial product of mixing is comprised of an agglomerate which is formed by the penetration of rubber in the voids between carbon black particles or other additives such as mineral fillers (e.g., clays) under the pressure and shear which builds up during Banbury mixing and on a roll mill. Once these voids are filled, the additive is incorporated but not yet dispersed. As soon as the agglomerates are formed, continuous mixing exposes them to shearing forces which break them up and eventually disperse them. In other words, both agglomerate formation and breakdown take place almost simultaneously. Figure 7 gives an example of mineral filler and carbon black dispersion in an EPDM, where the mix is exposed to different Banbury mixer speeds. The slides in this figure represent samples of the mix retrieved at different times during a continuous mix cycle, where the speed was stepped up to successively higher speeds at 2-minute intervals.

The composition of the carbon black agglomerates is established by the void volume per gram or per cubic centimeter carbon black (or filler). This is the volume determined during an oil adsorption experiment. This value is determinant for the carbon black concentration in the primary agglomerates.

Boonstra (1967) notes that blacks which are compatible in size and total surface area, but vary in structure and hence, void volume, require different mixing times to achieve the same level of dispersion. The reason for this is that since during mixing the voids between particles are filled, a greater void volume would require a longer time to be filled. Also, for most commercial blacks, increasing structure is associated with increasing surface activity. Lower adsorption activity in graphitized blacks promotes slippage of rubber molecules on the particle surface when filling interstices during black incorporation. This shortens the time needed for this process as compared to the original blacks. Easy incorporation however does not necessarily ensure good dispersibility. After a black has been completely incorporated, additional dispersibility depends on structure and adsorption activity. As a general rule, high-structure blacks tend to require more incorporation energy; however, once they are incorporated, less energy for dispersion is needed compared to lower-structure blacks.

An important consideration during the mixing and dispersion of carbon black and rubber is the degree of interaction between these two species which can lead to the formation of bound rubber. These bonds are both of a physical nature (i.e., adsorption onto highly active sites) and of a chemical nature. The latter is described as a reaction of free rubber radicals with reactive carbon black surface groups, promoted by high shearing energy. A parallel relationship exists between the formation of bound rubber and shearing forces in a mixer. Figure 8 shows a torque–time response curve obtained on a Haake rheocord mixer. It is observed that bound rubber can be formed almost immediately during mixing. The maxima observed in torque during mixing typically signify that degrading forces exist. It is not clear to what extent bound rubber influences vulcanizate properties. Usually with cross-linking in bound rubber, gel tends to be low. The formation of bound rubber is accompanied by an increase in viscosity. During bound rubber formation, shearing forces at constant rotor speeds increase because of this rise in viscosity. These shearing forces will affect both further dispersion of the carbon black and breakdown of the rubber. Vulcanizate properties eventually degrade after continued mixing because of a breakdown of the polymer.

The effect of fillers on cross-link density in vulcanizates can be characterized by either modulus or swelling measurements. A common problem in selecting the proper black is the difficulty of knowing whether the black interferes with

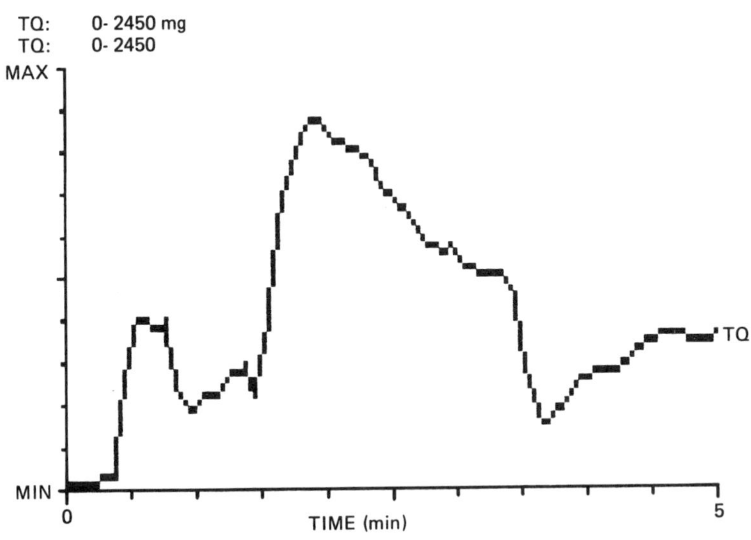

Figure 8 Torque–time response curve obtained on a rheocord, suggesting formation of bound rubber.

the vulcanization, causing higher cross-link density. It may also lower the degree of swelling for physical reasons, while the cross-link density in the rubber compound remains unchanged. A distinction can be made between adhering and nonadhering fillers depending on their degree of interaction with the polymer in the vulcanizate state. Adhering fillers (most carbon blacks) tend to reduce the swelling of their vulcanizate in a manner that is proportional to the filler content. In contrast, nonadhering fillers (e.g., $CaCO_3$, glass beads) tend to increase swell through the formation of voids surrounding the filler particles. The effect of carbon black on swell properties can be established by constructing a plot of the ratio of swelling values of pure gum and filled vulcanizates as shown in Figure 9. The parameter V_{ro}/V_r represents the reciprocal swelling values V_{ro} and V_r which are the volume fraction of rubber in the swollen pure gum vulcanizate and that in the swollen rubber matrix in the carbon black vulcanizate, respectively. Figure 9 illustrates that with increasing carbon black loadings in the vulcanizate, there is a proportional decrease of the V_{ro}/V_r ratio. The slope of this curve depends both on the type of filler and the properties of the swelling medium. The stronger the medium (i.e., the smaller V_{ro}) the greater the slope for a specific carbon black. Boonstra (1967) and others have shown that a plot of the slope versus V_{ro} for different solvents results in a straight line, and at $m = 0$, one observes the same V_{ro} value for all carbon blacks (again this depends on the particular solvent used for the swelling experiments). In other words, a critical V_{ro} value is observed for

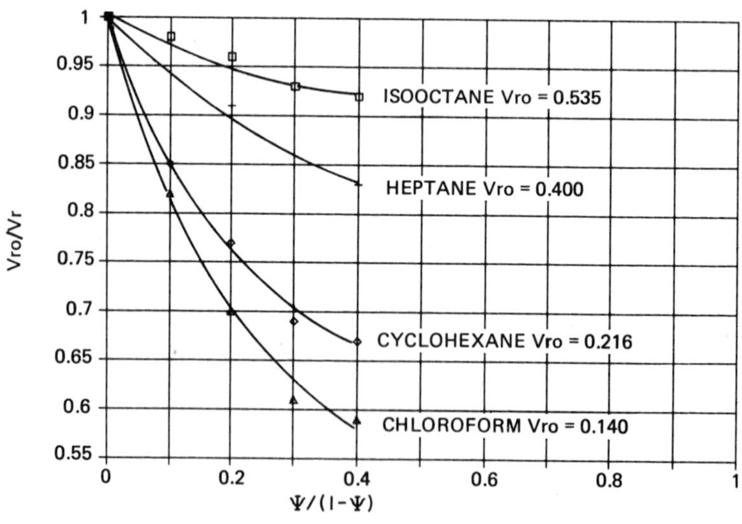

Figure 9 Swell index as a function of carbon black loading.

all carbon blacks and the value tends to vary only slightly with increasing cross-link density. The reason for this is that polymer molecule chains tend to cling to the black surface, which allows an extension that corresponds to some fraction of isometric swelling. In contrast, with nonadhering fillers, vacuoles filled with solvent are formed around the filler particles, resulting in an increase in swelling behavior with increasing filler levels. The behavior of graphitized carbon blacks on swell properties is very different. These typically have no influence on vulcanizate swelling in all solvents, basically because they are homogeneous owing to the heat treatment process.

The term *reinforcement* usually refers to improved tensile strength that is achieved by the addition of carbon black to noncrystallizing elastomers. In the case of butadiene–styrene copolymer, tensile strength can be improved by a factor of 10 or 12 over the pure gum vulcanizate through the addition of only 50 parts of reinforcing carbon black. The magnitude of this enhancement is somewhat unique and typifies the special nature of the forces existing between blacks and elastomers. Consider a noncrystallizing, unfilled elastomer. During elongation of the unfilled vulcanizate, the most highly elongated polymer chains reach their maximum elongation first and then break. When this occurs, those chains no longer carry any tension, and hence no longer contribute to tensile strength. As elongation continues, the next most highly elongated chains break. This process continues until eventually only a few chains contribute to the final tensile strength. It can be concluded therefore that those materials displaying low tensile strength experience unequal distribution of loading. This heterogeneous distribution can be improved by wetting or slippage. For example, if the most highly elongated molecular chains do not break down and retain their part in the elongation, a more uniform stress distribution will result and the final material will be stronger. The mechanism of mobile adsorption can contribute to homogeneity of stresses. Elastomer chains that are adsorbed on to the carbon black surface will remain in two-dimensional motion on the surface without being separated from it. At excessive elongation, they tend to slip along the surface. In this case the amount of energy per chain segment required for the process is lower than that needed for complete separation from the surface. Adsorptive energies are additive, and hence they can attain considerable values if a large number of successive segments are adsorbed. At the same time, while chain's slip along the surface during elongation, distributed segments line up sequentially on the surface building up tension that balances the elastic forces at a given elongation. The greatest contribution is derived from the points of high adsorptive activity.

The phenomenon described parallels that of crystallization, which occurs naturally in rubbers. Pure gum vulcanizates of crystallizing elastomers display greater tensile strength than elastomers that are noncrystallizing (amorphous). As such, crystalline polymers show only slight reinforcement through carbon black incorporation.

Carbon black vulcanizates display a permanent set after elongation. In contrast, pure gum vulcanizates show little, if any, permanent set. This permanent set tends to increase with increasing filler loading.

Mixing Practices and Effects of Degree of Mixing

As noted earlier, incorporation refers to the wetting of carbon black with rubber. During this operation, entrapped air is squeezed out between the voids of the rubber and carbon black particles. Early stages of mixing reveal that as the carbon black becomes incorporated, relatively large agglomerates (on the order of 10 ~ 100 μm) form. Cotten (1984, 1985) suggests that the time required for full carbon black incorporation can be determined by measuring the time required to reach the second power peak during a mixing cycle. Cotten (1985) performed mixing studies in a Banbury mixer, observing from microscopic examination of rubber–carbon black compounds the progression of black dispersion at different times. The rates of carbon black dispersion in this study were computed from the maximum torque data. A typical power curve generated by the author in reproducing Cotten's work is shown in Figure 10 using an oil-extended EPDM. Following Cotten, the rubber was first masticated for about 2 minutes and then the rotors were stopped. Carbon black was charged in the chute, the mixer ram inserted, and the rotors started again. Mixing times were

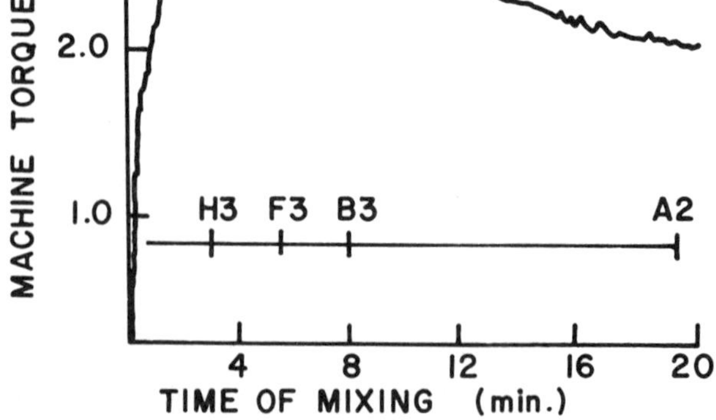

Figure 10 Typical torque response curve for rubber mixing in a Brabender.

measured from the instant when the rotors were restarted. The carbon black incorporation time was taken to be the time required to attain the second power peak shown in Figure 10.

Bound rubber in studies such as this can be measured by standard solvent extraction techniques. For example, toluene at room temperature or boiling heptane or hexane extractions can be used with a known aliquot of rubber sample. From a measurement of the residue weight, one can calculate the percentage of insoluble polymer.

The percentage (by volume) of unincorporated (nonwetted) carbon black can be estimated from density measurements. The volume of air in the batch (V) at time t can be computed from the difference in densities (ρ' and ρ) at time t and the time at the termination of mixing when carbon black dispersion is better. The formula used for this calculation is

$$V = \frac{B}{\rho'} - \frac{B}{\rho} \tag{4}$$

where B is the batch weight (phr).

The first power peak corresponds to the ingestion of the batch into the mixing chamber. This coincides with the instant that the ram in the loading chute reaches the bottom of its flight and thus removes any additional hydrostatic pressure from the mixing chamber. Cotten has found that the fraction of undispersed carbon black decreases linearly with the time of mixing at the point where the second power peak occurs. At shorter mixing times the compound can appear inhomogeneous and crumbly. Figure 11 illustrates some typical correlations for SBR and an EPDM. The mixing times can be observed to be inversely proportional to the rotor speed. When normalized, data tend to collapse to a single linear correlation. Cotten's work shows that there is a strong correlation between incorporation time and the rate of incorporation as computed from the slope of regression lines for change of densities with time for carbon black compounds in a single polymer.

The existence of a double power peak depends in part on the properties of the carbon black itself. Using pelletized carbon black will result in an absence of the first power peak because no additional hydrostatic pressure is applied when the mixer chute is lowered (hence the measured torque is no longer affected). The time to reach the second power peak, however, remains the same. Cotten notes that with fluffy carbon black, the time to reach the second power peak increases if no pressure is applied. Large agglomerates of carbon black form during the initial mixing stages regardless of the type of carbon black used.

Other studies by Turetsky et al. (1976) and Smith (1976) have shown good correlation between the measured black incorporation time (BIT) and weight average molecular weight, $\overline{M}_w$, of polymers. Incorporation times will increase

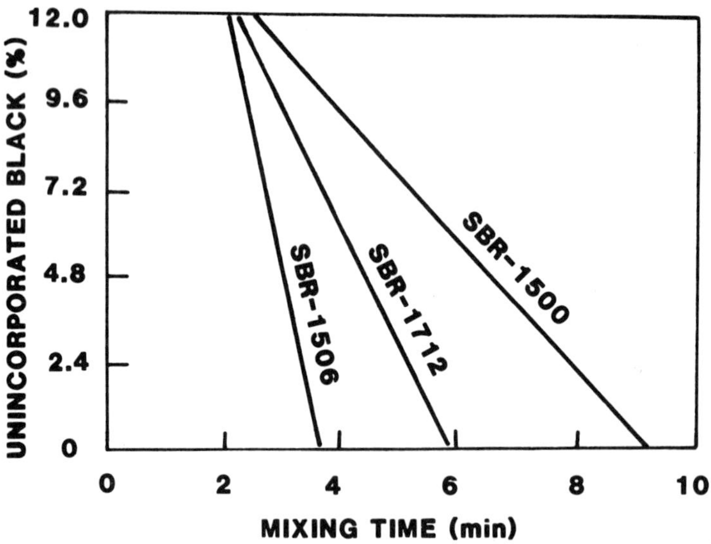

Figure 11 Correlation of unincorporated black with mixing time.

sharply with increasing polymer molecular weight. This effect tends to mask expected decreases in incorporation times when the Mooney viscosity is decreased through the addition of oil.

Decreasing concentrations of carbon black and increasing oil loading tend to reduce incorporation time. This is generally thought to be related to a lowering of the polymer viscosity.

Typical filler-reinforced elastomers contain in the rubber filler, curatives, plasticizers, and stabilizers. After mixing and curing a masterbatch, both physical and chemical cross-linking processes transform the system into a network structure. The structure is composed essentially of two networks: a stationary and a transient network. The stationary network is comprised of chemical cross-links that are formed by both curatives and fillers. This network also includes permanent crystalline structures from crystallizable polymers. In contrast, the transient network is comprised of trapped entanglements and transient ordered structures. Payne and Watson (1963) and Lee (1984) have proposed two mechanisms which explain the reinforcing effects of filler in elastomers. The first of these mechanisms can be described as a hydrodynamic interaction in which the filler particles are responsible for the reinforcement. The second mechanism is that of strong chemical interactions between the filler and the matrix. Both of these reinforcing interactions contribute to the mechanical properties of filler reinforced polymers. As noted earlier, the mechanical properties of engineering importance are the tensile modulus, the tensile strength, and the ultimate stretch ratio. Tensile modulus is mathematically defined as the slope of the stress–strain curve at zero strain for a given masterbatch. Tensile strength is defined as the force at break per unit area of the original sample. The ultimate stretch ratio is the uniaxially fractured gage length divided by the original gage length of the filler-reinforced polymer specimen.

A problem in designing polymer structures for a particular reinforced elastomeric application is a lack of quantitative physical relationships which define reinforcing effects of fillers on mechanical properties. Such constitutive expressions should ideally account for both the effects of fillers on the detailed structure and for the degree of mixing or dispersion of the masterbatch. Lee (1985) has tackled this problem somewhat by attempting to develop a mixing rule to correlate these effects. Lee's study focused on establishing a relationship for the tensile modulus and the ultimate stretch ratio of the filler-reinforced elastomer system.

Smallwood (1944) developed a simple relationship for the elastic modulus of filled rubber, in terms of the modulus of the gum E_1:

$$E = E_1(1 + 2.5\phi) \tag{5}$$

where ϕ is the volume fraction of filler. This expression is in fact the universal viscosity relation of Einstein. The relation accurately describes the modulus of filled rubber systems for dilute filler concentrations (up to 1%). A more accurate expression based on a polynomial relation is

$$E = E_1(1 + 2.5\phi + 14.1\phi^2) \tag{6}$$

This relation accounts for the increase in stiffness caused by spherical particles.

For particles that tend to agglomerate into chainlike clusters, a factor that accounts for the asymmetric nature of the cluster must be introduced:

$$E = E_1(1 + 0.67f\phi + 1.62f^2\phi^2) \tag{7}$$

where f is the geometric factor defined simply as the ratio of cluster length to width.

From the concepts of occluded volume and effective occluded volume, a slightly more complex relation was presented by Lee (1985):

$$E = E_1(1 + 2.5\phi'_{aggr} + 14.1\phi'^2_{aggr}) \tag{8}$$

where ϕ'_{aggr} is the effective volume fraction of the aggregates.

In the preceding expressions, the values of f and ϕ'_{aggr} can be deduced from a polynomial regression of the modulus data. The literature provides some working correlations for projecting compounding properties. For heat-treated and thermal blacks $1.3 < \phi'_{aggr}/\phi < 2.4$. For the ultimate stretch ratio of filled polymers λ_b, a ⅓ power law relation (i.e., $\lambda_b \sim \phi^{1/3}$) is applicable for some filled plastics.

From thermodynamics, the entropy of mixing defines the tensile modulus and the ultimate stretch ratio. For a given elastomer composition, different mixing histories give rise to different states of entropy. Therefore, a logical basis for defining tensile properties of filled reinforced elastomers is in terms of thermodynamic mixing rules.

As noted earlier, conventional mixing practices of fillers into different rubbers results in the formation of filler–rubber structures (e.g., bound rubber for carbon black filled elastomers). The specific filler–rubber structures formed affect the thermodynamic state of the elastomer system. This means that the reinforcement properties of the filler change because of the formation of these rubber-filled structures. Entropy of mixing for an ideal system can be computed from the Flory–Huggins equation:

$$\Delta S'_m = -k_B(N_1 \ln \phi_1 + N_2 \ln \phi_2) \tag{9}$$

where

k_B = Boltzmann constant
ϕ_1 = volume fraction of gum
ϕ_2 = volume fraction of filler
N_1, N_2 = numbers of effective polymer chain segments and effective filler particles, respectively

The entropy of mixing for a real system is

$$\Delta S^R_m = -k_B(N_1 \ln \phi_1 + N_2 \ln \phi'_2) \tag{10}$$

where

ϕ'_2 = volume fraction of filler actually wetted by polymer matrix

Figure 12 illustrates schematically the difference between ideal and real mixing in terms of a hard-spheres model. For real mixing, not all the fillers are incorporated into the polymer–filler wetting process because of the formation of filler agglomerates. Following Lee's (1985) development, the excess entropy

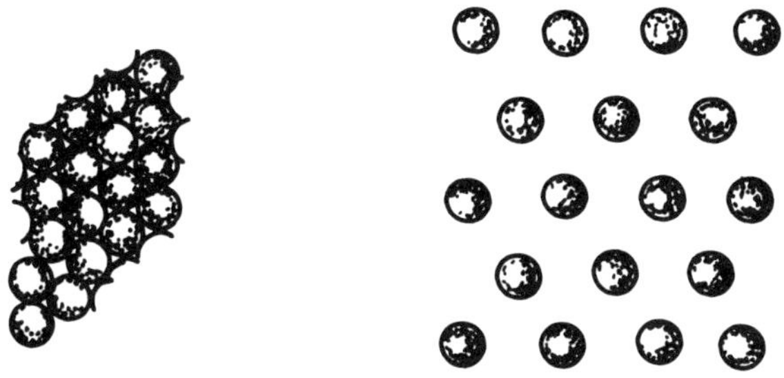

Figure 12 Hard sphere model illustrating good and poor mixing dispersion.

due to the heterogeneous distribution of the ith-type filler in the rubber matrix after mixing is

$$\overline{\Delta S}_{p-i} = \overline{\Delta S}_m^R - \overline{\Delta S}'_m; \qquad i = 2, 3, \ldots \tag{11}$$

And combining expressions, Lee obtains

$$\overline{\Delta S}_{p-2} = \frac{\Delta S_{p-2}}{N_2} = -k_B \ln\left(\frac{\phi_2'}{\phi_2}\right) \tag{12}$$

For an ith component system, $\phi_i' = \phi_i$, $\overline{\Delta S}_{p-i} = 0$, and hence

$$\overline{\Delta S}_{p-i} = -k_B \ln\left(\frac{\phi_i'}{\phi_i}\right) \tag{13}$$

Equation (13) defines the case of perfect mixing.

Using the effective volume fraction of the filler, Lee proposes a semiempirical mixing rule to describe the relative tensile modulus of filled elastomers.

$$\ln E^{\text{rel}} = \ln\left(\frac{E}{E_1}\right) = \phi' \ln K \tag{14}$$

where

E = modulus of filled elastomer
E_1 = modulus of gum
ϕ' = effective volume fraction of filler
$\ln K$ = reinforcement effectiveness factor

Equation (14) can also be written as

$$E^{\mathrm{rel}} = \exp\{(\ln K)\phi \exp[A(\phi - 1)]\} \tag{15}$$

where A is an index defining the degree of mixing.

The parameters ln K and A can be evaluated by a nonlinear regression of E^{rel}, ϕ data. Figure 13 shows the relationship between the effective volume fraction (ϕ_2') and the actual volume fraction (ϕ_2) of the filler computed for different mixing indices, A, and zero strain. Lee (1985) has shown that Eq. (12) provides a good fit of the modulus data for different rubbers and carbon blacks. Different rubbers evaluated in this study were chlorosulfonated polyethylene (CPSE), ethylene–propylene–diene rubbers (EPDM), and polychloroprene (CR).

The effective volume fraction of aggregate, ϕ'_{aggr} can be correlated with the effective volume fraction of filler:

$$\frac{\phi'_{\mathrm{aggr}}}{\phi} = \left(\frac{\ln K}{2.5}\right) \exp[A(\phi - 1)] \tag{16}$$

Using tensile modulus data, the relative mixing index for different elastomers can be determined (e.g., through A). This is a useful approach to assessing

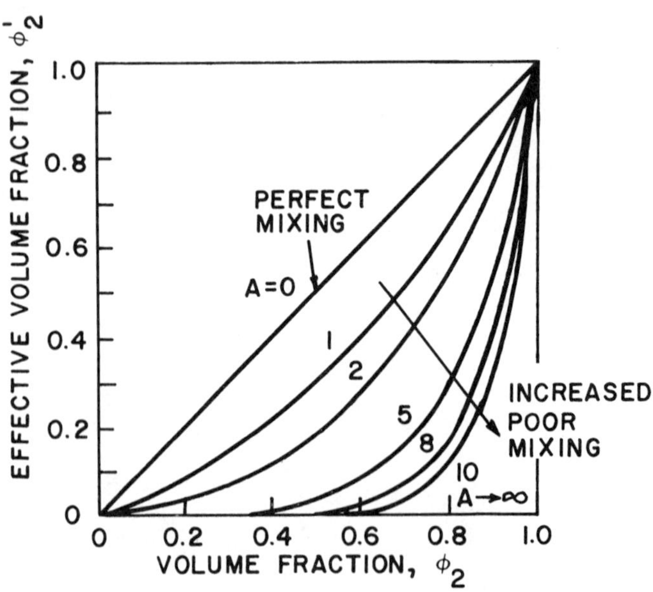

Figure 13 Relationship between A, ϕ_2 and ϕ_2'.

the degree of mixing of carbon black in different polymers, and hence is practical for the end user, developing compound formulations for different applications, as well as the polymer technologist, developing elastomeric products for customers. The mixing index *A*, based on tensile modulus data, is found to correlate qualitatively with morphological studies using the SEM (scanning electron microscope). Hence, the effective volume fraction of filler ϕ' can be used for characterizing the nonideal mixing behavior of filled elastomers.

Increasing the concentration of carbon black results in a decrease in the ultimate stretch ratio of the compound stock. Also, laboratory tests show that the ultimate stretch ratio of a good mixing system is higher than that of a poorer mixing elastomer system. The reason for this is that the poor degree of mixing of the filler produces weak spots which lead to premature failure (Lee, 1984). A relationship for the ultimate stretch ratio of filled elastomers based on a power-law expression is

$$\frac{\lambda_b}{\lambda_{b_1}} = \lambda_b^{\mathrm{rel}} = (1 - \phi')^{J/3} \tag{17}$$

where

λ_b = ultimate stretch ratio of filled elastomer
λ_{b_1} = ultimate stretch ratio of gum
J = fracture coefficient
ϕ' = effective volume fraction of filler

Nielson (1974) gives the following expression for the relative percent ultimate strain:

$$\frac{\epsilon_b}{\epsilon_{b_1}} = (1 - \phi^{1/3}) \tag{18}$$

where

ϵ_b = percent ultimate strain of filled elastomer (i.e., $\epsilon_b = (\lambda_b - 1) \times 100$)
ϵ_{b_1} = percent ultimate strain of the gum
ϕ = volume fraction of filler

The problem with this correlation is that it does not account for the effect of the degree of mixing on the ultimate stretch ratios.

Lee (1985) derived a power-law relationship between the mixing index *A* and the fracture coefficient *J* for conventionally mixed CSPE, CR, and EPDM:

$$J = 3.89A^{2.87}, \qquad \text{for } A \geq 1 \tag{19}$$

This correlation can be combined with Eq. (17) to give

$$\frac{\lambda_b}{\lambda_{b_1}} = \{1 - \phi \exp[A(\phi - 1)]\}^{1.30A^{2.87}} \tag{20}$$

This correlation is applicable for the range: $1 \leq A \leq 3$ (i.e., reasonably well mixed elastomer systems).

LATEX AND SOLUTION MASTERBATCHING

In latex–black masterbatching, polymer and black are brought together while both materials are in a finely divided colloidal suspension in water. SBR latex–black masterbatches were first introduced about 45 years ago. Early masterbatches used dispersing agents to keep the black in an aqueous suspension state. Today, a continuous dispersant-free latex masterbatching technique is used where the black is maintained in aqueous suspension under high turbulence while being brought into contact with the latex. Since the black has an inherently high surface area, it also serves as a creaming agent, which results in the production of a porous crumb that can be readily washed and dried.

In the dispersant-free latex production method, the parameters of importance are dry carbon black, oil, latex, and stabilizer emulsion. The initial procedure involves grinding the carbon black. The black is then slurried in water and blended with the latex, stabilizer emulsion, and oil emulsion under high-speed agitation. Oil is normally added separately to the mixer. From the mix tank, the material is sent into an acid coagulant step where rapid coagulation is promoted. The crumb then passes through a series of dewatering screens, filter presses, and driers prior to being baled and packaged for customer shipping.

Critical to producing a good latex masterbatch is the ability to achieve thorough wetting of the black by the polymer. Good wetting also results in a finer subdivision of the carbon black agglomerates. The presence of large, dry agglomerates (i.e., poor polymer penetration) of black leads to poor physicals and, subsequently, detrimental end-use performance.

It should be noted that although latex masterbatching conceptually results in very good filler dispersions, it does not necessarily provide a better-performing product. With adequate mixing, dry batch masterbatching procedures can produce compounds that are equivalent to latex masterbatching. One advantage of latex masterbatching, however, is the reduced need for costly mixing equipment.

Solution masterbatching differs from the latex process in that the polymer starts out in a hydrocarbon solution. Water is added to this solution to form small droplets, which are brought into contact with an aqueous slurry of carbon black. The aqueous carbon black slurry is prepared using the conventional latex procedure. The two suspensions are brought into contact under high-agitation conditions and rapid phase transfer occurs as the organophilic black aggregates interact with the polymer–solvent droplets. The finishing end of the process is essentially that of latex masterbatching (i.e., crumb formation, dewatering, and drying stages). This process does, however, require an additional step of solvent recovery. Solution masterbatching (also referred to as hydrosolution masterbatching) has been claimed to offer easy final processing with better product uniformity over the latex process. In comparison to latex masterbatching, solution mixing of black has some advantages over dry mixing in terms of operational maintenance and less costly processing equipment. Scott and Eckert (1966) and Burgess et al. (1965, 1974) describe the process in detail.

TEST METHODS FOR EVALUATING CARBON BLACK DISPERSION

As described above, the degree of dispersion of carbon black in rubber is one of the key factors affecting the properties of the final vulcanizate. The significance of good dispersion, the way in which the process of dispersion takes place, and the influence of dispersion on physical properties have already been discussed. It should be noted that no single method, nor even any combination of methods, gives all the information which would be desirable regarding the state of dispersion. The specific information needed includes:

The size and distribution of the carbon black and/or carbon black–rubber agglomerates, which may be as small as the smallest carbon black "primary aggregates," [0.05 micrometer (μm) in diameter] and may range in size up to about 500 μm or 0.5 mm.

The ratio of carbon black to rubber in these agglomerates.

The methods of laboratory dispersion measurement depend on the amount of information desired, the amount of time that can be devoted, and the nature of the samples. Cabot Corporation has published perhaps the most extensive procedures for evaluating carbon black dispersion in rubbers. The techniques are widely used throughout the rubber industry, although each laboratory has its own variation. The two most widely used methods are the "microtomed section" method and the "torn surface" method. The two methods give different information and should be regarded as complementary rather than exclusive.

The microtomed section method basically consists of sectioning frozen rubber samples, mounting the sections on a standard microscope slide, measuring or estimating (in a light microscope) the area of undispersed agglomerates, and calculating from this and other pertinent data the percentage of dispersed carbon black in the stock. The method of counting agglomerate area and calculating dispersion ratings is given by Leigh-Dugmore (1956) and Medalia (1961). Faster calculations, however, can be made for estimating the dispersion rating by comparison with a dispersion classification chart. This eliminates the time-consuming measurement of agglomerates by the counting method. This microscopic dispersion classification method is one of the methods described in ASTM test method D2663.

The equipment required for this evaluation includes the following:

Microscope Components *Oculars* (eyepieces): 10× Kellner. This ocular is easily divisible for inserting the graticule. The Kellner ocular provides a fairly wide field of view. It is desirable to have three oculars (whether for monocular or binocular microscopes): one clear, one with a standard graticule inserted, and one with a Leigh-Dugmore graticule inserted. This allows for examination in different ways by simply inserting the proper ocular without taking it apart to remove or replace the graticule.

Graticules: For Cabot dispersion classification described in ASTM, use a Cabot graticule for either 60× or 77× (depending on total magnification). The diameter of the graticules should be specified according to the recommendation of the eyepiece manufacturer.

Objectives: Either 6× or 8×; 6× Achromatic is generally supplied with the microscope.

Microscope A good-quality microscope should be used. It is preferable to obtain a microscope with a binocular or trinocular head which does not incorporate a complicating magnification factor, so that the total magnification is simply the product of the powers of the ocular and the objective. For photomicrography, a trinocular head is generally used. The total magnification must be kept close to either 77× or 60× (5×), depending on the graticule. Special care must be taken to incorporate the magnification factor (if any) of the binocular head into the magnification calculations. For example, with a Leitz Labolux microscope, which has a magnification factor of approximately 1.25 (depending on the interpupillary distance), a 6× objective and 10× ocular give an overall magnification of approximately 75×. Simple monocular microscopes are available at a lower cost and can be used for visual microscopy or even photomicrography, but they are somewhat inconvenient for routine measurements.

Microtome A sledge type of microtome with the required ruggedness and precision should be used. It is particularly convenient in that the sample remains fixed in place and therefore can be used with a liquid nitrogen freezing stage of simple design. In the past, freezing stages have been used which were cooled by expansion of liquid CO_2. This process is noisy and also somewhat hazardous owing to the high pressure of liquid CO_2, and, in addition, it does not provide the low temperatures which can be obtained with liquid nitrogen (particularly important for polybutadiene).

A knifemaker can be utilized to fabricate the glass knives used in the microtome. This instrument will produce glass knives to fit any commercially available microtome. The knifemaker is an instrument that provides alignment of the glass strips, as well as knife scoring and knife breaking. In operation, the scoring and breaking take place at the same point on the instrument without it being necessary to move the glass between these operations. The glass is pressed down firmly against the table by two studs located symmetrically on each side of the score. Two spherical-headed pins are then forced up against the lower surface of the glass directly under the score at a distance calculated to cause fracture of the glass. Movement of the pins is controlled by a lever arm which mechanically eliminates the small variations in the movement of the operator's hand from one break to another and produces a slow, easily controllable movement. Therefore, a knifemaker eliminates any uncontrollable torsion, such as occurs in manual breaking, and, by increasing the reproducibility of the forces brought to bear on the glass during breaking, produces knives of high quality with a minimum of waste.

The procedure recommended by Cabot Corporation for sectioning and mounting vulcanized rubber samples is as follows:

1. Cut a block approximately 3 × 3 × 5 mm from the rubber sample to be examined; a single-edge razor blade is a suitable cutting instrument for this purpose. Discard the blade after cutting four or five samples.

2. Select a glass knife and wipe it dry, being careful to avoid the sharp edge. Place the glass knife in the knife holder and tighten just enough to hold it in place.

3. Place the cut sample on the dry stage and position it so that it lies lengthwise on the axis of the cutting plane of the knife blade.

4. To cement the sample to the stage, surround it with mucilage solution from a medicine dropper. Take care to avoid covering the sample completely or the knife edge will be prematurely dulled in planing down the excess mucilage.

5. Freeze the sample by pouring liquid nitrogen into the stage reservoir. The mucilage will become creamy white when freezing is completed.

6. Set the stage beneath the knife holder and adjust the blade so that the cutting edge nearly touches the surface of the frozen sample at an angle of about 10° between the bottom of the knife and the cutting plane. Tighten the knife blade firmly in this position.

7. Plane down the sample with a few preliminary strokes until the surface of the sample is flat and the shape of the initial sections is rectangular. Operate the stage advance mechanism manually during the initial planing procedure. It is important to work quickly when planing and cutting because thermal expansion of the sample may cause the knife to cut too deeply, thus dislodging the sample.

Remove any excess mucilage and shavings from the knife edge (during planing) with a xylene wetted brush to avoid damage to the sample surface.

8. When planing has been completed, adjust the stage advance mechanism to 2 μm. Make one or two more preliminary cuts and discard these sections.

9. Working quickly, wet the knife edge with xylene and cut 8 to 12 sections. Remove the cut section from the knife edge after each stroke, using a xylene wetted No. 2 brush, and deposit all the sections in a beaker containing about 5 ml of xylene. Always remove sections while the knife is in the forward position (toward the operator).

10. Measure the sample dimensions on the stage. Record both dimensions to the nearest 0.1 mm. The area thus obtained is the ''before swelling'' factor. (This step is used in connection with the Leigh-Dugmore counting calculation but may be omitted for the Cabot dispersion classification procedure, unless the sample is so lightly cross-linked as to have an area swelling factor greater than 3.1).

11. Using a No. 2 sable brush, transfer a section from the beaker onto a clean glass slide and spread the section by holding one end with one brush and rolling it out with a second brush. If the section is too tightly rolled or stiff, hold it at one end with the point of a dissecting needle. The section should be mounted in the center portion of the slide. If it is too close to one end, the slide will not be supported properly on the microscope stage. A label should be affixed to one end of the slide and if two or three sections of the same sample are mounted on the same slide, a consistent reference scheme should be used, such as designating the section closest to the label as (A), the next as (B), and so on. When the section has been brushed out flat, allow the excess xylene to evaporate to near dryness and continue to brush out remaining wrinkles carefully so that the section adheres to the slide over its entire area. Avoid excessive stretching of the section.

12. Check the preparation microscopically for the following defects:

a. Wrinkles: Wrinkles will appear as long black streaks in varying patterns. These must be removed since they may be mistaken for carbon black or, in certain instances, will obscure smaller agglomerates of black.

b. Tearing: Samples should be free of tears. With some difficult samples, small tears cannot be avoided. Tearing occurs either during sectioning (by burred knives) or during spreading of the section. In some cases small tears or pinholes are formed when an agglomerate is pulled out during sectioning. The cause of excessive tearing must be found and eliminated wherever possible.

c. Proper freezing: The cutting temperature is important since improper freezing can affect the overall interpretation of a given sample. Underfreezing is usually characterized by the appearance of brownish, patchy areas throughout the section which follow no definite pattern. This problem may arise with unvulcanized samples which have been ''cold cured'' (see below) and with samples containing polybutadienes.

Although underfreezing itself cannot be corrected, the symptoms usually are indicative of other factors such as incomplete curing or poor mixing. Overfreezing to the point of extreme brittleness of the sample is characterized by chipping of the block or severe knife "chatter." This can be overcome by allowing the sample to warm until sections can be cut and then maintaining that temperature by using the coolant sparingly.

d. Knife marks: These occur in sections in two general categories: (1) knife "chatter" and (2) long streaks parallel to the cutting stroke. "Chatter" or alternate light and dark bands normal to the cutting stroke may be the result of an excessively brittle sample, a rough sample surface, an improper knife angle, or a loosely held knife. Serious knife "chatter" should be minimized, as it may obscure agglomerates, rendering the sections unacceptable. Long knife streaks which appear parallel to the cutting stroke are generally caused by a burred knife edge and except in extreme cases seldom interfere with the dispersion evaluation.

e. Transparency of transmittance: Sections must transmit enough light so that carbon black agglomerates can be distinguished from the background. Where carbon black loadings are very high, sections must be cut less than 2 μm thick because of the increased optical density of the sample. This can be done by setting the control to manual and making periodic strokes, relying on visual observation and trial and error to yield satisfactory sections. A bit of practice will be required to make good sections this way. Samples in which the carbon black is well dispersed and which are properly frozen will show a homogeneous brown background.

13. When a suitable section is obtained and mounted, cover it with a coverglass which has had a small drop of Duco cement placed near each corner. Avoid using excess cement, since it may react with some sections, discoloring them. Press the corners of the coverglass gently to the slide with a soft object such as the eraser end of a pencil.

14. Measure the section after mounting and record as the "after swelling" area. (This step is required only in conjunction with step 10 when the Leigh-Dugmore counting method is employed.)

The binocular microscope should be set up with the proper graticule inserted into one eyepiece. The light source and condenser should be arranged as in conventional microscopic work for low power observation. Many microscopes come equipped with built-in light sources which are adequate for the relatively low magnification used in this procedure. Condensers are frequently equipped with removable top lenses; any such lens should be removed or swung out for the low magnification used here, since if left in place it usually narrows the beam of light so that only the central portion of the field of view is illuminated. The brightness of illumination should be controlled by neutral density filters (included in certain microscope illuminators or obtainable as a set from microscope supply houses). Other methods which can be used for cutting down brightness are not recommended. These include (1) cutting down the aperture diaphragm on

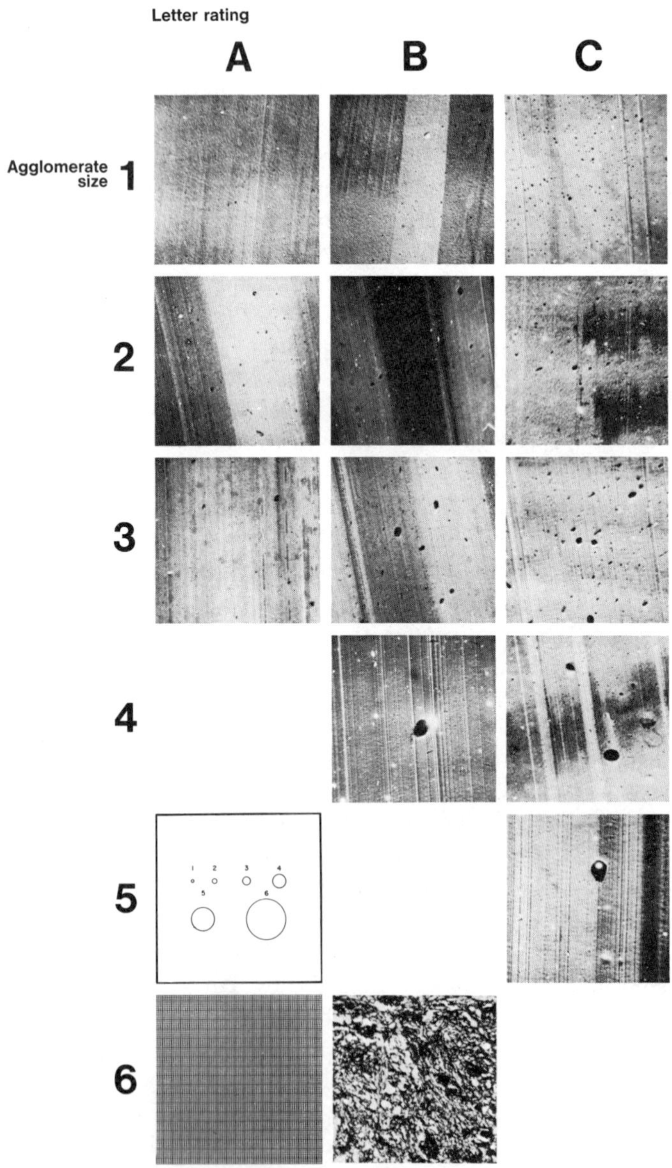

Related dispersion ratings Dunlop	100 – 98.6	98.6 – 96.1	96.1 - 91.8
Cabot corrections	100 – 99.7	99.7 – 99.1	99.1 – 98.2

Figure 14 Dispersion classification chart. (Courtesy of Cabot Corp., Boston, MA.)

D	E	F	G	H

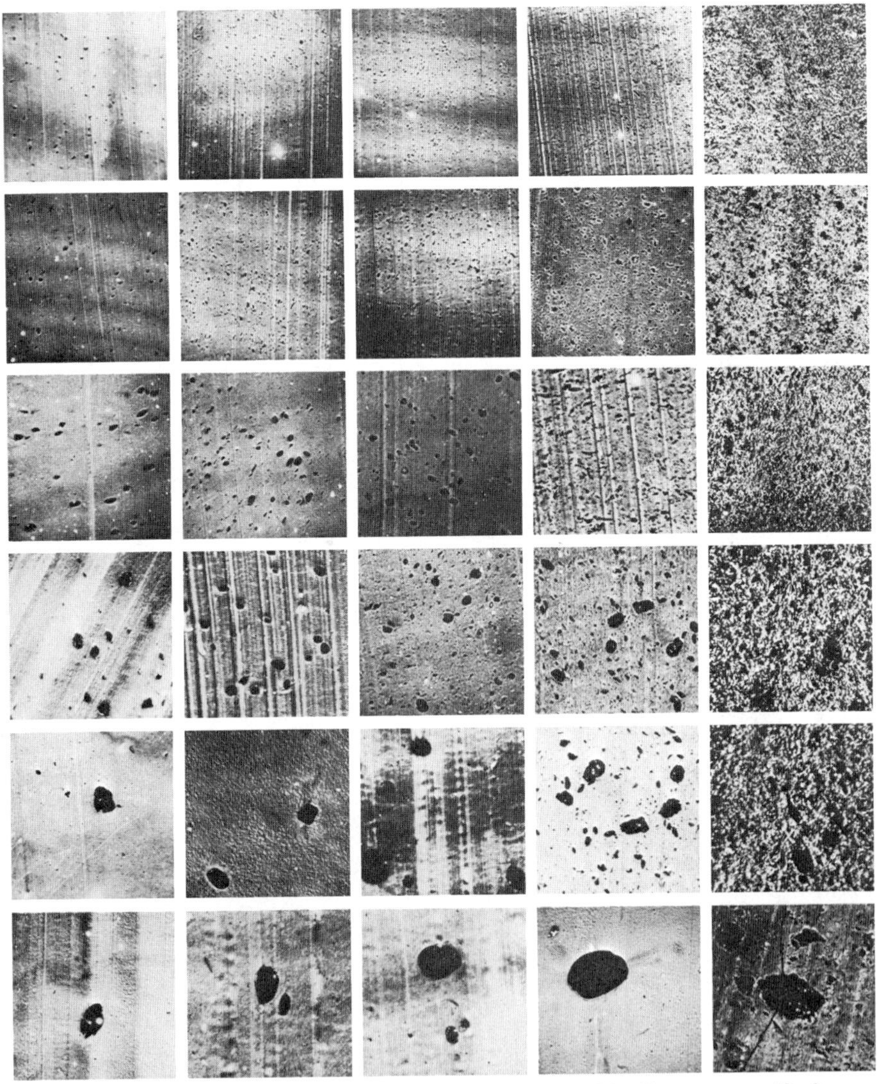

91.8 – 84.4	84.4 – 71.4	71.4 – 48.9	48.9 – 10	10 – 0
98.2 – 96.5	96.5 – 93.7	93.7 – 88.6	88.6 – 80.0	80.0 – 0

the condenser, which cuts down the resolution, and (2) decreasing the voltage of the lamp, which makes the light redder and thus interferes with estimation of the quality of dispersion in the matrix.

The dispersion classification chart is shown in Figure 14. It is comprised of six horizontal rows numbered from 1 to 6 and eight vertical columns lettered from A to H. The agglomerate size increases from 1 through 6 and the total number of agglomerates present increases from A through H. The information given by this method of presentation has a distinct advantage over the older methods where only the percent dispersed was available. In the present method not only percent dispersed can be made known but also the size of the undispersed agglomerates.

The specimen is examined by transmitted light, making sure that both the section and the graticule are in focus. Five fields are examined, one in the center and four close to the corners of the section, as shown in Figure 15. The total area of these fields is about one-third the area of a typical section, so this procedure provides good sampling without making any special effort to select "representative" fields. The best visual comparison is made between each field and the chart, focusing attention on the size and number of agglomerates and not on the appearance of the background.

To carry out the comparison, the size of the agglomerates in each field is estimated directly from the graticule, which appears superimposed on the field. The average size selected should correspond to that of the larger agglomerates,

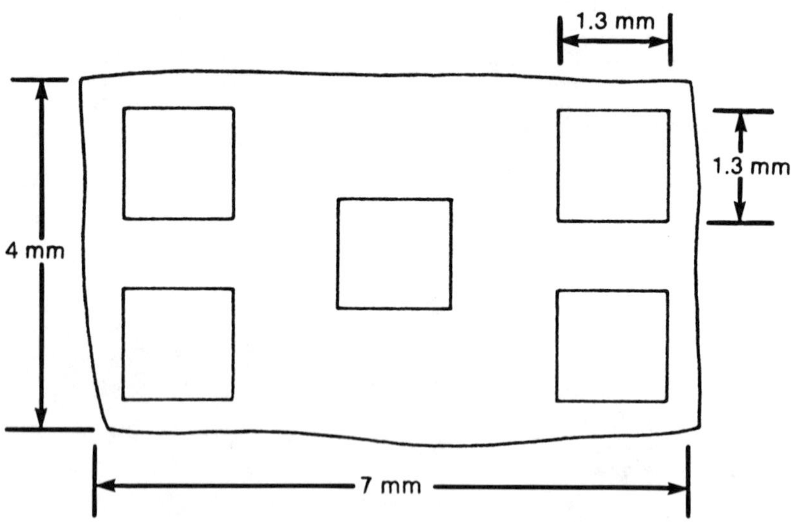

Figure 15 Location of five fields in measurements.

which appear to comprise a significant portion of the total agglomerate volume. Some small agglomerates, and an occasional larger one, will generally be present and should not be considered in the size category assignment. However, if the distribution of agglomerate sizes is unusually broad in comparison with the standard chart, this should be mentioned by giving the range of size categories within which most of the agglomerates fall.

The dispersion range (letter rating) is next chosen by comparison with the standard chart, in the row corresponding to the size category which has already been established; remembering that the square border shown in the graticule corresponds to the size of the photomicrographs in the standard chart. Some judgment is required in this step, particularly if the distribution of sizes is broad. It is helpful for the operator to check himself occasionally by carrying out an agglomerate area count on the same field (not merely the same section) on which a classification estimate has been made. If desired, the fields can be indexed for the purpose by means of the coordinate readings on the mechanical stage.

After recording the estimated ratings for each field, the ratings for the five fields must be combined. This is no problem if all five fields are in the same range; or if four are in one range and the fifth in an adjacent range; or if three are in one range and the other two in adjacent ranges on either side. In such cases the predominant range is the one reported. In more complex cases, the range letters cannot be simply averaged (e.g., A and E do not average to C, but to D).

For averaging the ranges of different fields, it is convenient to work with mean values of the percent of undispersed black in each range, $\overline{100\text{-D}}$. These are shown in Table 3, column 3. The ranges of several fields are found by averaging the values of $\overline{100\text{-D}}$, then placing the value in the proper range. For example, with five fields of ranges A, C, D, B, F, the average is $(0.15 + 1.35 + 2.65 + 0.6 + 8.8)/5 = 13.5/5 = 2.7$. Thus the average value of $\overline{100\text{-D}}$ is in range D (by reference to Table 3, column 2).

Since this averaging will most often be done with five fields, values of $\overline{100\text{-D}}/5$ are also given in Table 3, column 1: the sum of these values gives directly the average of $\overline{100\text{-D}}$ for five fields.

The size categories are reported simply by giving the extreme categories among the five fields, as, for example, C2-4.

For agglomerate area counting the counting procedure is less dependent on operator judgment, except when the sample contains many small agglomerates at the borderline of the "half-square" cutoff. The microscope must be set up as in the dispersion classification procedure, but the eyepiece with the graticule is replaced by an eyepiece containing the counting graticule. This graticule has a central ruled grid area which measures 1.0 cm^2. This grid area is divided into 10,000 small squares, each of which has an actual linear dimension of 0.1 mm. At 77× magnification, the small squares appear to be about 13 μm on an edge with an area of 169 μm^2.

Table 3 Averaging Procedure for Ranges[a]

Range	Percent undispersed black individual field, $\overline{100\text{-}D/5}$	Percent undispersed black range limits of $\overline{100\text{-}D}$	Mean values of percent of undispersed black, $\overline{100\text{-}D}$
A	0.03	0– 0.3	0.15
B	0.12	0.3– 0.9	0.60
C	0.27	0.9– 1.8	1.35
D	0.53	1.8– 3.5	2.65
E	0.98	3.5– 6.3	4.90
F	1.76	6.3– 11.4	8.80
G	3.14	11.4– 20.0	15.70
H	6.00	20.0–100.0	30.00

[a]Courtesy of Cabot Corporation, Boston, MA.

Measurements are made on five fields as described above; or, for checking a visual estimate, measurement must be made on the same field as that estimated visually. A count is made of all the small squares having their area more than half-filled by carbon black agglomerates, using a hand-operated mechanical counter to tally the number of squares counted. The number counted in each field of 10,000 squares is recorded as U. Thus one accounts only for those agglomerates which have an area of at least 84 μm^2 (average linear dimension approximately 9 μm), assuming black in smaller agglomerates to be dispersed. The theory of the subsequent calculation is based on the fact that for a random section, which is considerably thinner than the smallest dimension measured, the percent of a component by area is equal to its percent by volume in the material as a whole.

Calculate $\overline{U}$, the arithmetic average of U (the number of squares in each field).

Multiply $\overline{U}$ by 0.022. This gives the percentage of undispersed black, $\overline{100\text{-}D}$; that is, $\overline{100\text{-}D} = 0.022\overline{U}$.

If it is desired to use the agglomerate area count as a check on the estimation by the classification chart, then this value of $\overline{100\text{-}D}$ may be used by insertion in Table 3 to obtain the corresponding range.

Cabot Corporation recommends the following calculation method for an indication of the true amount of carbon black dispersed, *D*. The formula used is as follows:

$$D = 100 - \frac{v\bar{U}s}{AL} \tag{21}$$

where

v = volume fraction of carbon in the agglomerates
$\bar{U}$ = average of U
s = areal swelling factor of entire section
A = areal swelling factor of agglomerates
L = carbon black loading (volume percent) in the compound

As a general rule, $v = 0.40$ and $A = s$. The formula then becomes

$$D = 100 - \frac{0.40}{L}\bar{U} \tag{22}$$

and for a volume loading of $L = 18$, as assumed in the visual estimation chart:

$$D = 100 - \frac{0.40}{18}\bar{U} = 100 - 0.022\bar{U} \tag{23}$$

Note that

$$L = \frac{(\text{specific gravity of sample}) \times (\text{parts of carbon black})}{(1.86) \times (\text{total parts of compound})} \times 100 \tag{24}$$

If the specific gravity of the sample is not given, it may be calculated from the parts of black per hundred parts of rubber (including oil):

$$\text{Specific gravity} = \frac{100 + (\text{parts of black})}{100/\text{SGR} + (\text{parts of black})/1.86} \tag{25}$$

where SGR is the specific gravity of the unfilled rubber. For hydrocarbon rubbers, SGR may be taken as 0.93 with sufficient accuracy for this purpose.

It should be noted that the Leigh-Dugmore ratings have been used by the industry for a long period of time and have received wide acceptance. These ratings are given in Table 4.

Some unvulcanized samples can also be frozen and sectioned as described above. Others must be cured in the laboratory before they can be sectioned

Table 4 Comparison of Letter Ranges with Leigh-Dugmore Ratings

Letter range	Leigh-Dugmore rating	Cabot corrections of the Leigh-Dugmore ratings
A	100–98.6	100–99.7
B	98.6–96.0	99.7–99.1
C	96.0–91.8	99.1–98.2
D	91.8–84.4	98.2–96.5
E	84.4–71.6	96.5–93.7
F	71.6–48.4	93.7–88.6
G	48.4– 0	88.6–80.0
H	0	80.0– 0

successfully. The compound should be cured in a hot press if sulfur is present. If not, the following precautions may be used.

For Unsaturated Polymers:

1. Place a 10–15 g slug of the uncured sample between two sheets of 5 mil aluminum foil. Place this sandwich in a laboratory press and press out a sheet approximately 10–15 mils thick. Cut this sheet into 10 to 12 1 × 1 in. squares.
2. Place about 10 g of sulfur and 10 g of CBTS into a 4 oz jar; mix. Place several pieces of the prepared, uncured stock in this jar, cap, then agitate vigorously to assure uniform dusting of the sample surface. Dust all of the samples in this manner.
3. Ply up the dusted samples to a convenient height and in a "button mold" (a 1 × 1 × 0.10 in. mold is satisfactory) and cure for 45 min at 150°C.
4. The sample may now be handled as any vulcanized stock.

For Saturated Polymers (butyl, EPDM):

1. Prepare samples as described for unsaturated polymers above.
2. Melt approximately 50 g of dicumyl peroxide in a 100 ml beaker using a waterbath. The prepared samples are dipped into the peroxide, then allowed to cool.
3. Proceed to cure as outlined above.
4. The sample may now be handled as any vulcanized stock.

Masterbatch compounds (e.g., latex masterbatches) which have not been compacted on the mill often contain voids which must be minimized before sectioning can begin. This is best done in a pellet press. Treat the sample as an unvulcanized stock and proceed as above. Sections prepared from dispersant-free latex masterbatches typically contain black streaks in a colorless background.

Many polybutadiene compounds, particularly those based on 100% polybutadiene, have a very low hardening temperature and are difficult to section at the temperatures attainable with a liquid carbon dioxide freezing stage. Satisfactory results are generally obtained with the liquid nitrogen freezing stage, but in case of difficulty the samples may be hardened.

The *torn surface method* makes use of the fact that in a compound in which the ingredients are perfectly dispersed, light is reflected from a freshly torn surface revealing a smooth, unblemished texture. The presence of improperly dispersed ingredients is shown by irregularities which usually take the form of circular, convex "bumps" or concave pockmarks on the surface, and their presence indicates a less-than-perfect dispersion of the compounding ingredients. Their size and frequency may be used to judge the degree to which the compound falls short of perfect dispersion. A set of standards based on size and frequency of these irregularities may be set up to which numerical ratings can be attached. This then provides a means of evaluating a sample under investigation and permits a number to be quoted for degree of dispersion by reference to the set of standards.

The basic procedure for "torn surface" evaluations is outlined in ASTM D2663. This procedure lists five standards, with dispersion ratings ranging between 1 (very low) and 5 (high). Cabot Corporation uses a set of ten standards. To maintain consistency with ASTM, the same 1 to 5 scale is used, with decimals indicating intermediate ratings.

The torn surface method has been described as being capable of predicting SBR tread performance. It appears likely that the imperfections seen under the relatively low power magnification are of such a nature as to be potential sources of localized failure. The method is designed to be used as a rapid factory control on mixing and subsequent processing conditions to assure adequate dispersion. This procedure requires the use of a microscope and light source.

It has been found most convenient to compare the field under view with a photograph of established standards. It has been observed that over the range of very bad to excellent dispersion, ten different categories can be readily distinguished. Thus a photograph has been made of each category at a magnification, which, when printed and mounted in a montage, will closely correspond to the field under view at 20×. It then becomes a simple matter to scan the montage and select the one which most closely resembles the field under view. Figure 16 shows the torn surface dispersion rating chart developed by Cabot Corporation.

A tensile sheet remnant, after the stress–strain specimens have been re-

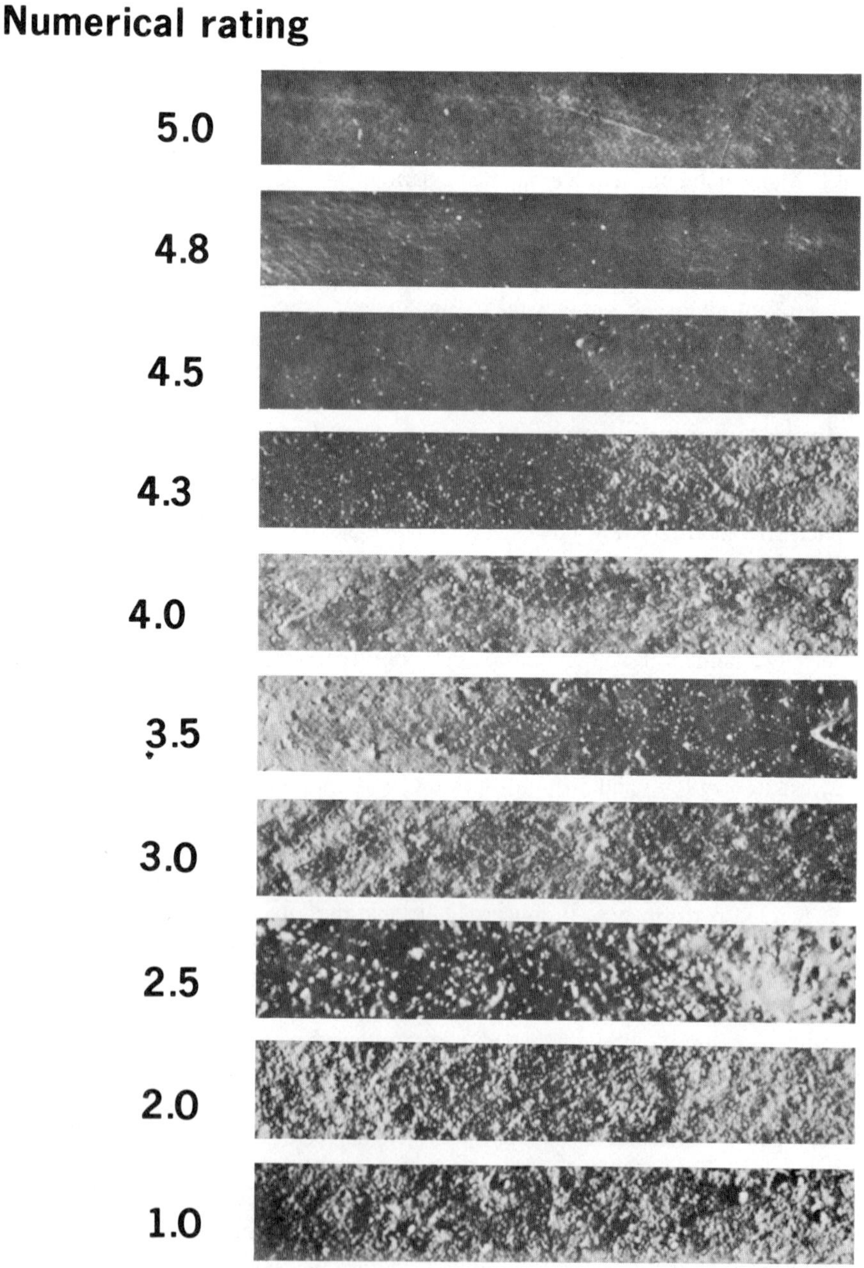

Figure 16 Torn surface rating chart. (Courtesy of Cabot Corp., Boston, MA.)

moved, is useful for this test. This type of specimen has the advantage of being readily available, and it is not necessary to prepare a special specimen.

A sharp cut about ⅜–½ in. long is made in the sheet with the shears. Grasping the sheet in both hands between the thumb and first finger with the edge of the sheet uppermost, a rapid movement of both hands away from each other and parallel to the line of the edge of the sheet will generally produce a smooth, even tear. After the tear has been induced, a specimen about ½–¾ in. may be cut from the sheet, which gives a size that can be viewed readily under the magnifier.

The specimen is held under the objective lens of the magnifier. Light is directed onto the torn surface with the angle adjusted to minimize glare from the shiny surface. A minimum of three fields are evaluated. Reference is made to the standard, and a rating given to each field. The median value of the three fields is reported.

Any comparison of methods for estimating degree of dispersion must take into account the uses to which the results will be put. In plant operations the main concern is for a rapid test by the control laboratory, capable of giving guidance to the mixing equipment operators. For this purpose examination of the cut edge of a cured, or even in some cases an uncured, specimen gives useful information by reference to the standards used in the Torn Surface Chart. Successful use of an uncured specimen would be gained with experience. The uncured surface in this case would be cut from the stock with shears or a sharp knife. The microtome procedure is more time-consuming, and its use is not justified for production control.

In technical service work it is likely that both methods are useful for investigating causes of failures or other areas in which the technical service man might become involved. The two methods described are complementary and give a picture of the gross dispersion on the relatively large surface plus the state of dispersion of the fine agglomerates, perhaps not visible as imperfections on the surface. In addition, the use of both methods would reveal surface irregularities which do not show up in a microtomed section. These irregularities can be caused by poor dispersion of sulfur or accelerator.

In research work one can distinguish between research on the compounding step itself and research where the step is only incidental to obtaining properties of the compound. In the latter case, if there is no reason to suspect poor dispersion, the torn surface method should suffice. Whenever poor dispersion may be suspected or whenever expensive research is involved, the microtome procedure should be employed as well. This would apply also in research on the compounding or mixing operation, or in any research on factors affecting the degree of dispersion.

Other useful techniques that can be combined with microtoming are x-ray mapping, polarized light microscopy and phase contrasting in which individual

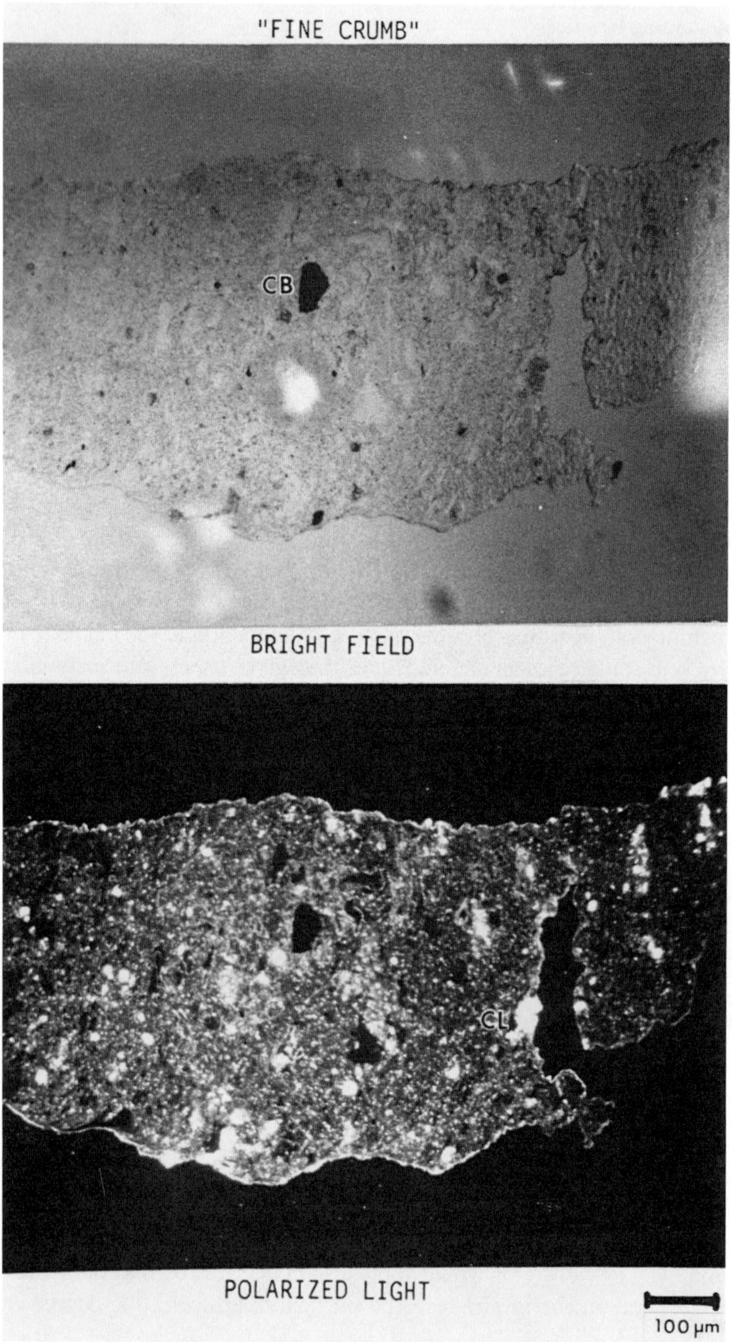

Figure 17 Photomicrographs for mixing/dispersion study.

"FINE CRUMB"

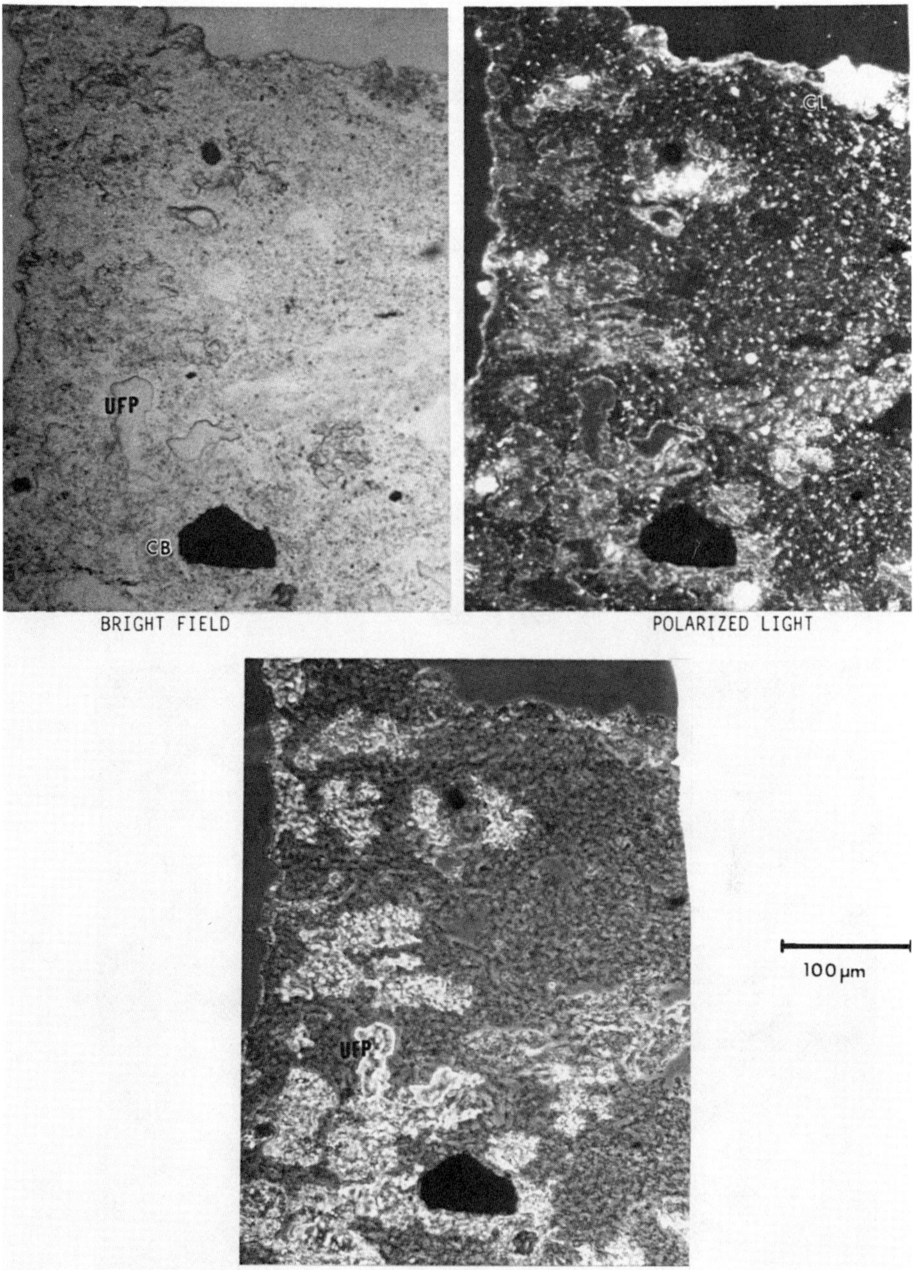
UFP
CB
BRIGHT FIELD
CL
POLARIZED LIGHT
UFP
PHASE CONTRAST
100 μm

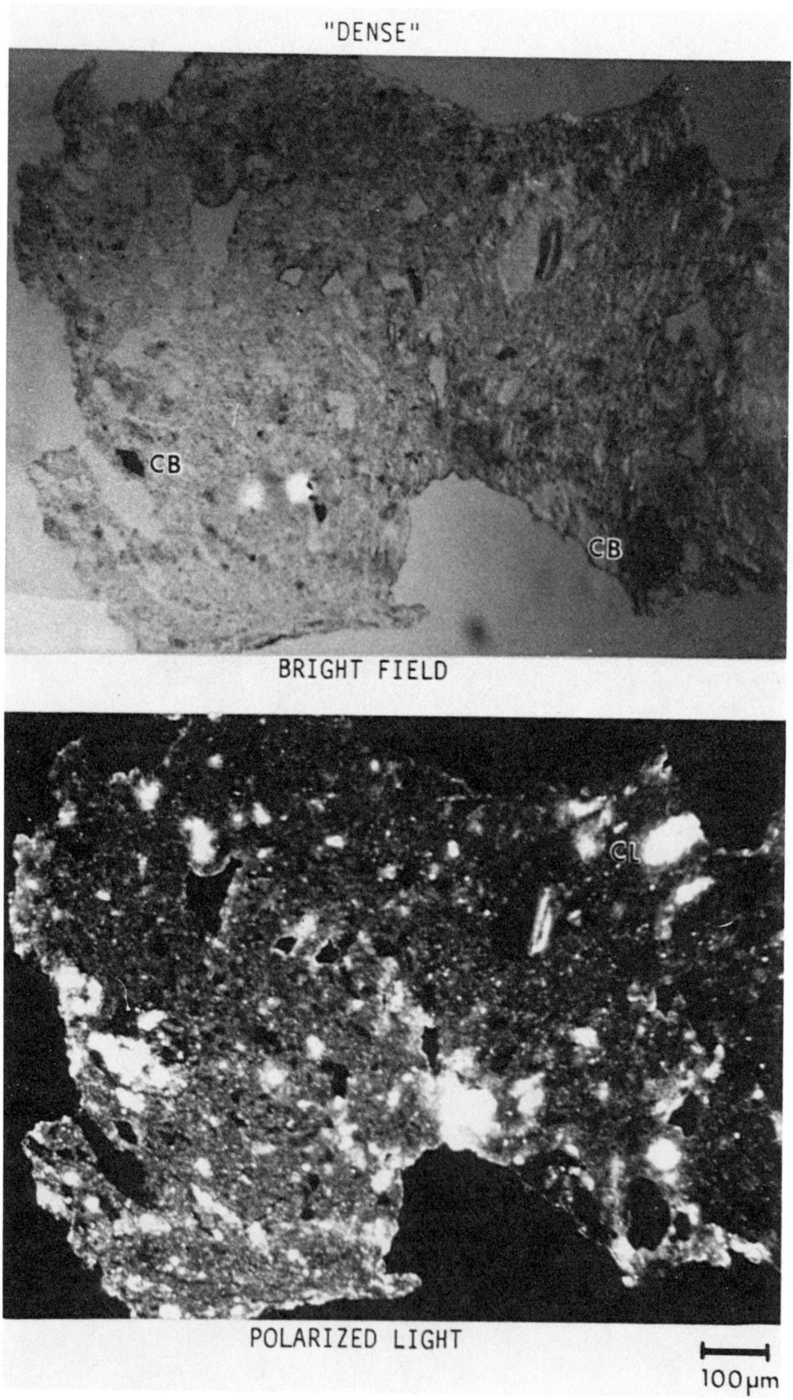

Figure 17 (Continued)

"DENSE"

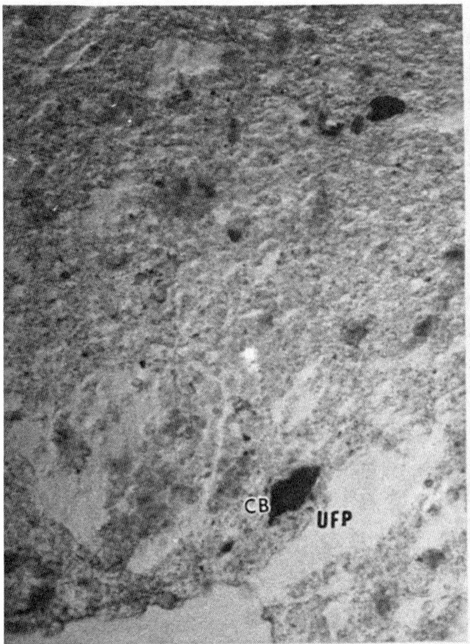

BRIGHT FIELD

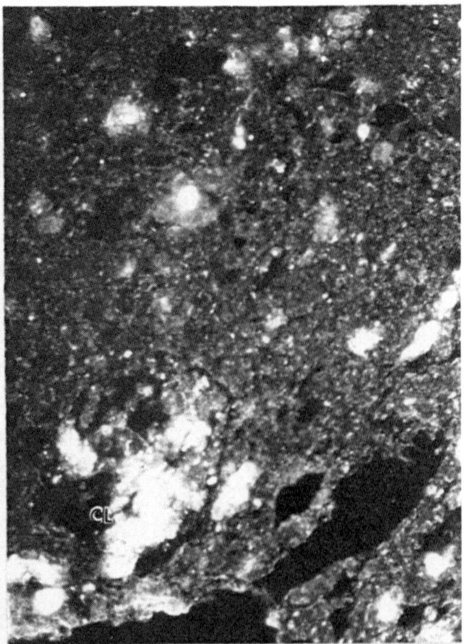

POLARIZED LIGHT

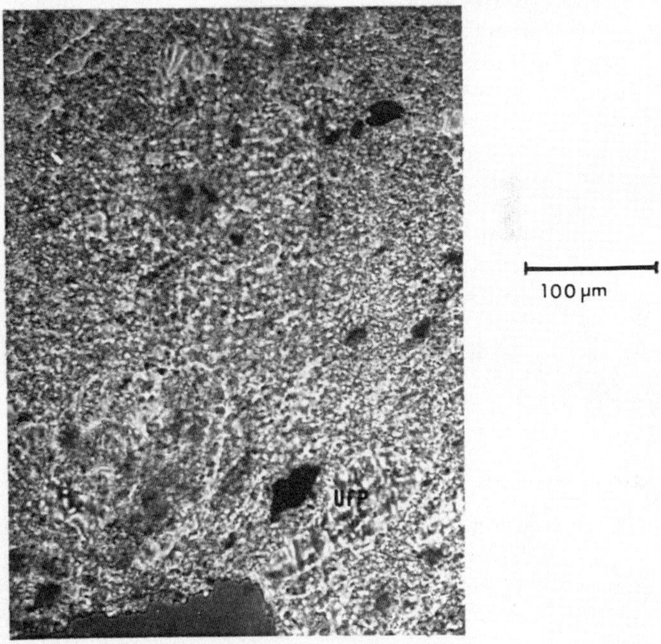

PHASE CONTRAST

filler constituents can be optically contrasted during microscopy to examine relative dispersions between samples. Figure 17 gives an example in which mineral filler and carbon black dispersion in two specimens (EPDMs) were examined. The specimens are the same EPDM mixed under identical conditions in a Banbury mixer. In one case the rubber was charged to the mixer in the form of fine crumb, and the other sample was introduced as dense chunks. Hence, the objective of this study is to determine whether the form of the rubber affects dispersion, given that mixing conditions are fixed. Because of the low loading of carbon black (2 phr) in these compounds, thick sections (~2 μm) were microtomed so that carbon black aggregation could be imaged. Examining the plates of Figure 17 reveals that both samples are very inhomogeneous, and a number of internal holes are present (see dark inclusions CB in the Bright Field images). In contrast, the clay dispersion appears to be finer in the Fine Crumb sample than in the Dense sample as shown by the polarized light images. The photomicrographs also reveal many large clear unfilled polymer zones present in both samples, as shown by the bright field and phase contrast images.

NOTATION

A	mixing index
B	batch weight
c	constant
D	degree of carbon black dispersion
D, d	particle diameter
E	elastic modulus
F	shape factor
f	ratio of cluster length to width
J	fracture coefficient
k_B	coefficient
L	carbon black loading, or length
M_w	weight average molecular weight
N_1, N_2	number of effective polymer chain segments filler particles, respectively
n	number of particles having diameter d
p	perimeter
S	entropy
s	swelling factor
V	volume
v	volume fraction of carbon in agglomerates
W	particle width

Greek Symbols

ϵ_b	percent ultimate strain of filled elastomer
θ	temperature
λ_b	stretch ratio
ρ, ρ'	density
ϕ	volume fraction of filler
ϕ_2'	volume fraction of filler wetted by polymer
ϕ'_{aggr}	effective volume fraction of aggregates

REFERENCES

Andrews, E. H., and A. J. Walsh, *J. Pol. Sci., 33,* 39 (1958).
Baranival, K. C., *Die Makromol. Chem., 100,* 242 (1967).
Beatty, J. R., *Rubber Chem. Technol., 37,* 1341 (1964).
Boonstra, B. B., *J. Appl. Polym. Sci., 11,* 389–416, 19.4 (1967).
Boonstra, B. B., in *Rubber Technology,* M. Morton, ed., Litton, New York, 1973.
Burgess, K. A., S. M. Hirshfield, and C. A. Stokes, *Rubber Age, 98,* 9, 85 (1965).
Burgess, K. A., S. S. Thune, and E. Palmese, *Rubber World, 149,* 4, 34 (1974).
Cheremisinoff, P. N., and F. Ellerbusch, *Carbon Adsorption,*
Corish, P. J., *Rubber Chem. Technol., 44,* 3, 814 (1971).
Corish, P. J., and B. D. W. Powell, *Rubber Chem. Technol., 47,* 3, 481 (1974).
Cotten, G. R., Paper No. 1, Rubber Division Meeting, ACS, Cleveland, OH, May 1975.
Cotten, G. R., Paper presented at Rubber Division Meeting, ACS, Los Angeles, CA, April 23–26, 1985.
Cotten, G. R., *Rubber Chem. Technol., 57,* 118 (1984).
Gessler, A. M., *Rubber Chem. Technol., 43,* 5, 943 (1970).
Gessler, A. M., W. M. Hess, and A. I. Medalia, Reinforcement of Elastomers with Carbon Black, Plastics and Rubber Processing Paper 1164 (1978).
Hawkins, W. L., *Rubber Plastics Weekly,* London, *142,* 291 (1962).
Hawkins, W. L., *Rubber Plastics Weekly*, London, *132*, 289 (1966).
Lee, M. C. H., *J. Appl. Polym. Sci., 29,* 499 (1984).
Leigh-Dugmore, C. H., *Rubber Chem. Technol., 29,* 1303 (1956).
Medalia, A. I., *Rubber Chem. Technol., 34,* 1134 (1961).
Nielson, L. E., in *Mechanical Properties of Polymers and Composites,* Vol. 2, Marcel Dekker, New York, 1974.
Payne, A. R., and W. F. Watson, *Rubber Chem. Technol., 36,* 147 (1963).
Scott, C. E., and F. J. Eckert, *Rubber Chem. Technol., 39,* 3, 533 (1966).
Smallwood, H., *J. Appl. Phys., 15,* 758 (1944).
Smith, B. R., *Rubber Chem. Technol., 49*, 278 (1964).
Smith, W. R., and G. A. Kasten, *Rubber Chem. Technol., 44,* 1287 (1970).
Turetsky, S. B., P. R. Van Bushirk, and P. F. Gurberg, *Rubber Chem. Technol., 49,* 1 (1976).
Walters, M. H., and D. N. Keyte, *Trans. IRI, 38,* 40 (1962).

6

Extrusion of Elastomers and Plastics

INTRODUCTION

An extruder is a standard piece of equipment used in rubber and plastic processing for such operations as injection and blow molding, for processing thermosetting resins, and for the hybrid process of injection blow molding. The apparatus is fed room temperature resin in the form of beads, pellets, or powders, or if rubbers are being processed, the feed material may be in the form of particulates or strips. The unit converts the feedstock into a molten polymer at sufficiently high pressure to enable the highly viscous melt to be forced through a nozzle into the mold cavity (injection-type molding) or through a die (e.g., blow molding or continuous extrusion of articles). In the initial portion of the extruder the polymer is conveyed along the extruder barrel and is compressed. The material is then heated until soft, eventually reaching a molten state. As fresh feed material enters, heat transfer takes place between the molten fluid and solid polymer. Once in the molten state, the extruder acts like a pump, transferring the molten polymer through the extruder channel building up pressure prior to flow through the discharge nozzle or die. The principal components of a single-screw extruder are illustrated in Figure 1. The machine has a motor drive, a gear train, and a screw that is keyed into the gear reducing train. The fluid layers between the screw flights and barrel wall maintain the screw balanced and centered.

Modern units are equipped with continuous variable speeds and barrels are usually electrically heated, usually by band heaters with either "on-off" or proportional control. The barrel can be "zoned" according to the number of

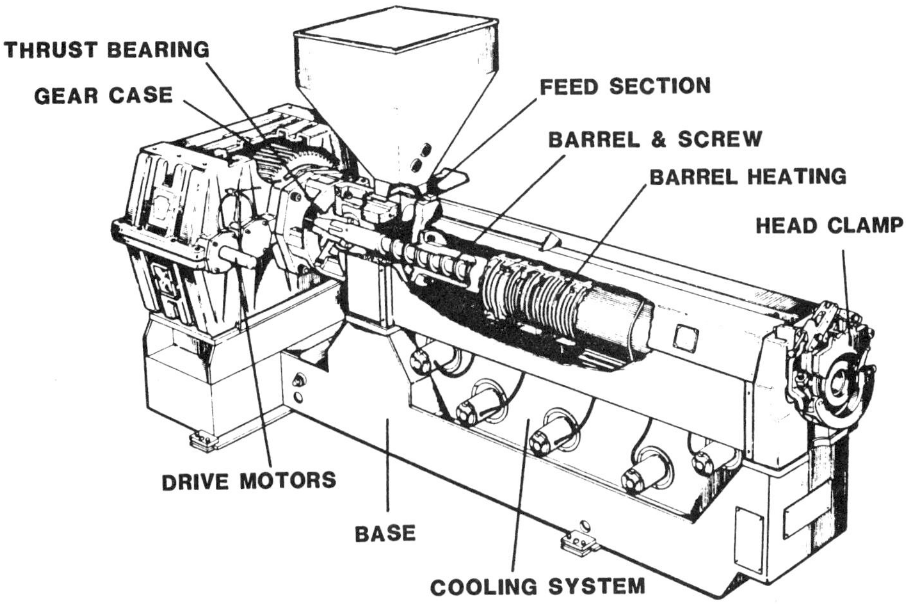

Figure 1 Principal design features of a single-screw extruder.

controllers on the heater bands. Depending on the application and type of service, the screw may be cored to allow for heating or cooling.

A die can be attached at the end of the extruder. Die designs range from very simple geometries such as an annulus for pipe and tubing profiles, to very complex shapes such as the rubber seals used as glass run channels around the windows of automobiles.

Older extruders used in manufacturing various consumer articles are equipped with minimal instrumentation. Standard instrumentation usually consists of a pressure gauge at one point along the barrel (usually at the head) and a thermocouple in the hot melt region. In noncritical operations, the operator will monitor pressure, temperature, and screw speed. Mass flow rates are typically monitored by the sample weight–time method.

In operations requiring close tolerances on extruded articles, a greater degree of instrumentation is employed. Usually, several pressure transducers and thermocouples along the barrel are used to ensure uniform extrusion and to control barrel and stock temperatures. Some designs may include thermocouples on the screw to monitor and control conveying flights.

There are numerous types of extruders, the most common of which is screw extrusion, which can be of the single or twin type. Another common type is the plunger or ram-type extruder. Materials such as TFE Teflon and ultrahigh-molecular-weight polyethylene are normally handled by ram extrusion. The melting temperatures of these polymers are very high; hence, these materials cannot be pumped readily as in screw extrusion. In this chapter discussions largely cover screw extrusion of low- to moderate-molecular-weight plastics and elastomers.

SINGLE-SCREW EXTRUSION

Many commercial extruders plasticate and pump materials in the range of 10 ~ 15 lb/hr-hp. However, pumping capacity is a relative quantity that depends on the material. In adiabatic mixing, machine capacities can be as low as 3 ~ 5 lb/hr-hp. Also, a machine which handles a thermoplastic elastomer could show as much as a 3 ~ 4 times increase in mass throughput when switched to a low melt viscosity material such as nylon.

Screw configuration depends to a large extent on the properties of the material being processed. Figure 2 illustrates several common arrangements. A constant-pitch metering screw is usually employed in applications not requiring intensive mixing. Where mixing is important, say for color dispersion, a two-stage screw equipped with a let-down zone in the center of the screw is appropriate. Turbulence promoters can also be included at or near the tip of the screw. In some applications with two-stage screws, venting at the let-down section may be needed.

The screw section immediately ahead of the gear train acts as a solids metering or feed zone. It is characterized by a deep channel between the root of the screw and the barrel wall. The plasticating zone is ahead of the solids metering zone. This is a transition region where the channel narrows. The purpose of this zone is to provide intense friction between solids and a region for melting of the polymer to take place. Near the tip of the screw is the melt metering zone, where pressure builds up. In this region the polymer melt is essentially homogenized and raised to the proper temperature for extrusion of the article.

The action of an extruder is analogous to that of a positive-displacement pump. The flight depth along the screw (i.e., the ratio of the solids metering channel depth to the melt pump channel depth) is known as the compression ratio. The purpose of screw flights is to enable the screw to transport polymer down the barrel. The pitch angle of the flights again depends on the type of material handled. Many elastomer applications employ a general-purpose screw which is of a constant pitch (i.e., flight equals the diameter). The pitch angle of this single-flighted screw is usually 17.61°. Typical extruder specifications are as follows:

Compression ratio: 2:1 ~ 6:1 (for materials ranging from LDPE to some nylons)

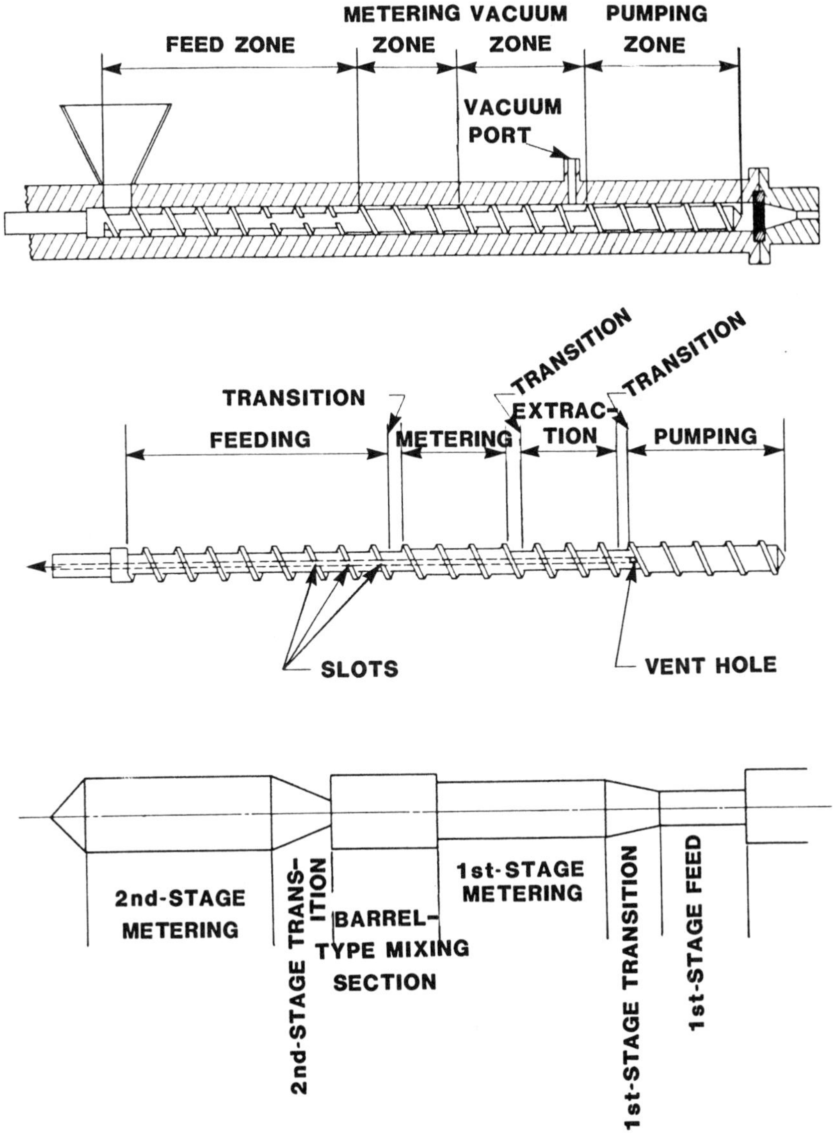

Figure 2 Screw profiles and cross section of vented barrel extruder.

Pitch angle: 12 ~ 20°
Length-to-diameter ratio (L/D): 16:1 ~ 36:1 (low 40's typically for easily melting/flowing polymers requiring high mixing and venting)
Extrusion pressures: 10000 ~ 30000 psi

For extrusion pressures, low ratings (~10000 psi) are usually sufficient for many thermoplastic materials. The upper limit typifies FEP Teflon.

The *lead* on a screw is defined as the distance between the flights. As an approximation it is equal to the ID of the barrel for a single-flighted screw. The radial clearance between the flight tip and the barrel is tight (usually ~0.001 in./in. of barrel ID). The reason for such a tight clearance is that if the gap is too great, the material may flow back along the barrel, resulting in a loss of melt pumping zone capacity.

The initial region of the extruder plays an important role in the machine's overall operation. It consists of a hopper or feed arrangement and a solid feedstock conveying region. Its purpose is to transfer the cold polymer feed from the feed hopper into the barrel, where it is initially compressed. This compression forces air out between the interstices of resin pellets or rubber chunks (air being expelled back through the hopper) and breaks up lumps and polymer agglomerates. This action creates a more homogeneous feedstock that can be readily melted.

Important features of the screw are shown in Figure 3. The relative motion of the extruder barrel to the screw is at an angle to the flights, where the helix angle of the flight is ϕ. From a vector diagram of the relative motion, the forward velocity of the plug is the velocity down the flights divided by ϕ. The volumetric capacity of the screw is

$$Q_s = U_z \rho \int_{R_s}^{R_b} \left(2\pi R - \frac{pe}{\sin \phi}\right) dR \tag{1a}$$

or

$$Q_s = \rho U_z \left[\frac{\pi(D_b^2 - D_s^2)}{4} - \frac{peh}{\sin \phi}\right] \tag{1b}$$

where

U_z = forward plug velocity
P = number of flights in parallel
e = width of flight
h = channel depth
R_s, R_b = radii of the screw root and inside barrel, respectively

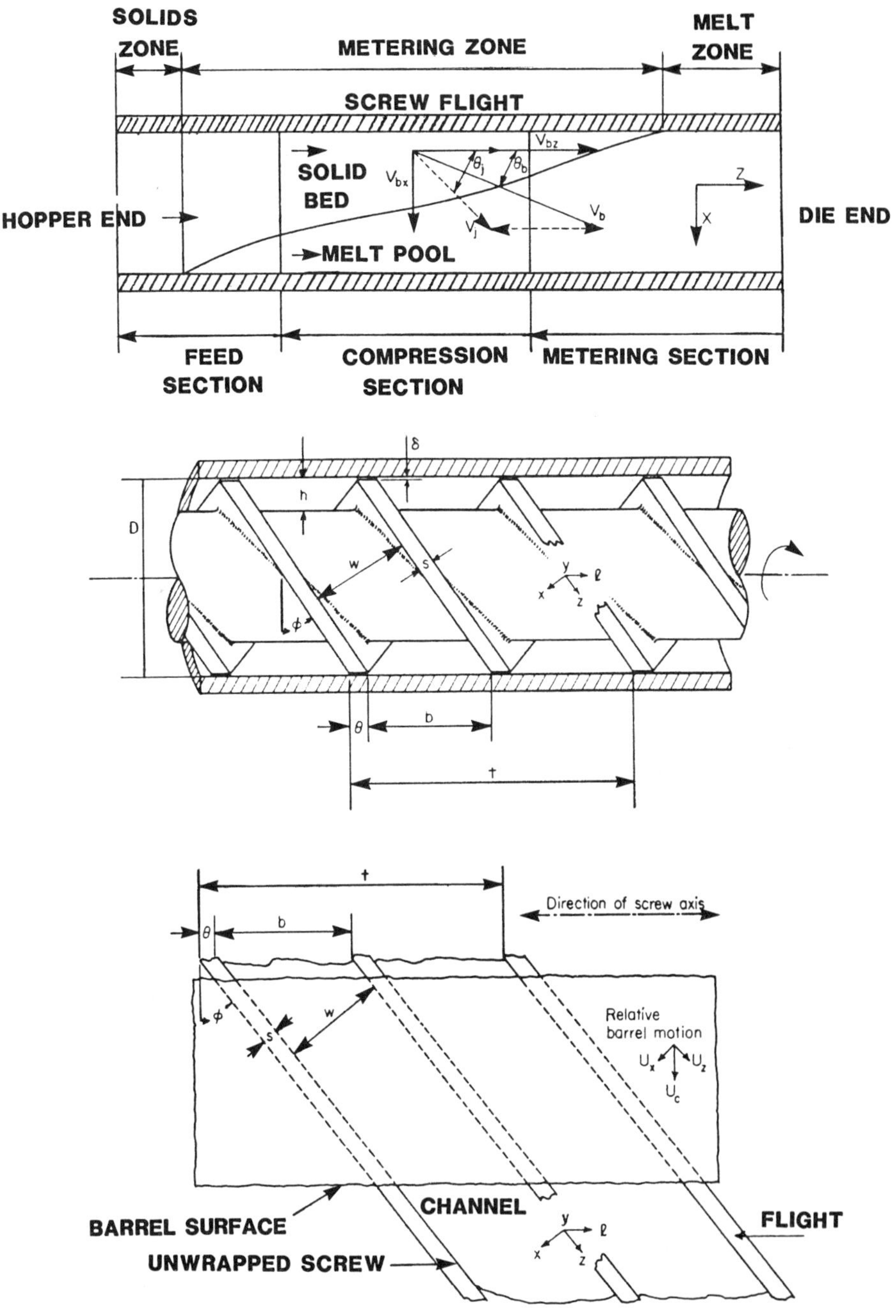

Figure 3 Configuration of screw.

Throne (1979) notes that the plug velocity at the barrel surface is:

$$U_{z\rho} = \frac{U_z}{\sin \theta} \tag{2}$$

where θ is the angle of advance of the plug relative to the screw axis (typically 10 ~ 25°).

Frictional forces between the plug and barrel depend on angle θ, which in turn is a function of the force and torque exerted on the polymer plug.

The preceding relationship is however an oversimplification. A more detailed force and torque balance on the plug shows that φ is a function of the forces and pressures:

$$Q_s = \rho\pi^2 NhD_b(D_b - h)\left(\frac{W}{W + e}\right)\left(\frac{\tan \phi \tan \theta}{\tan \phi + \tan \theta}\right) \tag{3}$$

where W is the average channel width (note that for most designs $W \gg e$). This expression is based on the following relationships:

$$U_{z\rho} = \frac{U_b \tan \phi \tan \theta}{\tan \phi + \tan \theta} \tag{4}$$

where

$$U_b = ND_b$$
N = screw speed (rpm)

$$W = \left(\frac{\pi}{p}\right)(D_b - h) \sin \phi - e \tag{5}$$

Following Throne (1979) and Tadmor and Klein (1969), the following values are introduced, which typify a more or less conventional single-screw design:

$\phi = 17.61°$
$\theta = 17.61°$
Negligible pressure drop along solids transport zone

Hence, $\tan \phi = \text{cotan}\ \theta$, and for no friction between the screw and solid plug:

$$Q'_s = \rho\pi^2 NhD_b(D_b - h) \sin \phi \cos \phi \tag{6}$$

and

$$\sin \phi \cos \phi = 0.2884$$

This provides a basis for defining a screw's discharge coefficient in terms of the solids conveying capacity as follows:

$$Q^* = \frac{Q_s}{Q_s'} = 0.2884\ (\tan\phi\ \tan\theta)/(\tan\phi + \tan\theta) \tag{7}$$

This ratio provides a relative rating of the solids flow rate. For practical ranges of θ, Eq. (7) indicates that the actual solids throughput can be of the order of 6% less than the maximum theoretical throughput.

Defining ψ_0 as the friction between the plug and screw channels; then

$$\psi = F\pi D_b L_s \cos(\phi + \theta) - \psi_0 \tag{8}$$

where

ψ = frictional force or push that a self-containing solid plug can exert on an obstruction
F = frictional force per unit area between the barrel and polymer solids
L_s = channel length filled with polymer

The resistance force that would obstruct the solid plug is

$$\psi_n = 2\pi D_b \xi \tag{9}$$

where ξ is the non-self-conveying friction per unit surface.

For solids to be conveyed, the condition that $\psi > \psi_n$ must exist. This can be stated more explicitly as:

$$\left[\frac{FL_s \cos(\phi + \theta)}{2\xi} - \frac{\psi_0}{2\pi D_b \xi}\right] \gg 1 \tag{10}$$

The form of the feed material can have a dramatic effect on the solids-conveying efficiency of the extruder. For example, in plastic extrusion applications, when the resin is fed as pellets, the coefficient of friction generally tends to be lower than in powder form; hence, pellets can tend to show higher conveying rates. In general, Eq. (10) indicates that increasing F (e.g., increasing the coefficient of friction between the material and the barrel) results in more effective conveying.

Mechanical heat is generated in the solids-conveying zone, which can be described in terms of the amount of heat added per pound of polymer conveyed:

$$\frac{H_s}{Q_s} = \frac{FD_b L_s(\tan\phi + \tan\theta)}{\rho h(D_b - h)[W/(W + e)]}\ (\tan\phi\ \tan\theta) \tag{11}$$

This thermal energy is the heat generated due to friction. This relationship shows that to generate frictional heat, a large frictional force between the barrel and solids is needed. This in part can be accomplished by using an extruder having a relatively shallow channel height and a long feed section.

Following Tadmor and Klein (1969), the pressure in the solids-conveying zone increases exponentially with distance down the flight and can be described by

$$P = P_1 \exp\left[\frac{(B_1 - A_1 k)Z_s}{B_2 + A_2 K}\right] \tag{12}$$

where

P_1 = pressure at the extruder inlet
Z_s = distance down the channel
$A_{1,2}$, $B_{1,2}$, K = functions of the solids properties and screw geometry

The inlet of the extruder usually has a hopper immediately above it when handling plastic resins. The pressure at the extruder inlet therefore follows the Janssen equation, which shows that the pressure exerted by solids on the floor of a vessel is independent of the height or head of material, that is

$$P_{v\infty} \simeq \frac{D\rho_b}{4\phi' K} \tag{13}$$

where ϕ' is the friction coefficient between the solid pellets or powder and the wall.

The value of $\phi' K$ can be determined experimentally. More accurate predictions of the pressure exerted at the base of the hopper can be obtained by considering the pressure profiles of the solids through the bin (Bridgwater and Scott, 1983).

An important limitation of the Janssen equation is that it is based on a static bin analysis. Extruder hoppers in fact are usually conical. In addition, they are not operated like static hopper-bins or freely discharging hopper-bins. Flow rates through extruder hoppers are controlled by the feed rate to the extruder. The operator must ensure that the feed rate from the hopper matches that of the extruder output. Pressure fluctuations due to variations of feed material levels in the hopper can cause surging and uneven extrudate profiles.

Once the polymer feed has been compacted, the solid resin particles or rubber chunks must be transformed into a homogeneous melt. Two major requirements are needed for proper operation:

1. There must be adequate heat conduction from the barrel to the resin or elastomer.
2. There must be sufficient mixing of the melt with the unmelted resin.

Shear heating between the resin and barrel surface may be necessary for the latter to occur. In general, the larger the amount of viscous dissipation within the melt–solid mixture, the less amount of energy need be supplied to the barrel, which translates into lower operating costs. Materials such as ABS can be heated from its resin state to a melt without supplying heat along the barrel. In this manner the extruder is operated adiabatically. In processing different elastomers the extruder operation can be isothermal; that is, the barrel, screw, and rubber are maintained at constant temperature and we rely on shearing energy to convert the polymer from a solid to a melt state.

Considering the melt as a Newtonian fluid and y as the direction from the top of the solid bed to the barrel surface, the velocity profile in the melt is

$$\frac{d^2U_z}{dy^2} = 0 \tag{14}$$

Imposing the no-slip condition at the wall, $U_z(0) = 0$, and the velocity component of the barrel in the down-channel direction is $U_z(\delta) = U_z'$ (where δ is the thickness of the melt layer), then

$$U_z = \left(\frac{y}{\delta}\right) U_z' \tag{15}$$

Equation (15) describes the velocity profile across the melt. Now the heat transfer which includes viscous dissipation is

$$k_m \frac{d^2T}{dy^2} + \eta \left(\frac{dU_z}{dy}\right)^2 = 0 \tag{16}$$

where

k_m = melt thermal conductivity
η = melt viscosity

Conduction is the primary mechanism for heat transfer in laminar flow. The melt temperature is $T(y = 0) = T_m$ and at $y = \delta$, $T = T_b$, which is the barrel temperature. The preceding relations combine to provide the temperature profile through the melt:

$$\frac{T - T_m}{T_b - T_m} = \text{Br}\left(\frac{y}{\delta}\right)\left(\frac{1 - y}{\delta}\right) + \frac{y}{\delta} \tag{17}$$

where Br is the dimensionless Brinkman number. The Brinkman number provides a measure of the relative importance of viscous heat generation to heat conduction due to the imposed temperature differential, $T_b - T_m$. It is defined as

$$\text{Br} = \frac{\eta U_z'^2}{2k_m(T_b - T_m)} \tag{18}$$

Differentiating Eq. (17) and evaluating the expression $y = 0$ gives the heat flow from the melt into the solid plug of polymer preceding it

$$-q = k_m(T_b - T_m)\delta + \frac{\eta U_z'^2}{2\delta} \tag{19}$$

In the solid bed the material's surface temperature is $T(0) = T_m$, and the temperature at some sufficiently large distance from the inlet is T_s. If the solids move in a plug flow fashion, the heat conduction for this material is

$$\rho_s c_s V_{sy} \frac{dT}{dy} = k_s \frac{d^2T}{dy^2} \tag{20}$$

where

V_{sy} = bed velocity in y-direction
c_s = specific heat
k_s = thermal conductivity

Solving Eq. (20) gives

$$\frac{T - T_s}{T_m - T_s} = \exp\left(\frac{V_{sy}y}{\alpha_s}\right) \tag{21}$$

where α_s = thermal diffusivity of the bed.

Equation (21) states that the temperature profile in the solid exponentially decreases from its melt temperature at the surface to the screw temperature T_s (if the extruder is sufficiently long) or some other reference temperature, in the interior of the bed.

The rate at which the melt enters the region above the solid bed is approximately the mass rate of flow leaving the melt pool; hence,

$$Q' = V_{sy}\rho_s x = \frac{V_{bx}\rho_m\delta}{2} \tag{22}$$

where

Q' = rate of melting per unit down-channel distance
ρ_s = solid's density
V_{bx} = velocity of barrel in the screw axis direction relative to the screw
ρ_m = melt density
x = average solids bed width

The heat flux into the solid is obtained from the heat conduction equation (Eq. (20)). From an overall heat balance

$$\left[\frac{k_m(T_b - T_m)}{\delta + \eta U_z'^2/2\delta}\right] - \rho_s c_s V_{sy}(T_m - T_s) = V_{sy}\rho_s H' \tag{23}$$

where

H' = heat of fusion of solids

Throne (1979) gives the following two expressions for the thickness of the melt and the velocity of the bed:

$$\delta = \left\{\frac{[2k_m(T_b - T_m) + \eta U_z'^2]x}{V_{bx}\rho_m c_s(T_m - T_s) + H'}\right\}^{1/2} \tag{24}$$

$$\nu = \left\{\frac{[V_{bx}\rho_m k_m(T_b - T_m) + \eta U_z'^2/2]}{2[c_s(T_m - T_s) + H'}\right\}^{1/2} \tag{25}$$

where ν is the speed of rotation and thus the velocity of the solid bed.

The rate of melting is proportional to the square root of the width of the solid bed, and the solid's volumetric flow rate is related to the rate of melting:

$$-\rho_s V_{sz} \frac{d(hx)}{dz} = \nu \tag{26}$$

Throne (1979) imposes the following model assumptions:

1. The melt film thickness is small and constant during the melt phase,
2. The height of the bed h_s is equal to the height in the channel.

3. The channel is tapered and therefore the height can be described by $h = h_0 - az$, where h_0 = channel height at beginning of melt zone and a = slope of the channel.
4. At the beginning of melting (i.e., at $z = 0$), $X = W$, the channel width.

By applying these assumptions, substituting Eq. (25) into Eq. (26), and simplifying terms, the following relation is obtained

$$(h_0 - az)\frac{dx}{dz} = ax + \left(\frac{\nu\phi}{\rho_s V_{sz}}\right)x^{1/2} \tag{27}$$

The solution to this equation is:

$$\frac{x}{W} = \left\{\frac{\psi}{a} - \left(\frac{\psi}{a} - 1\right)\left[\frac{h_0}{(h_0 - az)}\right]^{1/2}\right\}^2 \tag{28}$$

where ψ is a transformation variable defined in terms of the mass flow rate G:

$$\psi = \frac{\phi W^{1/2}}{G/h_0} \tag{29}$$

and

$$G = V_{sz}h_0 W\rho_s \tag{30}$$

A plot can be prepared of X/W versus z/Z, for different a/ψ values [where $Z = (h_0/\psi)(2 - a/\psi)$, the total melting length], which shows the effect of screw flight tapering on the solid bed profile. The relationship for Z shows that by increasing the extruder channel width, the length of melting required decreases. Further, increasing barrel temperature, melt viscosity, and screw speed decrease for melting length.

After the polymer has been completely melted it enters the metering or pumping zone of the extruder. This portion of the machine is designed for a certain length so that sufficient pressure builds up for injection or extrusion through dies. The most widely referenced model of this portion of extrusion is that of Squires (1964). The model was later modified by Carley (1971). This is a drag flow model. The relationship between pressure drop in the flow channel and the velocity profile is

$$\frac{dP}{dz} = \eta\frac{d^2V_z}{dy^2} \tag{31}$$

where

η = melt viscosity (assume Newtonian for immediate discussions)
V_z = velocity component along Z-direction
y = distance above the root of the screw

By integration we obtain:

$$V_z = \frac{yV_z}{h} - \frac{y(h-y)(dP/dz)}{2\eta} \tag{32}$$

Integration of the velocity profile across the channel height gives the volumetric flow rate

$$Q = W \int_0^h V_z \, dy$$

$$Q = \frac{V_z Wh}{2} - \frac{Wh^3(dP/dz)}{12\eta} \tag{33}$$

By incorporating terms describing the channel geometry, Eq. (33) transforms to the following:

$$Q = F_d\left[\left(\frac{\pi^2}{2}\right)(\sin\phi\cos\phi)D^3hN\right] - F_p\left[\left(\frac{\pi}{12}\right)\sin^3\phi h^3\left(\frac{D}{L}\right)\left(\frac{P}{\eta}\right)\right] \tag{34}$$

where

ϕ = helix angle
D = barrel diameter
N = screw speed
L = length of melt zone of screw in the axial direction
F_d, F_p = geometric functions of W/h; drag flow and pressure flow shape factors, respectively

The first term on the right-hand side of Eq. (34) represents the drag flow induced by the moving barrel surface. The second term represents the pressure flow induced by the axial pressure gradient in the extruder. Note that for melt flow in very narrow channels, $F_p = F_d = 1$.

INFLUENCE OF RHEOLOGY ON EXTRUSION

Proper scale-up and use of polymer processing equipment such as extruders, as well as an understanding of the physical processes taking place in the apparatus, require a knowledge of the polymer melt and its dependence upon the shear, temperature and molecular properties of the polymer. This information is perhaps most important in processes where the polymer melt undergoes high shear and heat treatment and/or where degradation of the polymer may occur. In some applications, polymer degradation may actually be desirable in order to achieve the proper characteristics of the final polymer melt.

There have been a number of attempts to relate rheological properties of polymers to screw extruder design. Zavadsky et al. (1982) developed mathematical descriptions of flow curves of a degradable type of polypropylene (TE430) and applied these relationships to the design of plasticating screw extruders for the production of polypropylene fibers. In this study the flow curves were measured on a Gottfert laboratory extruder over a temperature range of 190–310°C and shear rates of 1–300 s^{-1}. A Takara plastometer was used to obtain the final melt index for each measurement. The effective viscosity of the melt η was computed as the ratio of the wall shear stress $\tau = \Delta Pd/4L$ to the nominal shear rate Γ at the wall, calculated without any correction as $\Gamma = 8u/d$ (where d = die diameter, L = length of die, u = mean flow velocity). Munstedt (1978) describes a method to generate a master curve for all temperatures based on an Arrhenius-type relationship.

$$a_T = \exp\left(\frac{E}{R}\frac{T_0 - T}{T_0 T}\right) \tag{35}$$

where

a_T = shift factor
E = activation energy of flow
R = universal gas constant
T_0 = absolute reference temperature
T = absolute temperature of measurement

The master curve has the general form:

$$\frac{\eta}{a_T} = f(a_T\Gamma) \tag{36}$$

Zavadsky et al. (1982) redefined the shift factor in terms of the melt index (MI) such that

$$a_{T_m} = \frac{a_T}{\text{MI}} \tag{37}$$

A logarithmic plot of $\eta \text{MI}/a_T$ versus $\Gamma a_T/\text{MI}$ for measured data of the flow velocity u and pressure difference ΔP converted to η and Γ over the entire range of temperatures for the study was found to lie on a master linear correlation (see Figure 4). This type of relationship can be readily defined in terms of a power series:

$$\ln \eta = \ln a_{T_m} + \sum_{i=1}^{5} a_i (\ln a_{T_m} \Gamma)^{i-1} \tag{38}$$

where the temperature shift factor takes the form

$$a_{T_m} = (\text{MI})^{a_8} \exp\left(a_6 + \frac{a_7}{T}\right) \tag{39}$$

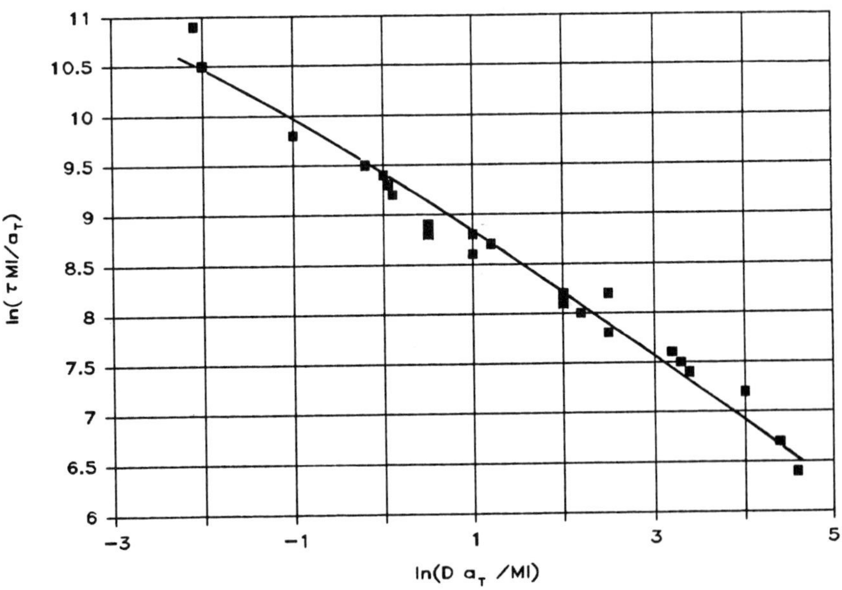

Figure 4 Master flow curve developed by Zavadsky et al. (1982) for a TE430 polypropylene for a temperature range of 190–310°C.

The coefficients in the power series can be evaluated by the method of least squares and for the data of Zavadsky, these values are as follows:

$a_1 = 10.306$ $a_2 = 0.52327$ $a_3 = -0.019027$
$a_4 = 0$ $a_5 = 0$ $a_6 = -10.718$
$a_7 = 4263$ $a_8 = -0.87968$

Melt index is a function of the temperature and residence time of the melt in the apparatus. Residence time can be quantified from the volumetric throughput of the extruder and a mathematical expression such as that derived by Zavadsky et al. can be derived from a multivariable regression analysis:

$$MI = 5.1 + t_c \times 1.0725^{(t-321.5)} + t_c^2 \times 1.2082^{(t-338.1)} \tag{40}$$

where

t = temperature
t_c = residence time

A more convenient relationship for describing the flow curves of polypropylene, and indeed many other polymers, is a power law model expression:

$$\eta = m_0 \exp(-kt)\Gamma^{n-1} \tag{41}$$

For calculation purposes, this relation is more convenient than Eq. (38). The power law coefficients k and n can be derived by the comparison of the partial derivatives of Eqs. (41) and (38) to give

$$n = 1 + a_2 + 2a_3 \ln(a_{T_m}\Gamma) \tag{42a}$$

$$k = \frac{na_7}{T^2} \tag{42b}$$

And the flow consistency index m_0 is simply

$$m_0 = \eta \exp(kt)\Gamma^{1-n} \tag{43}$$

Zavadsky points out that the parameters of the power law model change because of the degradation of the polymer even under conditions of constant shear rate and temperature. The importance of the melt index on these parameters becomes evident by performing a simple series of parametric calculations using the preceding relationships. Figure 5 illustrates the effect of MI on k, n, m_0 and η.

In the case of polypropylene degradation of the polymer can have a significant effect on the extrusion process. Zavadsky et al. applied their model to the evaluation of several single-screw extruder configurations. Figure 6 shows model predictions of the influence of polypropylene degradation on the energy dissipation, temperature, and pressure down-channel profiles, accounting for progressive degradation as given by the empirical relation of Eq. (40). The value of such a model is that it can guide the user in selecting an extruder geometry suitable for the particular polymer characteristics. The study shows that the influence of shear rate and MI on the rheological parameters k and n is distinct but diminishes with increasing values of both quantities. In general, the influence of shear rate on parameter m_0 is negligible but this parameter is a strong function of the melt index.

There are a number of mathematical treatments of the extrusion process based on various methods. For example, Marshall et al. (1965) and Klein et al. (1965) used regression analysis to fit a system of empirical equations to data obtained from an experimental extruder. These relations, however, can only be used to describe the performance of other machines of similar size working with similar materials. More universal methods are based on principles of continuum mechanics. In this case the extruder is conceptually divided into several zones in order to develop a mathematical description of the extrusion process as described earlier.

Zavadsky (1978) developed a mathematical model of an extruder by dividing the apparatus into the following zones:

The zone under the hopper
The solids-conveying zone
The delay zone
The melting zone
The melting/conveying zone

The mathematical description of a particular zone and its exact solution depend on several simplifying assumptions. The model equations can be solved numerically only by using a digital computer. Therefore, in choosing suitable mathematical models for the various zones a compromise must also be made between cost and accuracy.

For the zone under the hopper, the length of the zone is assumed to be equal to the inside diameter of the hopper neck. The barrel is usually cooled here. It can be assumed that the solid polymer in this zone is heated by hot air from the extruder to an average of the temperatures of the barrel, screw, and solids in the hopper. The pressure can be estimated by the solid polymer hydrostatic pressure.

In the solids-conveying zone the solids movement strongly depends on the friction coefficient in the solids-barrel and solids-screw regions. The mathematical model developed by Zavadsky is based on the work of Broyer (1972) for two-dimensional, nonisothermal, nonisotropic plug flow of solid material described

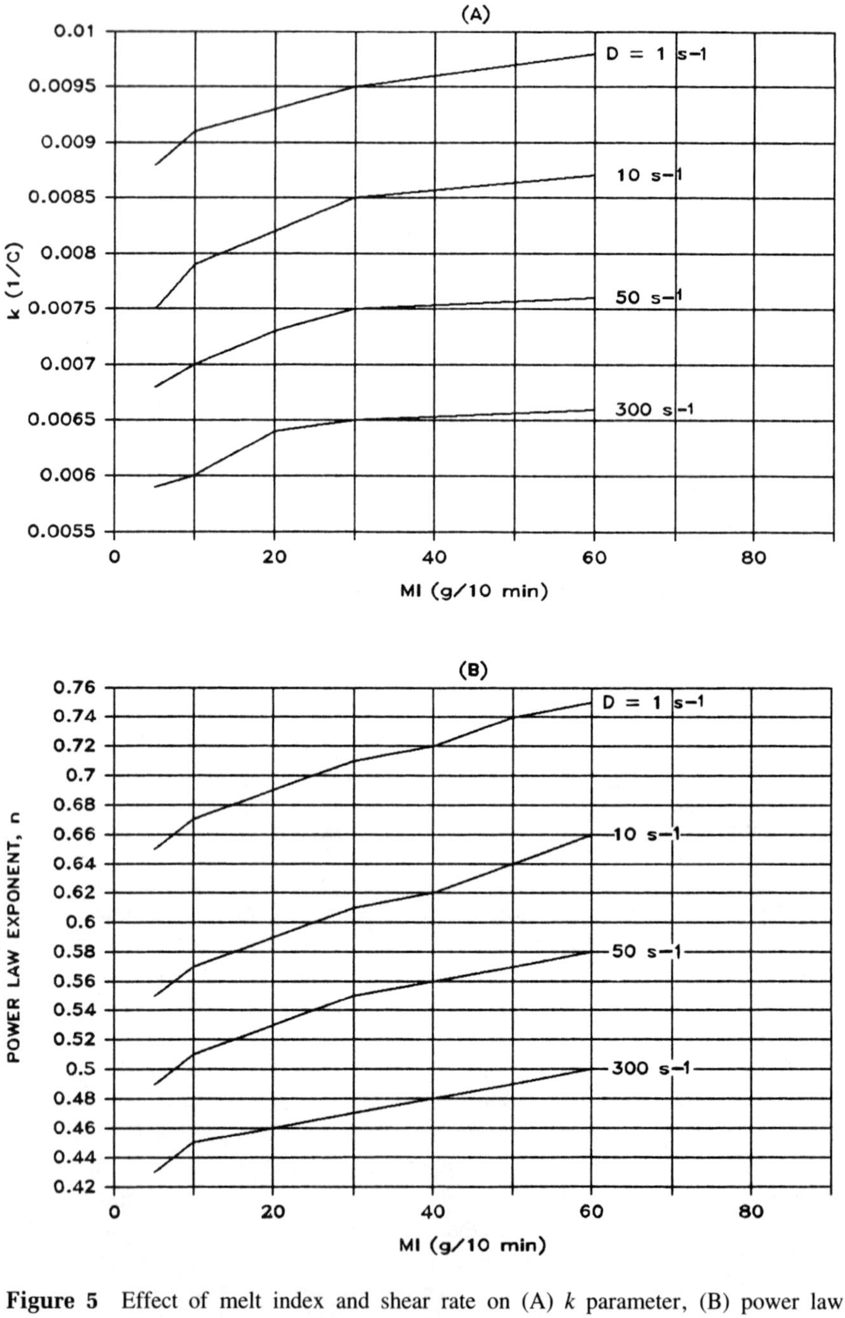

Figure 5 Effect of melt index and shear rate on (A) k parameter, (B) power law exponent n, (C) parameter m_0, (D) apparent viscosity η.

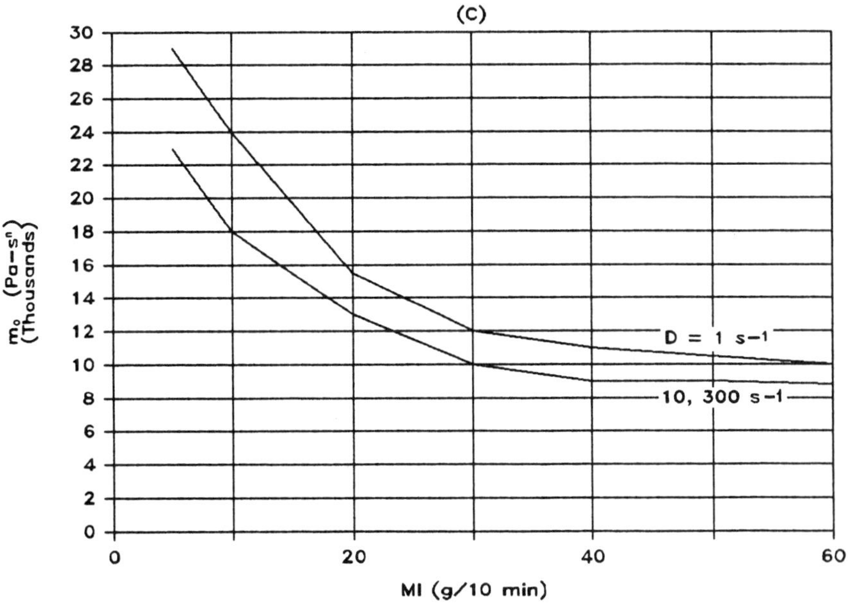
(C)
m_o (Pa–s^n)
(Thousands)
30
28
26
24
22
20
18
16
14
12
10
8
6
4
2
0
0
20
40
60
MI (g/10 min)
D = 1 s−1
10, 300 s−1

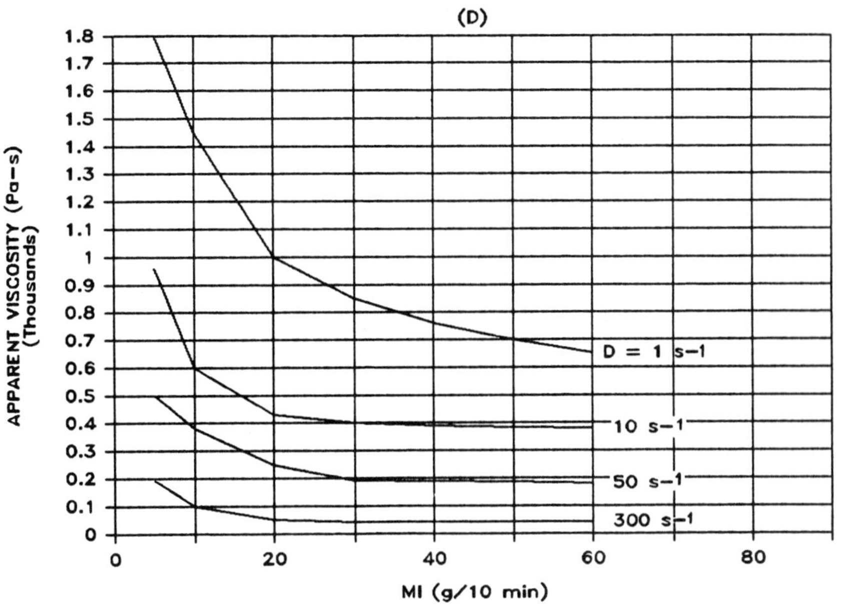
(D)
APPARENT VISCOSITY (Pa–s)
(Thousands)
1.8
1.7
1.6
1.5
1.4
1.3
1.2
1.1
1
0.9
0.8
0.7
0.6
0.5
0.4
0.3
0.2
0.1
0
0
20
40
60
80
MI (g/10 min)
D = 1 s−1
10 s−1
50 s−1
300 s−1

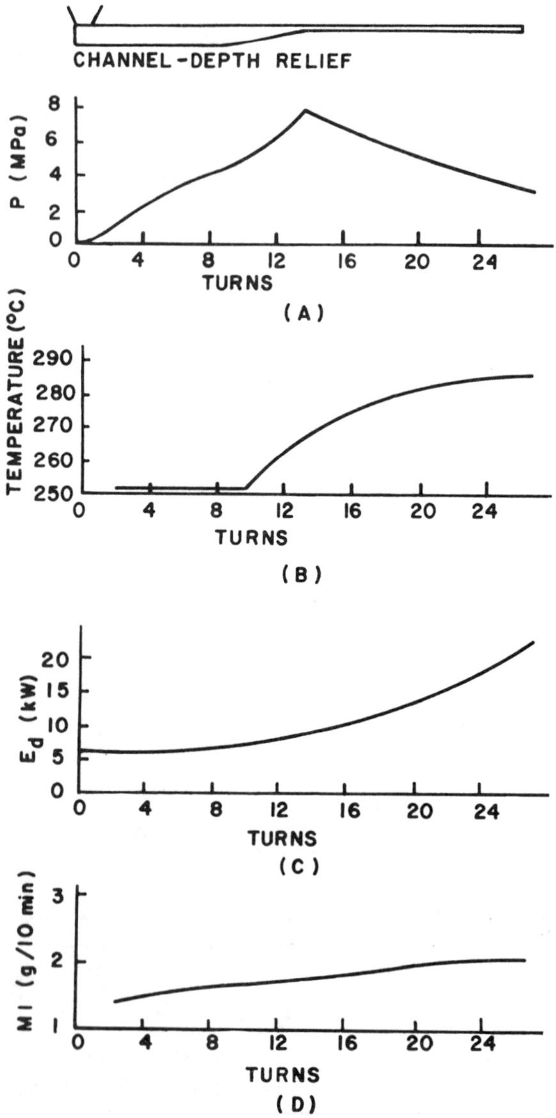

Figure 6 Computer prediction of (A) down-channel pressure profiles; (B) down-channel melt temperature profiles; (C) down-channel melt index profile; and (D) down-channel dissipated energy profiles, for one extruder configuration.

earlier. Examples of typical temperature and pressure profiles are shown in Figure 7. Friction coefficient–temperature relations are given by Chung et al. (1977).

As noted in an earlier discussion, there is a transition region in the extruder which can be thought of as a delaying zone. Once the solid resin has been conveyed a certain distance along the extruder it becomes important that it reach its melting temperature. The friction forces in the barrel are replaced by shear stress in the film. In the delay zone, heat is conducted from the hot barrel through the melt film to the melt–solid interface. Heat is also generated by viscous

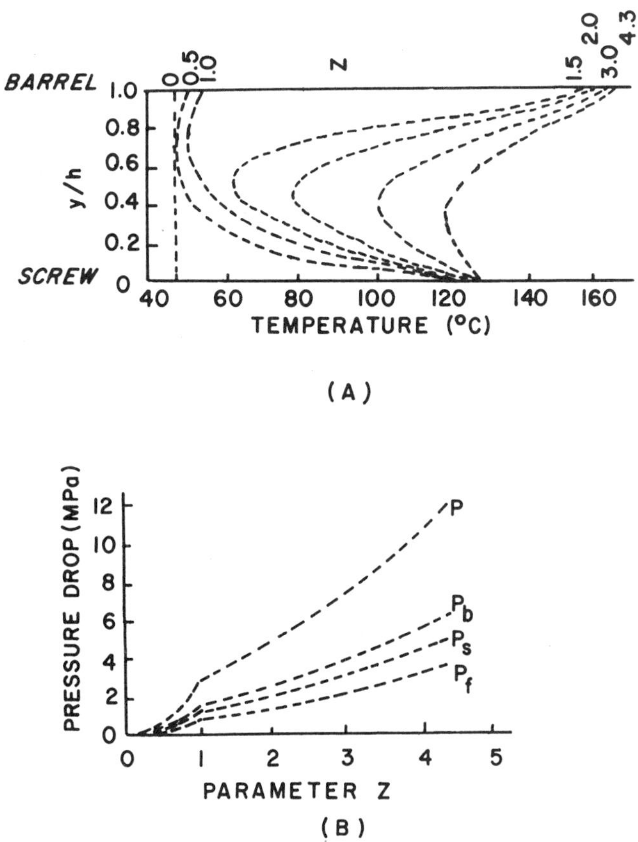

Figure 7 (A) Model predictions of the solid bed temperature profiles in the solids-conveying zone reported by Zavadsky and Karnis (1985) for one-extruder configuration, (B) Model predictions of the axial pressure distribution in the solids-conveying zone.

dissipation in the film. The heat energy melts the solid polymer causing the thickness of the melt film to grow. A mathematical model of this zone for constant depth of the screw channel and constant solid bed velocity was given by Kacir and Tadmor (1972). The average value of the film thickness at the end of the delaying zone can be five times larger than the flight clearance. The flow of polymer between the barrel and melt–solid interface can be modeled as the flow of a power law fluid between moving plates with a linear temperature profile across the film thickness. This mathematical model can be extended to account for changes in the channel depth. Typical pressure and film thickness profiles are shown in Figure 8. The end of the zone is given when the film thickness has grown to the initial value of film thickness in the melting zone.

Tadmor et al. (1966, 1970) provide a model of the melting zone, where the mechanism of heat transfer is similar to that described in the delaying zone. Here, however, the melt film is continuously scraped by the advancing flight and accumulates in the rear of the screw channel. Melting exists only at the melt–solid interface along the barrel. The model assumes a constant solid bed velocity along the axis of the channel. The flow of the melt in the film can be simulated in the same way as the flow in the delaying zone, that is, fully developed drag flow of a power-law fluid with linear temperature profile across the film thickness, with heat generated in the flow being conducted to either the solid bed or barrel surface. The rate of melting is given by the balance of heat fluxes on the melt–solid interface using an effective latent heat of fusion, and the change of the solid bed width is given by a mass balance. In a simplified analysis the solid bed can be treated as a semi-infinite plate with heated surface. Therefore, the boundary condition given by the screw temperature has no influence on the temperature of

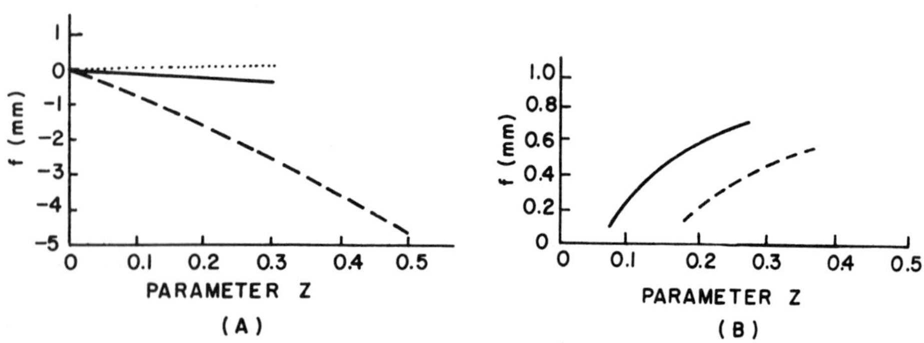

Figure 8 Model predictions of (A) axial pressure profiles in the delaying zone and (B) axial film thickness profiles reported by Zavadsky and Karnis (1985) for two-extruder configurations.

the solid bed and the bed temperature profile is independent of the downstream position. Relations for simulating the down-channel pressure profile were derived for the model of Zavadsky et al. (1982).

For the melt conveying zone, Zavadsky employs two different mathematical models. For optimizing the channel depth (maximum mass flow rate) with respect to strength and calculating the screw core and flight, a two-dimensional isothermal Newtonian model is used. The velocity is assumed to be fully developed in the axial direction and the channel rectangular with one moving plane, namely, the barrel. The computed mass flow is modified with the aid of correction factors for channel curvature and for flow in the flight clearance. The pressure gradient is given by the difference between the initial and required end pressures divided by the melt conveying zone length. The melt viscosity is calculated for the required melt properties at the end of the extruder. For computing the temperature and pressure profiles along the channel axis a non-Newtonian, nonisothermal two-dimensional model with heat dissipation is used.

The heat balance in an infinitesimal element of the length is made with the aid of barrel-melt screw heat transfer coefficients. The relation between shear stress and shear rate is given by a power law model:

$$\tau = m_0 \, e^{-kt} \, \dot{\gamma}^n \tag{44}$$

The parameters m_0, k, and n can be given as a function of temperature, shear rate, and varying properties of the melt, due, for example, to its degradation. In each zone the barrel temperature is constant, whereas the screw temperature grows linearly along its axis.

EXTRUSION OF ETHYLENE–PROPYLENE ELASTOMERS

Ethylene–propylene elastomers have been used widely in extruded parts for over a quarter of a century. This is principally due to the recognized ease of processability of E-P compounds in extrusion operations and the broad range of physical properties that are obtained to satisfy the myriad of applications. Major uses include hoses, dual durometer weatherseals, glass run channels for automobiles, building profiles, and electrical insulation for the automotive, architectural, and construction industries. The reader should refer to the work of Johnson (1983), which provides a detailed review of extrusion technology. In this section an overview of the characteristics of ethylene–propylene polymers is presented to show how these parameters affect the extrusion process. Specific structural factors, namely, molecular weight and breadth of molecular-weight distribution, are then related in more detail to processability in the extruder. In a more general sense, these principles are equally applicable to other types of polymers, such as plastics. It is important to keep in perspective that regardless of the particular polymer in use, the rheological properties of that material play a dominant role in the extrusion or mixing process. In the previous section the concepts of extruder modeling were reviewed, with emphasis placed more on apparatus scale-up and design. However, the consumer or end-use manufacturer most often does not have the freedom to alter equipment configuration. He must in fact work with fixed equipment and is therefore faced with the problem of selecting a polymer of suitable properties for manufacturing the article. To achieve the proper rheology for processing and, of course, the correct physical, chemical, and cure properties of the final product, the polymer manufacturer must synthesize a suitable molecular structure. Rational selection of engineering polymers must therefore be based on our ability to relate molecular properties to physical, chemical, and rheological characteristics, which in turn, must be related to the materials' processability and end usage.

Ethylene–propylene elastomers are uniquely suited for many extruded automotive and industrial applications noted above because of their unlimited ozone resistance, outstanding weatherability, good heat resistance, and low compression set. EP polymers have characteristically long, flexible, and mobile chain segments, which permit easy incorporation of fillers, plasticizers, and other rubber additives. From a processability viewpoint, this is an important feature because extrusion throughput is greatly increased as the viscoelastic effects of the rubber are reduced. In addition, compound cost savings also accrue from increased loadability of the E-P rubber. Thus, the finished article has the potential for attractive manufacturing economics.

Ethylene–propylene elastomers are described in terms of both structural and compositional factors for complete definition of any product. For a very thorough discussion of ethylene–propylene elastomers, the reader should consult Ver Strate in the *Encyclopedia of Polymer Science and Engineering* and the paper by Keller (1985). These parameters are as follows:

- EP molecular parameters for grade definition
  - Structural factors
    - Molecular weight (or Mooney viscosity)
    - Molecular-weight distribution
    - Long-chain branching
    - Monomer sequence distribution
  - Compositional factors
    - Ethylene–propylene ratio
    - Diene type and concentration

Ethylene and propylene are the basic monomeric units of this elastomer as illustrated by the structures shown in Figure 9. Polymerization proceeds in the presence of transition metal halide/aluminum alkyl chloride catalyst systems to yield high-molecular-weight elastic products. The copolymer may be combined

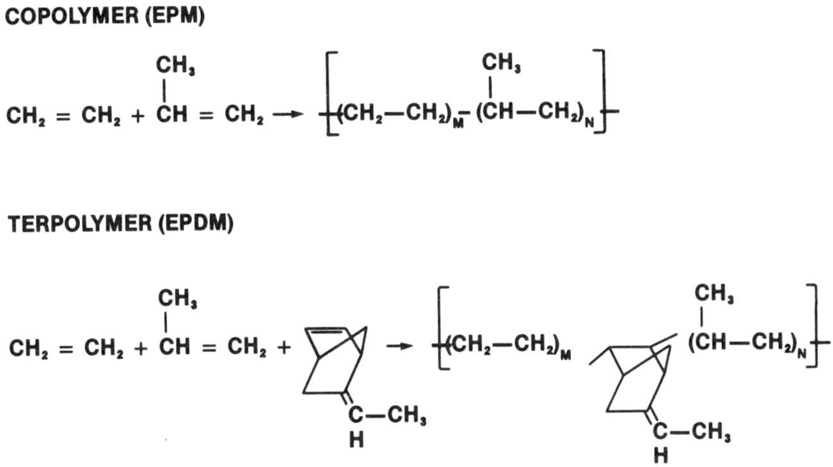

Figure 9 Structure of EP elastomers.

with a third or in some instances even a fourth monomer to provide unsaturated functionality along the polymer chain to increase vulcanization response with peroxides or to make the rubber sulfur curable. These structures are somewhat oversimplified. Rather than alternating monomers even at a 50/50 weight ratio of C_2/C_3, there are more or less random distributions of ethylene and propylene units which are determined by catalyst system and other polymerization conditions. In general, the Mooney viscosity at 125°C of commercial EP elastomers varies from less than 20 to well over 100 for oil-extended grades. Typical extrusion grades are found in the medium-to-high Mooney series, although blends of polymers from all groups may be used.

In addition to the sequence distribution of monomers in any given polymer chain, molecular-weight distribution also varies greatly among commercial EP elastomers. MWD typically spans the entire range from narrow to broad, with some bimodal and even multimodal distributions. A generalized schematic of narrow and broad distributions is given in Figure 10. In the extrusion process, narrow distribution grades tend to have higher throughput rates and extrudates are usually smoother. However, because of the absence of high-molecular-weight ends, stocks generally feed poorly. On the other hand, broad MWD polymers have better extruder feeding characteristics but also yield higher die swell. Another manifestation of MWD is the effect on state of cure. Figure 11 illustrates a curemeter response for two EP polymers that differ only in MWD (Mooney, ethylene, and diene content are similar). The narrow elastomer cures at a faster rate and to a higher cure state, because there are fewer chain ends in these polymers,

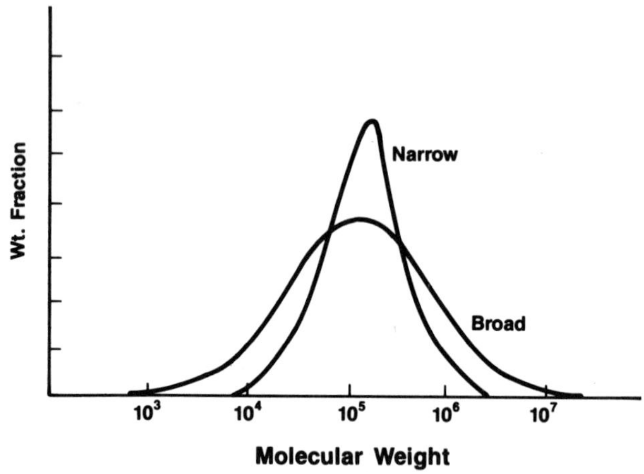

Figure 10 General molecular-weight distribution of EP elastomers.

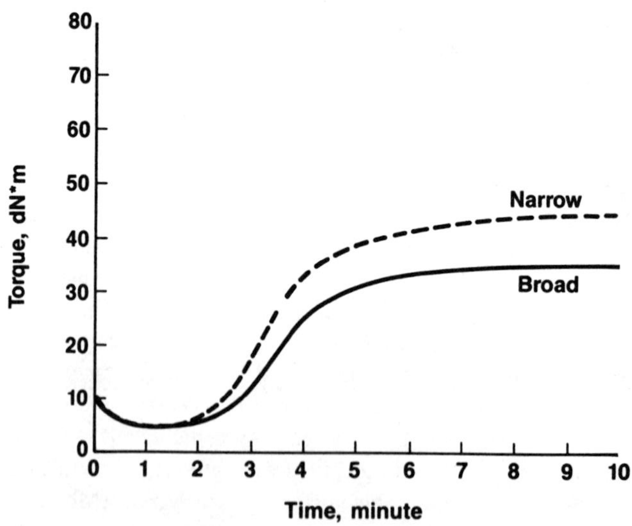

Figure 11 A comparison of narrow and broad MWD polymers in cure response.

which results in a more efficiently formed network. In general, as the average weight-to-number molecular weight (M_w/M_n) decreases, cure state of the compound increases. This is illustrated by the data in Figure 12, which is a plot of the difference between the maximum and minimum readings in the curemeter versus the weight to number average ratio as obtained from GPC techniques.

When the ethylene content in the polymer is typically above 55 wt %, measurable levels of crystallinity can be detected using differential scanning calorimetry (DSC) techniques. There are several important aspects for the use of high ethylene products in extruded articles. Toughness and green strength are imparted to the compound, essential in a cold feed process; thus, the poor feed qualities of narrow MWD polymers are overcome via the semicrystalline route. Further, stocks employing high ethylene grades tend to be more thermoplastic, which enhances mass flow. Finally, the profile should have better size control and dimensional stability. Figure 13 shows the effect of ethylene content on the green strength of an extrusion compound. Depending on the catalyst and reactor conditions employed, a 1–2 MPa increase in green strength is observed as low as 57 wt % in one type of polymer while at 64 wt % in another. Thus, at any given

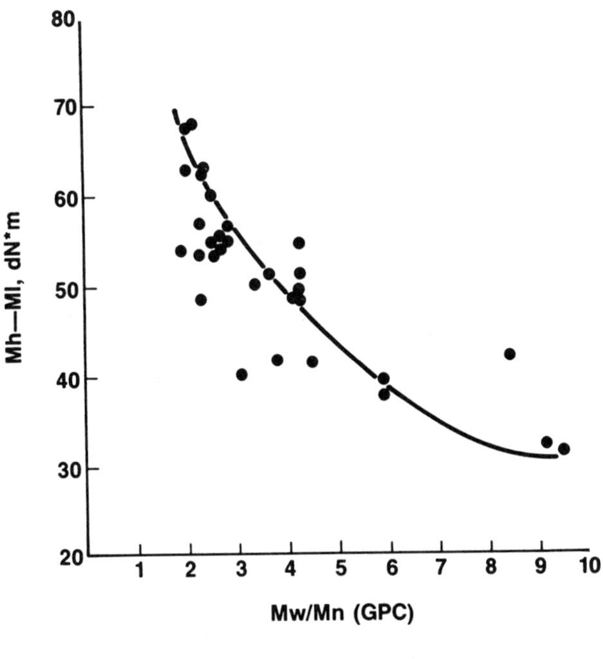

Figure 12 The effect of MWD on the state of cure.

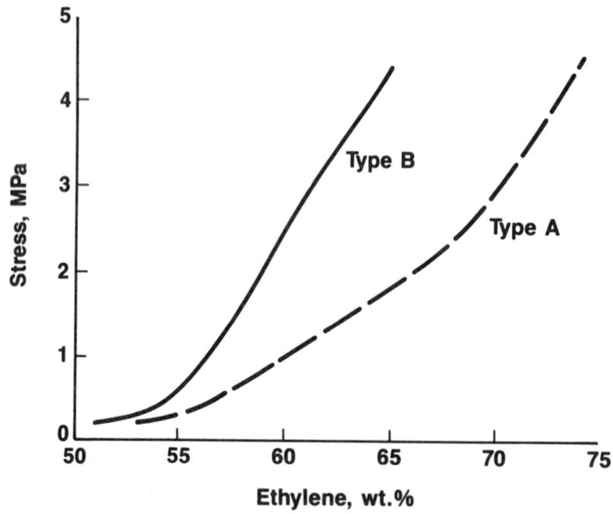

Figure 13 The effect of ethylene content on rubber green strength.

ethylene content, type A polymer is more amorphous than type B. This is noteworthy because ethylene content has a major influence on tensile strength. For the same level of cold green strength or toughness as might be encountered in the case of mandrel forming a radiator hose, the type A polymer is expected to yield higher tensile properties in the finished product.

The green strength of the rubber compound is critical to cold feed extruders and generally depends on the level of crystallinity in the polymer. A measure of crystallinity is the polymer's heat of fusion, as measured by DSC techniques. A typical DSC curve is shown in Figure 14. It is important to note that the presence of monomers such as nonconjugated dienes can affect the level of crystallinity. As shown in Figure 14B, the addition of a small amount of diene [ethylidene norbornene (ENB)] significantly reduces the heat of fusion. The effect of the monomer is to disrupt the ethylene sequence distribution so that the true ethylene content of the polymer is approximately defined as follows:

$$\text{Corrected ethylene content (wt \%)} = (1\text{-diene content}/100) \times \text{ethylene level} \tag{45}$$

Figure 15A shows crystallinity data for a group of polymers made using the same catalyst system. As shown, below about 55%, the polymers are amorphous. Figure 15B shows the dependence of compound green strength (polymers compounded in a sulfur cure system with carbon black) on the level of crystallinity.

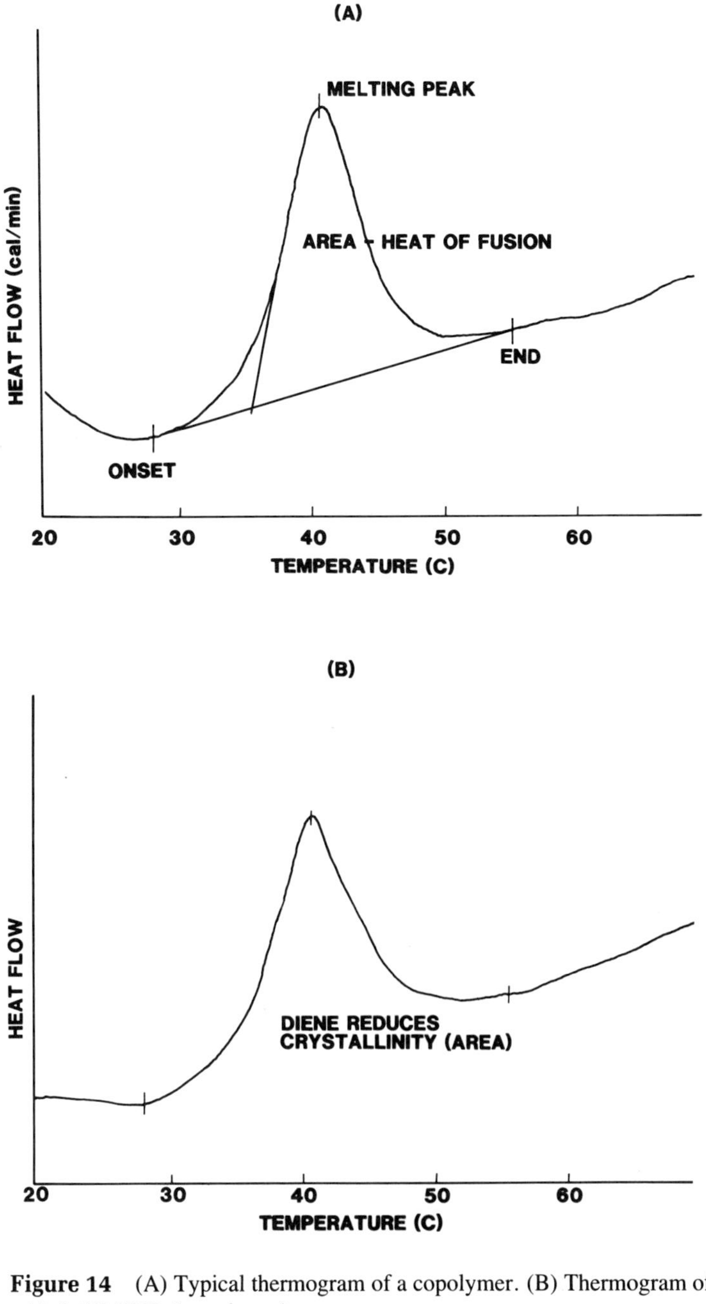

Figure 14 (A) Typical thermogram of a copolymer. (B) Thermogram of same polymer with 0.5% ENB (terpolymer).

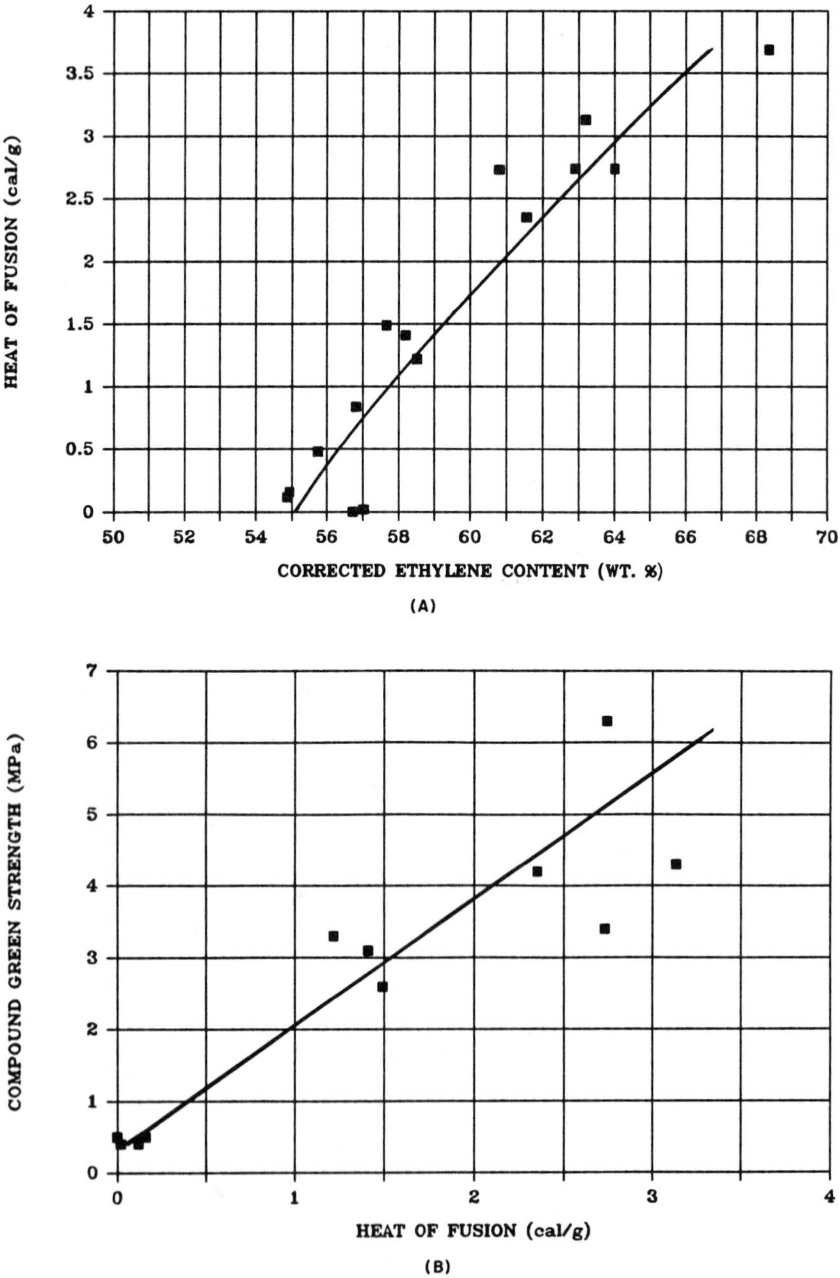

Figure 15 (A) Dependence of crystallinity on ethylene content of polymer. (B) Relationship between green strength and crystallinity.

In cold feed extruders, high green strength values tend to produce smoother feeding and compaction within the extruder with the net result of increased throughput. Figure 16 shows the effect of green strength on mass extrusion rates for a series of polymers extruded in a single-screw extruder.

As noted above, the monomers employed to introduce olefinic sites for sulfur curing are nonconjugated dienes. The most widely used is ENB. This diene polymerizes through the strained ring, resulting in a carbon–carbon double bond pendant to the main chain. The unique feature is its fast cure response, which is an important consideration in extruded profiles, particularly for sponge as well as other applications. Other dienes, illustrated in Figure 17, are also utilized to provide some distinctive characteristic. 1,4-Hexadiene (HD) is the most difficult to incorporate into the main chain because it does not involve a strained ring as does ENB. However, terpolymers containing HD generally impart somewhat better heat resistance than other EPDM grades. Dicyclopentadiene (DCPD) polymerizes into the main chain, but the second or five-membered ring double bond is also accessible to Ziegler catalysis, which can lead to long chain branching. Norbornadiene (NBD) has two equivalent double bonds that are reactive in polymerization. In some E-P rubber grades, a small amount of NBD is utilized to develop controlled branching. In general, reaction of the residual double bond results in tetrafunctional long-chain branch points in EPDM. In

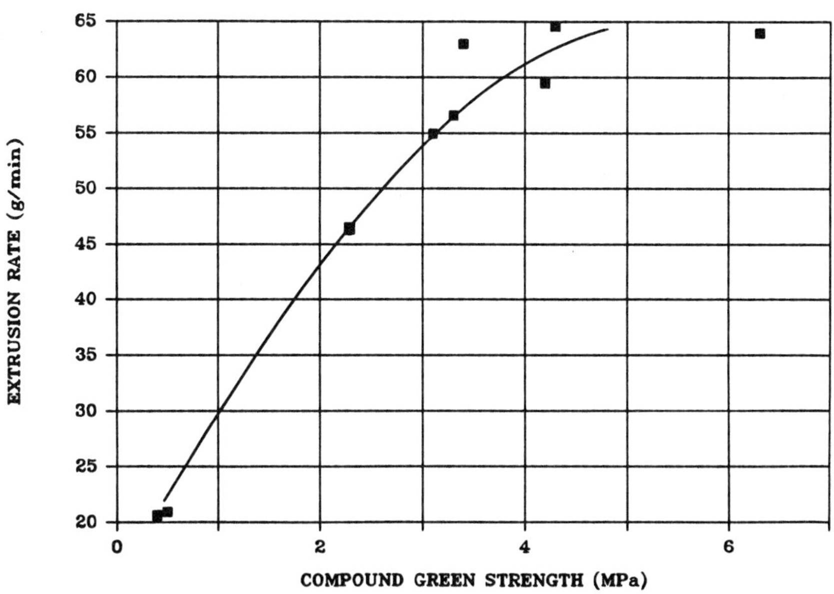

Figure 16 Extrusion throughput increased with compound green strength.

Name & Structure		Features	Branching
Ethylidene Norbornene	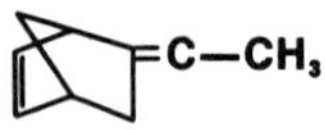	• Fast Cure • Good Tensile • Good Set	Nil
1,4 - Hexadiene	C = C - C - C = C - C	• Best Compression Set • Normal Cure Rate • Scorch Safe	Nil
Dicyclopentadiene		• Low Cost • Very Slow Cure • Good Set	High
Norbornadiene		• Branching	Very High

Figure 17 Chemical structures of different dienes used in EPDM elastomers.

some extruded applications a small amount of branching may be desirable as a process aid and to maintain profile definition prior to vulcanization. However, if the branching reactions are left uncontrolled, gel formation occurs, which is obviously undesirable.

As noted earlier, the processing characteristics of EPDM and other polymers depend on compositional and molecular-weight properties. In designing extrusion products and/or assessing different polymers in new or different extrusion applications formulas, the ability to project small-scale processing tests to commercial extrusion is essential. Laboratory rheological testing can assist in defining the proper balance of molecular-weight distribution to enhance extrusion operations. The molecular-weight distribution (MWD) is defined by various molecular-weight averages or moments of the MWD. Figure 18 shows a typical MWD curve with the important moments defined as follows:

$$\overline{M}_n = \frac{\Sigma\, N_i M_i}{\Sigma\, N_i} \tag{46}$$

$$\overline{M}_w = \frac{\Sigma\, W_i N_i}{\Sigma\, W_i} \tag{47}$$

where N_i, M_i, and W_i are the number molecular weight and weight fraction of the ith species.

The high ends of the distribution are characterized by the z and $z + 1$ averages:

$$\overline{M}_z = \frac{\Sigma\, N_i M_i^3}{\Sigma\, N_i M_i^2} \tag{48a}$$

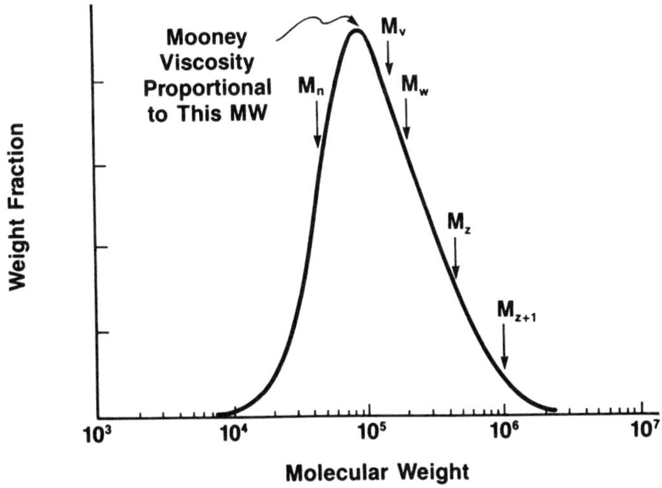

Figure 18 Important features of a typical molecular-weight distribution curve for a polymer.

$$\overline{M}_{z+1} = \frac{\Sigma N_i M_i^4}{\Sigma N_i M_i^3} \tag{48b}$$

$$[\dot{\eta}] = KM_v^{\alpha} \tag{49}$$

The Mark–Houwink relationship of viscosity to molecular weight states that where $[\dot{\eta}]$ is the limiting viscosity number (or inherent viscosity), M is the molecular weight, and K and α are constants which depend on the polymer–solvent system in use. $\overline{M}_v$ is defined as the viscosity average molecular weight, which can be approximated by (Slade, 1975):

$$\overline{M}_v = \alpha \frac{\Sigma N_i M_i^{1+\alpha}}{\Sigma N_i M_i} \tag{50}$$

with α a constant that is temperature and solvent dependent. From Figure 18 it is seen that:

$$M_n < M_v < M_w < M_z < M_{z+1} \tag{51}$$

Indexes of MWD such as M_w/M_n, M_z/M_w can be used to compare different polymers on a relative basis.

The importance of these moments or indexes to extrusion can be studied with small-scale rheological testing of compounded stock. In the elastomers industry the Mooney viscometer and the Monsanto processability tester (MPT) are widely used to characterize both neat rubber and compound rheological properties.

The Mooney viscometer measures torsional flow between a stationary and rotating disk to determine the shear stress at the outer radius of the disk, and, as normally applied in product quality control, provides a single-point viscosity measurement at one shear rate and temperature. In general, Mooney viscosity can be used to show the overall effect of MWD on the bulk polymer viscosity. Figure 19 shows a plot of compound Mooney viscosity versus gum Mooney viscosity. Curve A polymers have M_w/M_n averages of about 2.0, whereas the lower compound viscosity polymers (curve B) have $3 < M_w/M_n < 4$. Figure 19 implies that broader-MWD polymers process (i.e., mix, mill, extrude) better than narrow-MWD polymers. This is a generality, however, and can be better defined in terms of the moments of the MWD.

The relationship between compound viscosity and the higher moments of the distribution are illustrated by a plot of the compound Mooney viscosity versus M_z/M_n in Figure 20. This plot illustrates that the high ends of the distribution can play an important role in compound rheology.

A capillary rheometer also provides a means of studying MWD effects on rheology, over a wide range of shear rate conditions, which more closely approximate extrusion operations. The Monsanto processability tester (MPT) is widely

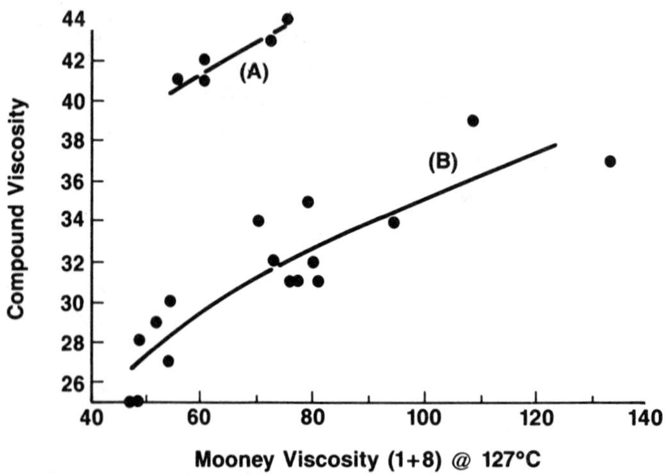

Figure 19 A plot of compound versus gum polymer Mooney viscosity.

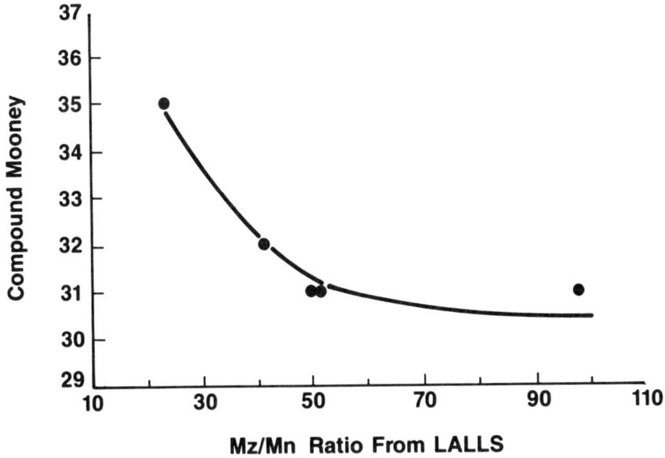

Figure 20 Effect of the high ends of the molecular-weight distribution on the compound Mooney viscosity. All polymers have approximately the same neat Mooney viscosity and were compounded in the same carbon black formulation. (LALLS, low-angle laser light scattering.)

used to assess the flow and elastic response of various compound stocks. This instrument is an automated capillary rheometer equipped with a data acquisition system which can record shear stress and running/relaxed die swells at shear rates ranging from 1 to $10^4\ s^{-1}$. Shear rates in commercial extruders are typically in the region from 30 to 1000 s^{-1}. Shear stress can be correlated in the usual manner with a power law expression

$$\tau = k\dot{\gamma}^n \tag{52}$$

where k and n are the so-called consistency and pseudoplasticity indexes. The consistency index is calculated from the intercept on the τ axis at unit shear rate. From Newton's law of viscosity,

$$\tau = \eta_a\dot{\gamma} \tag{53}$$

The apparent viscosity can be related to coefficients k and n by combining Eqs. (52) and (53):

$$\eta_a = k\dot{\gamma}^{n-1} \tag{54}$$

Figure 21 shows typical shear rate-viscosity flow curves on a log-log plot for two different polymers in both gum and compounded stocks at 100°C. The broader MWD polymer is shown to have a lower compound viscosity over the entire range of shear rates than the narrower MWD sample, despite the fact that in the gum state, the polymer is much more viscous than the narrow MWD sample. At very high shear rates, the rheological flow curves tend to converge as illustrated by the compound viscosity data in Figure 21. Nonetheless, over the shear rate range of interest for extrusion applications, the broader polymer with a high molecular weight end is lower in viscosity, which directionally should translate into faster extrusion.

Sezna (1984) has shown that the k and n constants for compound stocks as measured in the MPT properly describe the rheological characteristics of extrusion. By relating k and n to MWD, some generalizations for selecting polymers for certain extrusion applications can be made. Table 1 reports typical k and n values for both gum and compound samples along with their MWDs. Data are shown for four samples in the same compound formulation, and one sample in two widely different compounds. In scanning the table, it is found that n is nearly constant despite significant variation in MWD and that it appears to be more of a function of the specific compound. The k-values, however, vary, implying a k-MWD and M_w dependency.

Figure 22 shows a plot of compound k-values versus M_z/M_w averages for

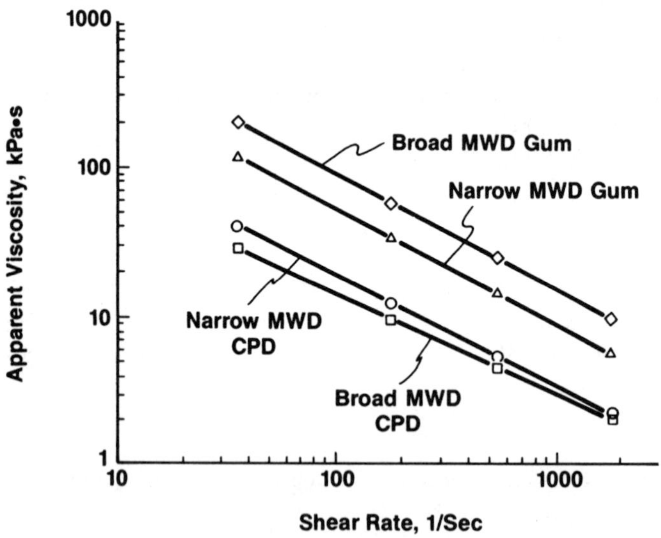

Figure 21 Effect of shear rate on viscosity of extrusion polymers at 100°C.

Table 1 Values of k and n for Gums and Compound Polymers

Mooney viscosity	MWD	$\overline{M}_z/\overline{M}_w$	Gum k	Gum n	Compound k	Compound n
53	3.1	16	2223	0.26	402.9	0.29
72	3.8	30	2296	0.26	804.8	0.24
73	3.4	12.4	3125.9	0.23	323.9	0.31
60	2.2	3.1	1844	0.23	571.2	0.26
60	2.2	3.1	1844	0.23	148.6	0.16

several polymers with approximately the same M_w/M_n. This illustrates that compound viscosity can be significantly altered by varying the moments of the MWD. In this example the importance of the high end of the distribution is illustrated. The lower k-value signifies lower compound viscosity, which should result in a faster extrusion for the compound formulation studied.

As discussed, the processing characteristics of EPDM and other elastomers depend on contributions from three main areas: compositional properties, molecular-weight properties, and the presence of long-chain branching. In terms of composition, when the ethylene content in the polymer is generally above 55 wt %, measurable levels of crystallinity can be detected using differential scan-

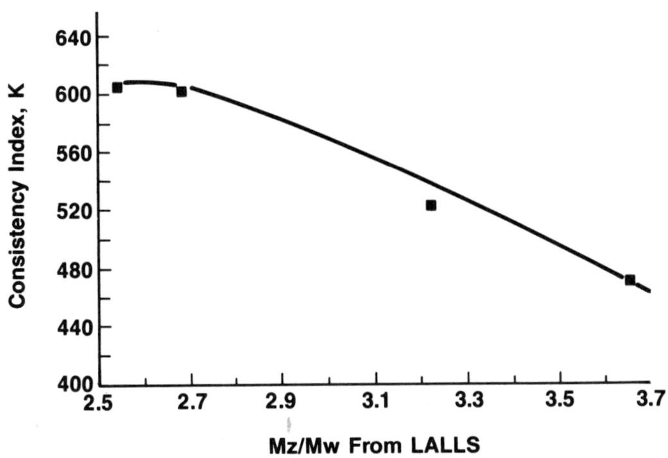

Figure 22 Dependency of polymer flow index on M_z/M_w ratio.

ning calorimetry (DSC) techniques. Crystallinity can play an important role in extruded articles. Toughness and green strength in compounds are generally derived from crystallinity, which can establish good feeding characteristics in cold feed extruder operations. In addition, compound stocks derived from crystalline-like polymers tend to be more thermoplastic, which enhances mass flow. Finally, extruded profiles typically have better size control and dimensional stability.

Narrow-MWD polymers tend to demonstrate higher throughput rates in extrusions, with extrudates typically having smooth surfaces. Note, however, that the absence of high-molecular-weight ends results in a compound stock having little guts; hence these materials generally feed poorly. Narrow-MWD polymers are sometimes preferred because they cure faster than broader polymers; hence the poor feed qualities of the narrow MWD polymers are overcome somewhat by introducing semicrystallinity. In contrast, broad-MWD polymers have better extruder feeding characteristics; however, they tend to show higher die swell.

The chemistry of long-chain branching (LCB) is complex and theoretical discussions are given by Flory (1953). During polymerization, propagated by free radicals, a chain may occasionally enter into a chain transfer reaction with a polymer molecule. The growing chain is terminated and the reactivated polymer molecule proceeds to grow a new chain attached to the unit involved in the chain transfer. The resulting branched polymer may subsequently undergo activation at another of its units, and hence another chain is added to the molecule at this point. Repetition of this process may yield highly branched molecules. Polymers with low levels of LCB tend to be more shear-thinning, and hence will exhibit a lower viscosity when subjected to stress. The net result is faster extrusion. Hence, from a processing standpoint a controlled amount of branching can be favorable.

The presence of branching can be detected by coupling the techniques of gel permeation chromatography (GPC) with laser light scattering in a multidetector apparatus for molecular-weight characterization measurements. This technique enables comparison of linear model molecules with their branched counterparts. At a given elution time, the quantity

$$g = \left(\frac{\mathrm{MW}_{\mathrm{linear}}}{\mathrm{MW}_{\mathrm{branched}}}\right)^{1+\alpha} \tag{55}$$

where α = Mark–Houwink exponent for the solvent used, can be related to the degree of branching. The technique is described by Baldwin and Ver Strate (1972), Krevelen (1972), Kresge et al. (1984), and Scholte (1983). A crude branching index can therefore be defined from the ratio of weight average molecular weights as measured by GPC and low-angle laser light scattering (LALLS).

Table 2 Summary of Data for Extrusion Study

Polymer	K	n	M_w/M_n	M_z/M_w	$M_w \times 10^{-5}$	Branch index
A	589.5	0.25	2.22	1.8	1.7	1.17
B	605.9	0.29	2.69	2.53	1.94	1.01
C	602.73	0.25	2.78	2.67	1.895	1.04
D	661.53	0.25	2.87	2.69	2.002	1.16
E	983.24	0.25	3.51	2.73	2.525	1.3
F	822.49	0.23	2.87	2.77	2.005	1.07
G	832.6	0.24	4.04	2.92	3.402	0.96
H	957.4	0.22	3.83	3.11	3.1	0.94
I	523.46	0.26	2.7	3.22	1.587	0.83
J	607.88	0.25	2.85	3.53	1.899	0.92
K	842.9	0.21	3.64	3.64	2.774	0.98
L	472.2	0.3	2.67	3.65	1.837	0.78
M	323.9	0.32	3.35	4.17	2.086	0.9
N	415.2	0.27	3.12	4.7	1.661	0.94
O	458.8	0.27	3.24	5.76	2.059	0.84

To illustrate the coupling of molecular properties to rheology and then to the ultimate extrusion process, 15 polymers were synthesized in the Exxon laboratories and a capillary rheometer was used to determine the flow and elastic responses of their compounded stocks over a range of shear rates anticipated in the extruder. All stocks were prepared in the same applications formula using IRB#5 carbon black at about 30%. Shear rates ranged from 30 to 180 s^{-1} and measurements were made at three temperatures (75°C, 100°C, 125°C). These temperatures covered the wall temperature ranges anticipated for the extruder. Table 2 provides a summary of the power law model rheological coefficients obtained at 100°C along with the molecular weight and branching characteristics of each polymer. A stepwise regression of k-values (using log values) as a function of M_w/M_n, M_z/M_w, M_w, and branching index was performed and the results are tabulated in Table 3. As shown, k correlates with the weight average molecular-weight distribution (M_z/M_w). Usually when branching indices are larger than or equal to 0.9, it is believed that no branching exists; hence the lack of k-dependency is not surprising. The correlation derived from this empirical fit of the data is

$$k = \exp(6.386)(M_z/M_w)^{-0.558}(M_w \times 10^{-5})^{0.923} \tag{56}$$

Table 3 Spreadsheet Regression Results: Run Summary

Proc #	Proc Name	Result
1	STEP	ok

Proc01: Step

Dep.Var:	LLN(K)	R-Square:	67.037
File :	REGRESS	adj R-Sq:	61.543
Range:		St.Error:	0.199
3.00	Variables	F-Stat :	12.202
15.00	Observations	Dur/Wat :	1.811
		Det. :	1.000

Stepping Report:		adj R-Sq.
Step 1:	Add LN(Mw)	38.7333
Step 2:	Add LN(Mz/Mw)	61.5430
Step 3:	Add LN(INDEX)	61.4801
Step 4:	Del LN(INDEX)	61.5430

Var Name	Coeff	t-Stat	St.Err	Beta
Intercept	6.386			
LN(Mz/Mw)	-0.558	-2.951	0.189	-0.489
LN(Mw)	0.923	3.936	0.234	0.652

Actual K-Value	Calcltd. K-Value	Residual	Percent Deviation
589.50	697.70	-108.20	-18.35
605.90	651.78	-45.88	-7.57
602.73	618.93	-16.20	-2.69
661.53	648.42	13.11	1.98
983.24	796.73	186.51	18.97
822.49	638.78	183.71	22.34
832.60	1010.44	-177.84	-21.36
957.40	895.31	62.09	6.49
523.46	473.32	50.14	9.58
607.88	530.67	77.21	12.70
842.90	740.11	102.79	12.19
472.20	505.15	-32.95	-6.98
323.90	527.35	-203.45	-62.81
415.20	399.74	15.46	3.72
458.80	435.10	23.70	5.16

Figure 23 shows a parity plot of the calculated versus the measured k-value, indicating that the correlation is good to within $\pm 20\%$.

The power law exponent n in this example appears to be nearly constant (refer to Table 2). For EPDMs it appears to be a strong function of branching.

The consistency index varies with temperature according to the Arrhenius-type relationship:

$$k = \alpha \exp\left(\frac{E}{T - T_0}\right) \tag{57}$$

where

T, T_0 = measurement and reference absolute temperatures, respectively
E = activation energy of flow

To summarize to this point, we can relate the MWD of a polymer through an empirical relation, Eq. (56), to its power law coefficients, and the rheology–temperature relation can be determined by Eq. (57).

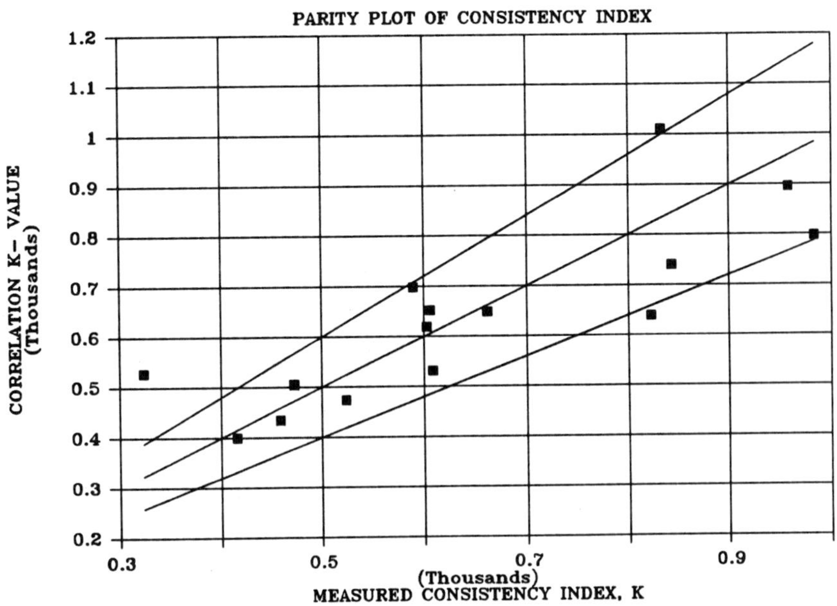

Figure 23 Parity plot of consistency index k-correlation.

The elastic response of the test stocks in a capillary rheometer must also be measured in terms of a Bagley correction experiment. Using the method described in Chapter 2, the Bagley correction is calculated by comparing capillary pressures at the same shear rate for two or more dies with different L/D ratios. The entrance correction term g can be computed from

$$g = \frac{[P_2(L/D)_1 - P_1(L/D)_2]}{P_1 - P_2} \tag{58}$$

where

P_1 = pressure at one shear rate using a die with $(L/D)_1$
P_2 = pressure at same shear rate using a die with $(L/D)_2$

This correction accounts for the additional pressure developed as the stock is accelerated into a smaller die opening. Using this entrance correction, the true shear stress at a die opening is

$$\tau_w = \frac{P}{4(L/D + g)} \tag{59}$$

Figure 24 shows a plot of g versus the effective shear rate $\dot{\gamma}$, generated in the capillary die.

We may now relate the preceding information to extruder operation by restating the one-dimensional extruder expressions in terms of the apparatus's configuration. The extrusion data reported in the discussions that follow were generated in a 2.5-in.-diameter, single-screw extruder that was 69 in. in length and equipped with an adjustable slit die. The extruder was equipped with three-zone heating and although conditions were set to maintain close to isothermal operation, conduction-loss calculations indicated that the barrel was typically 10 ~ 15°C cooler than the die. Head pressures and temperatures were measured with dual thermocouple–transducer sensors. Linear mass rates were measured using the weight–time method at near steady-state conditions as indicated by machine torque measurements. Relaxed die swell measurements were made on extrudate samples.

The shear rate through the slit die can be approximated from the analysis of flow between parallel plates to give

$$\dot{\gamma} = \frac{Q}{WH^2} \tag{60}$$

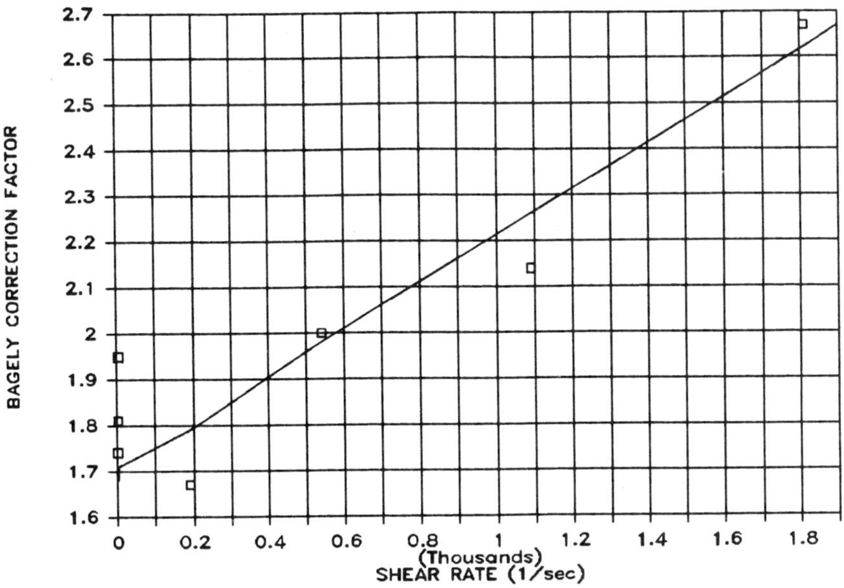

Figure 24 Bagley correction term as a function of shear rate.

where

Q = volumetric flow rate
W = die width opening
H = die height opening

The shear stress at the die is

$$\tau = \frac{P}{4(L/D + g)} \tag{61}$$

Using the definition that $\tau = k\dot{\gamma}^n$ in Eq. (61) and solving for the extruder head pressure gives

$$P = 4\left(\frac{L}{D} + g\right)k\left(\frac{Q}{WH^2}\right)^n \tag{62}$$

To solve Eq. (62) for P and Q, a second expression is needed. Considering the cold feed extruder as a melt conveying screw of simple geometry, the apparent flow through the apparatus then reflects two types of flow, namely, drag flow (i.e., no restriction at the die, Q_d) and pressure flow, Q_p. Q_d represents the maximum theoretical output of the extruder, which is reduced by the head pressure developed at the die (Q_p). Therefore, the actual throughput is

$$Q = Q_d - Q_p \tag{63}$$

As discussed earlier in this chapter, the drag flow contribution can be approximated simply by

$$Q_d = AN \tag{64}$$

where

$$A = [\pi^2 D^2 H'/120] \sin \phi \cos \phi \tag{65}$$

N = screw speed
D = barrel diameter
H' = minimum flight depth
ϕ = helix angle of screw (17.6°)

For the extruder used in the experiments, $A = 2.87 \times 10^{-4}\ m^3$-min/s.

The pressure flow contribution can be estimated from

$$Q_p = B\left(\frac{P}{\eta_s}\right) \tag{66}$$

where

$$B = [\pi D H'^3/12L] \sin^2 \phi \tag{67}$$

P = head pressure
η_s = viscosity of the stock in the barrel

For the extruder, $B = 2.62 \times 10^{-6}\ m^3$.

The viscosity of the stock in the barrel is based on the channel apparent shear rate:

$$\dot{\gamma}_a = \frac{\pi DN}{60H'} \tag{68}$$

which at the mean barrel wall temperature is

$$\eta_s = k\, \dot{\gamma}_a^{n-1} \tag{69}$$

where k depends on temperature and the molecular weight and molecular weight distribution of the polymer.

The output of the extruder with a die can now be estimated by

$$Q = AN - B\left[\frac{P}{\eta_s}\right] \tag{70}$$

Equations (70) and (62) can be solved by the method of successive approximations. The iterative calculation procedure is as follows:

1. Input data are the MW (molecular weight) and MWD (molecular-weight distribution), extruder conditions (barrel and screw temperature, screw speed). The rheological coefficient k can be estimated from the empirical equations (56) and (57). The power law exponent n in this example is assumed a constant, reflecting the average value for the samples.
2. Compute the channel shear rate from Eq. (68) using the specified N. Using the appropriate k-value, obtain an estimate of η_s from Eq. (69).
3. Make an initial estimate of Q and
   a. Compute the die shear rate, Eq. (60).
   b. Compute the head pressure at the die, Eq. (62).
   c. Use Eq. (70) to compute the head pressure based on N, Q_{est}, and η_s for the barrel section.
4. If the computed head pressures agree to within some specified tolerance (e.g., 2%), then the estimate of Q is correct. If not, then select a new Q and repeat the calculations until convergence is achieved.

Model predictions of the pressure drop for polymer A (Table 2) are compared to measurements in Figure 25 in terms of a plot of head pressure versus screw speed. As shown, the model predictions are in good agreement with experimental data for the majority of experiments. The scatter in measured head pressure and the low values at higher screw speeds suggest problems with the transducer accuracy.

Model predictions of the linear extrudate rate for three of the polymers listed in Table 2 are shown in the parity plot of Figure 26. As shown, the model provides good predictions for polymers widely different in molecular-weight distribution. In fact, the more accurately the molecular-weight distribution is characterized, the more precise are model predictions. Figure 27 shows model predictions of the volumetric throughput as a function of screw speed for two polymers. The comparison to measured values in this figure illustrates that the simple model is capable of quantitative predictions.

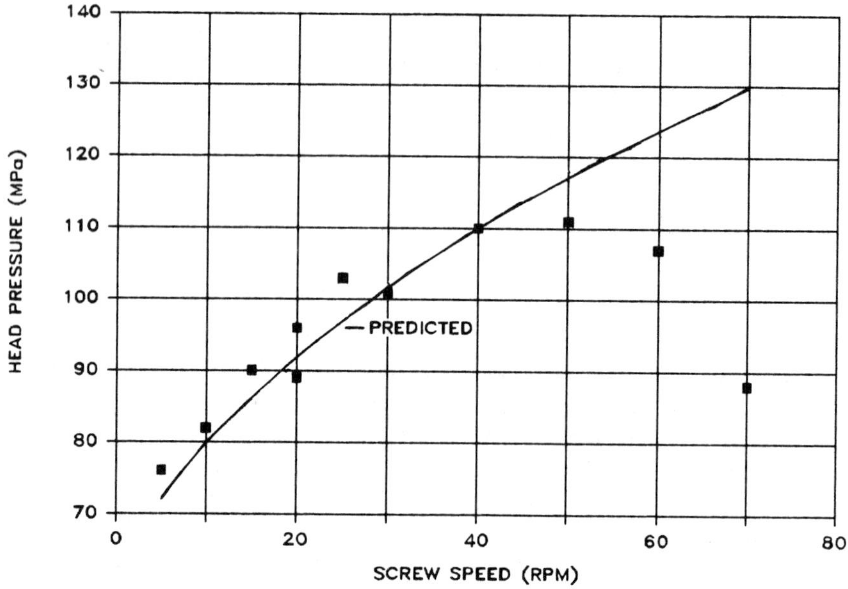

Figure 25 Model predictions (solid line) of head pressure are compared to measured values.

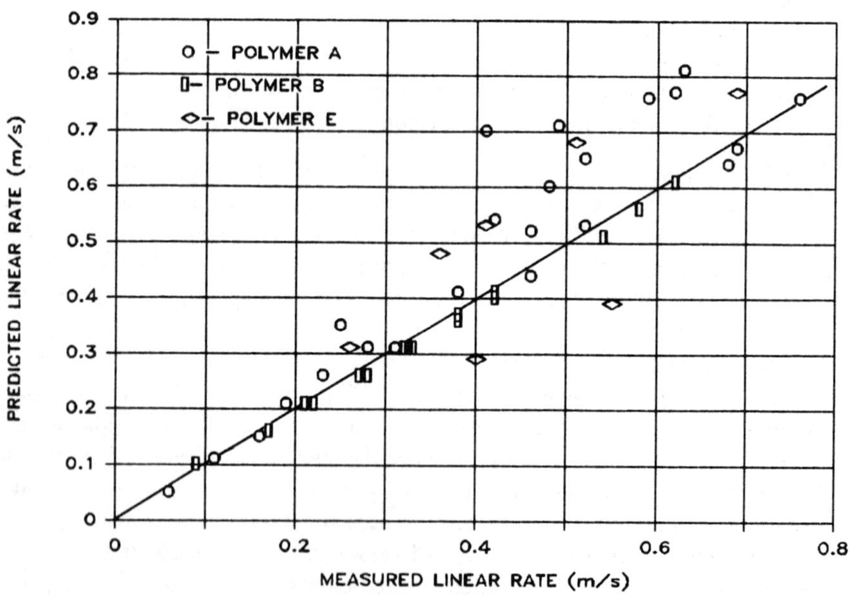

Figure 26 Parity plot showing model properly predicts linear throughput.

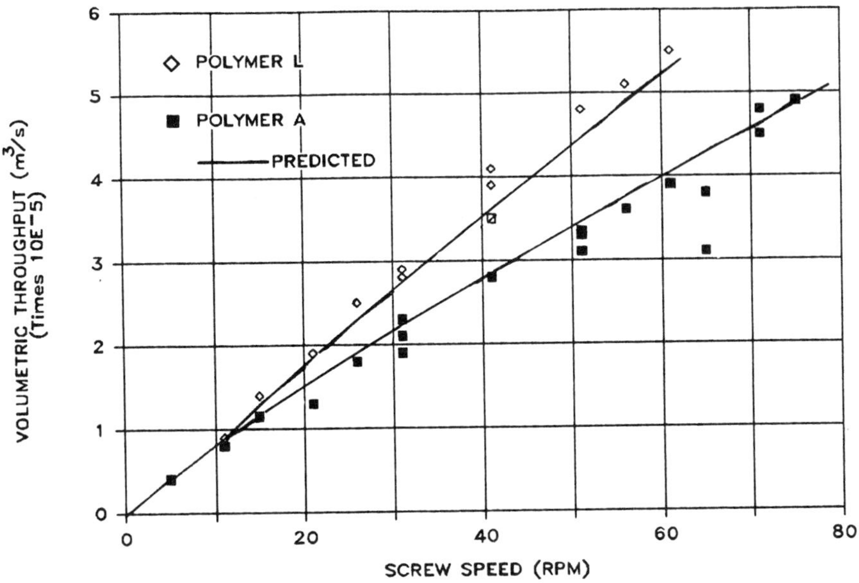

Figure 27 Model predictions of volumetric throughput for two polymers.

The manufacturer of extruded articles not only must define and control his throughput capacity, but must also control the article's swell. In rubber applications dies are usually cut to accommodate the article's tolerances based on the characteristics of a single polymer supply or fixed blend production, the reason for this being largely cost; that is, dies, particularly complex geometries such as cross-head dies, are simply too expensive to retool to accommodate different polymer supplies. Consequently, a common problem among elastomer manufacturers in attempting to qualify their products as a second source of supply or even in introducing new products to extruded article manufacturers is to provide a polymer having proper swell characteristics. Again, capillary rheometer data can provide some guidance as a predictive tool in designing polymer structures and/or selecting polymers for a particular application. A very simple, empirical approach is to correlate capillary rheometer swell measurements with a power law relation:

$$\text{Swell} = k'\dot{\gamma}^{n'} \tag{71}$$

Table 4 provides tabulated data on the swell indices k' and n' along with molecular-weight properties of the various polymers tested. An additional pa-

Table 4 Summary of Swell Index Data Generated in Capillary Rheometer

Polymer	n'	K'	M_w/M_n	M_z/M_w	M_w	Index	IV
A	0.08	11.44	2.22	1.80	1.70	1.17	2.28
B	0.12	14.13	2.69	2.53	1.94	1.01	
C	0.06	12.67	2.78	2.67	1.90	1.04	2.46
D	0.05	13.26	2.87	2.69	2.00	1.16	2.47
E	0.08	15.20	3.51	2.73	2.53	1.30	3.03
F	0.05	15.56	2.87	2.77	2.01	1.07	2.70
G	0.05	16.23	4.04	2.92	3.40	0.96	2.98
H	0.05	16.49	3.83	3.11	3.10	0.94	2.78
I	0.13	10.74	2.70	3.22	1.59	0.83	
J	0.05	12.28	2.85	3.53	1.90	0.92	2.30
K	0.06	15.57	3.64	3.64	2.77	0.98	2.92
L	0.09	14.58	2.67	3.65	1.84	0.78	
M	0.10	13.30	3.35	4.17	2.09	0.90	2.39
N	0.10	9.15	3.12	4.70	1.66	0.94	1.99
O	0.11	9.43	3.24	5.76	2.06	0.84	2.06

rameter, IV (intrinsic viscosity), is also tabulated. The intrinsic viscosity is a solution viscosity normally measured in a solvent such as decolin and can be viewed as a crude measure of molecular weight. A multivariable regression of the data in Table 4 reveals that k' is directly proportional to IV:

$$k' \sim \mathrm{IV}^{1.357} \tag{72}$$

The results of the regression are summarized in Table 5, and the correlation is shown in Figure 28. The correlation offers no surprise; as expected, higher-molecular-weight polymers generally tend to swell more (i.e., they have higher k'-values). Index n' does not show any dependency to any of the parameters tabulated in Table 4; hence an average value representative of the polymers studied can be used.

Using the effective shear rate for the extruder die, Eq. (72) was applied directly to estimating the extrudate swell, and comparisons to experiments are shown in Figure 29. The simple relationship provides a good qualitative prediction of the extruder operation and for polymer A actually provides quantitative predictions. Figure 30 shows a parity plot of predicted and measured extrudate

Table 5 Spreadsheet Regression Results

Proc01: Step

Dep.Var:	LLN(K')	R-Square:	90.341
File :	REGRESS	adj R-Sq:	89.376
Range:		St.Error:	0.065
2.00	Variables	F-Stat :	93.534
12.00	Observations	Dur/Wat :	2.858
		Det. :	1.000

Stepping Report:		adj R-Sq.
Step 1:	Add LN(Mw)	52.8098
Step 2:	Add LN(Mz/Mw)	71.0405
Step 3:	Add LN(IV)	87.0506
Step 4:	Del LN(Mw)	88.4285
Step 5:	Del LN(Mz/Mw)	89.3755
Step 6:	Add LN(INDEX)	89.2526
Step 7:	Del LN(INDEX)	89.3755

Var Name	Coeff	t-Stat	St.Err	Beta
Intercept	1.329			
LN(IV)	1.357	9.671	0.140	0.950

ACTUAL K'	CALCLTD. K'	RESIDUAL	PERCENT DEVIATION
11.44	11.56	-0.12	-1.03
12.67	12.81	-0.14	-1.13
13.26	12.88	0.38	2.83
15.20	17.00	-1.80	-11.85
15.56	14.54	1.02	6.56
16.23	16.62	-0.39	-2.42
16.49	15.13	1.36	8.27
12.28	11.70	0.58	4.75
15.57	16.17	-0.60	-3.85
13.30	12.32	0.98	7.36
9.15	9.61	-0.46	-5.03
9.43	10.07	-0.64	-6.80

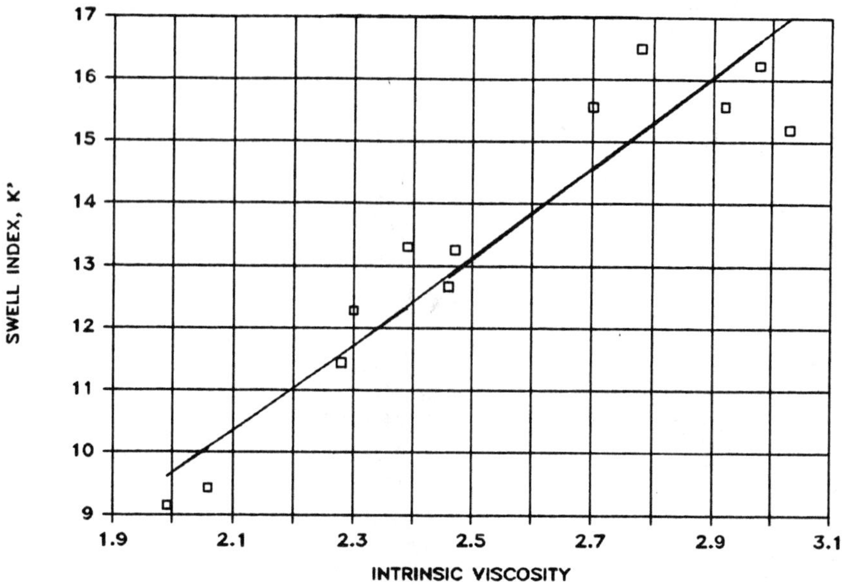

Figure 28 Relationship of swell index to intrinsic viscosity for 100°C data.

swells for several of the polymers listed in Table 4. Although the relationship of Eq. (72) does not provide highly accurate predictions, it in fact provides a good relative comparison between different polymers. As a first-pass approach to selecting a polymer for an intended extrusion application, Eq. (72) certainly provides some guidance.

Sezna (1984) provides a more elegant approach to predicting die swell using data generated on the MPT (Monsanto processability tester). As noted by Sezna, the extruder die may have a different die geometry, and hence a different stress history is applied to the stock. Also, the stock extruded through a capillary is much smaller than the extrudate discharging from an extruder; thus the thermal histories of the stocks can be very different. These differences in cooling rates lead to differences in stresses, resulting in different die swell levels. Sezna regressed the MPT die swell data to a general quadratic equation with temperature, shear rate, and die L/D ratio as the independent variables. The shear rate used for predicting die swell was based on the effective shear rate in the extruder, which was estimated as $\dot{\gamma} = 32Q/\pi d^3$. The general form of the die swell expression obtained is

$$\text{Swell} = k_0 + k_1\dot{\gamma} + k_2T + k_3(L/D) + k_4\dot{\gamma}^2 + k_5T^2 + k_6(L/D)^2 + k_7(\dot{\gamma}T) + k_8[\dot{\gamma}(L/D)] + k_9[T(L/D)] \quad (73)$$

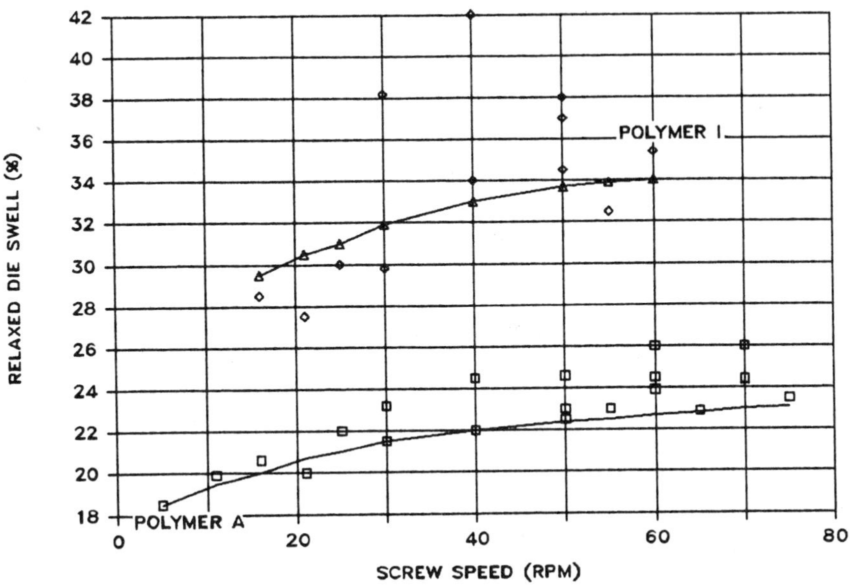

Figure 29 Extrudate swell increases with screw speed (i.e., shear rate). Solid lines are predictions from Eq. (72); symbols are measured swell.

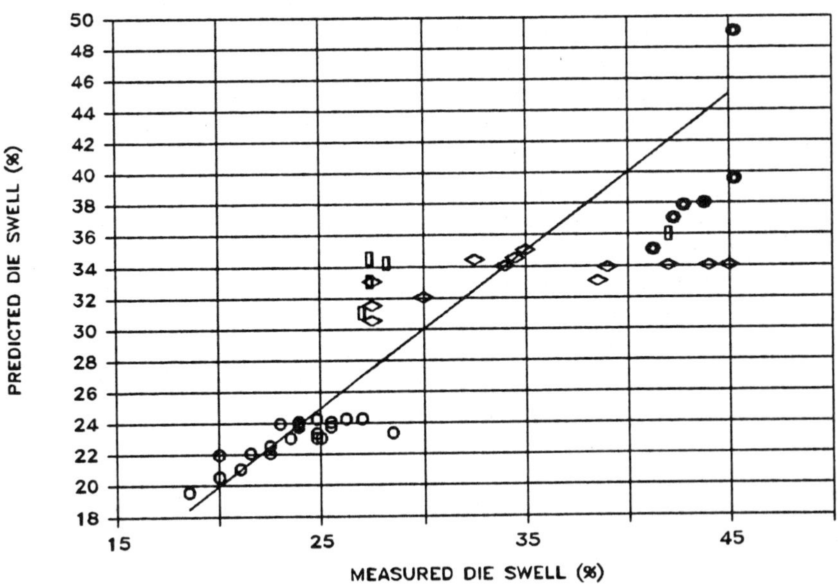

Figure 30 Parity plot comparing predicted and measured extrudate swells for several polymers.

TWIN-SCREW EXTRUSION

There are a variety of twin-screw extruder designs employed throughout the polymers industry, with each type having distinct operating principles and applications in processing. Designs can be generally categorized as corotating and counterrotating twin-screw extruders. Eise et al. (1981) has classified twin-screw extruders in terms of the mechanisms of operation, and the same terminology is adopted for the present discussion.

Different arrangements of twin-screw systems are illustrated in Figure 31. The figure illustrates differences between corotating and counterrotating as well as between intermeshing and nonintermeshing screw arrangements. Note the difference between fully intermeshing and partially intermeshing systems and between open- and closed-chamber types. For example, a counterrotating screw shown in Figure 31 (type 1) is an axially closed, single-chamber pumping system, whereas a corotating screw (type 4) is an axially open mixing system relying on drag forces.

A screw system open in axial direction is actually open in the longitudinal direction of the screw channel since it has a passage from the inlet to the outlet of the apparatus. This means that material exchange can take place lengthwise along the channel. In a closed arrangement the screw flights in the longitudinal direction are closed at intervals. It is important that the cross section of the screw channel be open in order for material exchange to take place from one flight to the other in a direction normal to the screw channel. There is usually some leakage over the screw crests and through the areas required for mechanical clearances. Whether the screws are open lengthwise or crosswise, or have a closed geometry, has a direct effect on conveying conditions, mixing action, and the pressure buildup capacity of the system. Nonintermeshing systems, for example, are open lengthwise and crosswise (refer to Figure 31). Fully intermeshing, counterrotating systems can be closed lengthwise and crosswise, and regardless of mechanical clearances they can develop closed chambers. This is the case with screw pumps. Lengthwise and crosswise closed, fully intermeshing corotating systems (Figure 31, type 2) are theoretically impossible, as is a lengthwise open and crosswise closed counterrotating system.

Fully intermeshing, corotating screws are open lengthwise. When normal screw flights are employed they are closed crosswise, and when staggered screw discs are used they are open crosswise. This design is illustrated by types 4 and 6 in Figure 31.

There are several types of partially intermeshing screws, e.g., lengthwise open and crosswise closed systems and lengthwise and crosswise open systems. Most commercial intermeshing counterrotating screw extruders are combinations of Systems 1, 9a and 9b, shown in Figure 31. Commercial nonintermeshing, counterrotating screws correspond to System 11, and are used along with a single-screw discharge extruder.

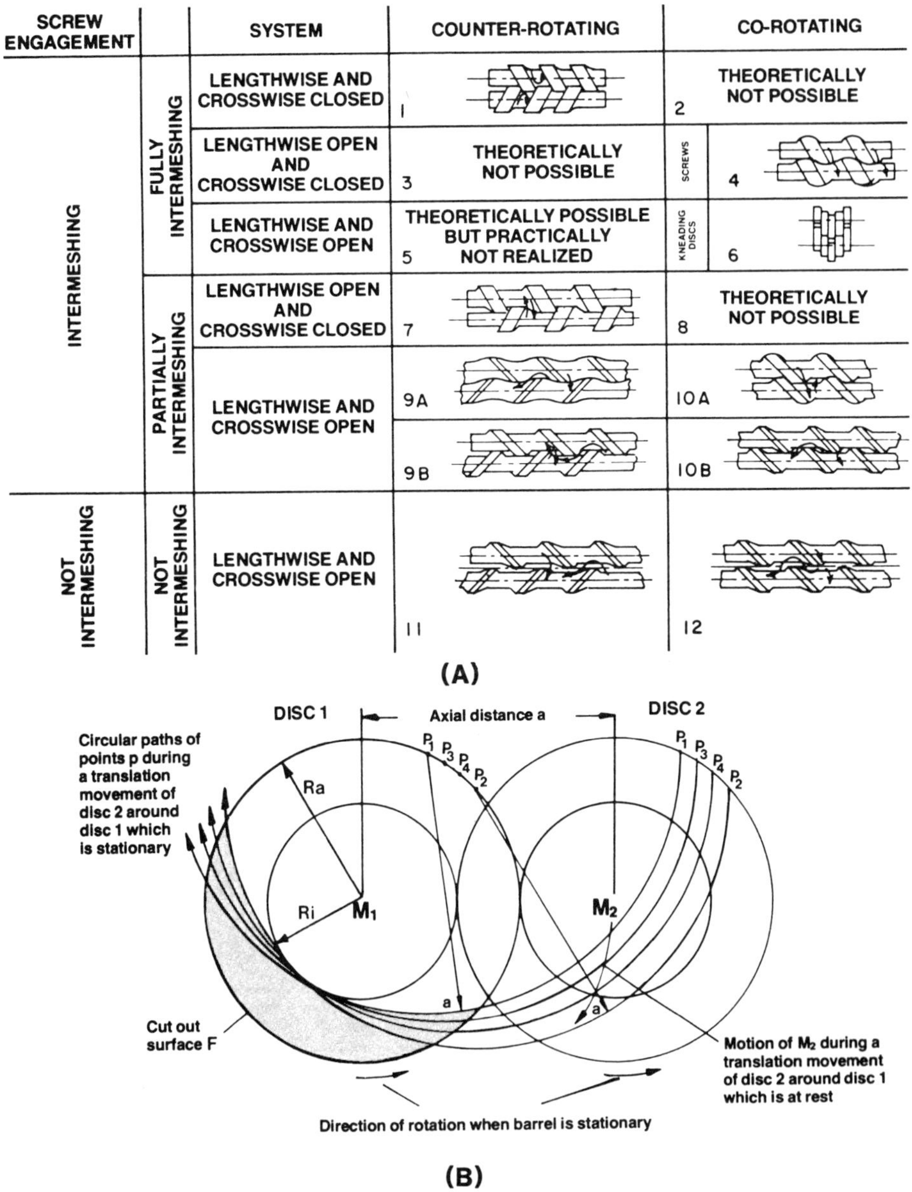

Figure 31 (A) Different twin-screw arrangements. (B) Movement principle of a co-rotating screw.

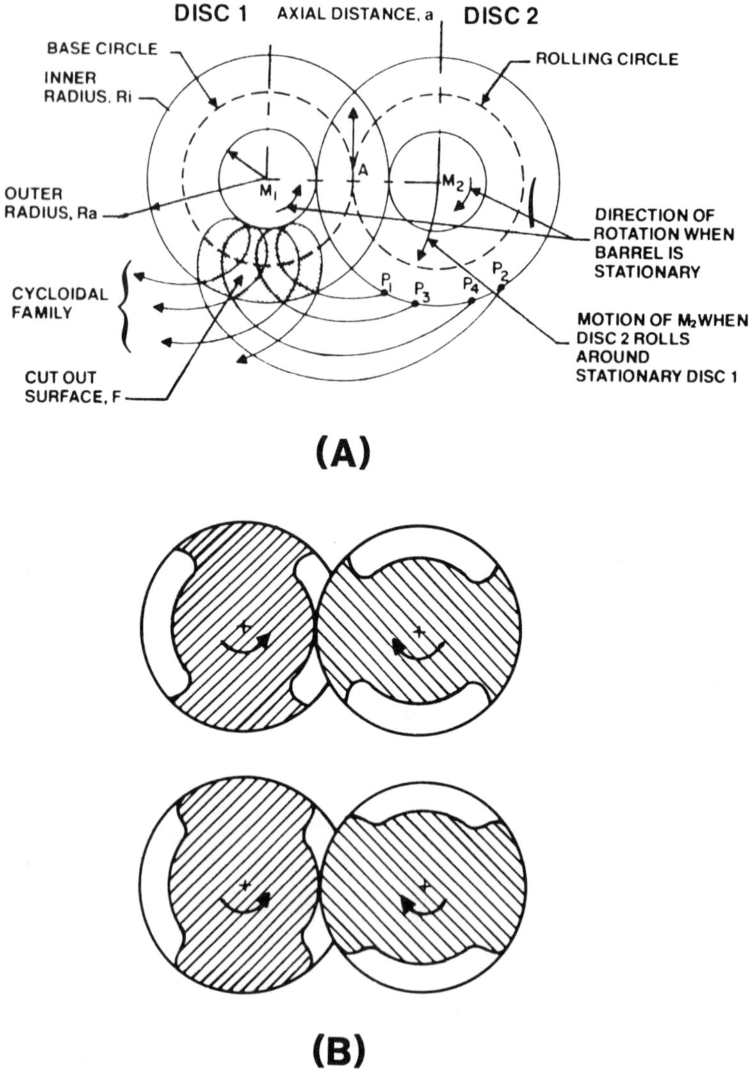

Figure 32 (A) Vector diagram of a counterrotating screw. (B) Cross section of counterrotating screws.

Counterrotating screws operate with both shafts turning with equal angular velocity in a counteracting rotation. Figure 32 shows a vector diagram illustrating the velocity fields and forces of a counterrotating screw operation. In contrast, with corotating screws as shown in Figure 31, when screw 1 is held stationary, screw 2 will not roll off, but will make a circular translation motion around screw 1. The center point M_2 moves in this fashion on a circular path around M_1, with the radius M_{1MPV2}; screw 2 itself produces no rotation of its own. Each point describes a circular path with the same radius. If the arc P1–P2 on the circumference of disc 2 is a screw crest, the points of this arc describe (as shown with two intermediary points P3 and P4, Figure 31), a family of arcs which cut out a halfmoon-shaped surface area F from screw 1.

Each point along the axial screw surface is tangentially wiped by the other screw with a constant relative velocity. For simplicity, consider the screw simply as a flat disc as shown in Figure 32. If disc 1 is held stationary, disc 2 with rolling circle radius AM_2 rolls off the base circle with the radius M_1A of disc 1. Each point of disc 2 describes an elongated or abbreviated epicycloid depending on its distance from M_2. Let the arc P1–P2 of the circumference of disc 2 define a screw crest; then the points of this arc represented with two intermediary points (P3 and P4) describe a family of cycloids which "cut out" the surface area F from disc 1. The shape of the cutout area F corresponds to the crosscuts of the screw channel perpendicular to the turning axis. Figure 32B illustrates two examples of these screw crosscuts. The forward rotation of the thin discs in a direction lengthwise to the shaft axis results in screws with positive and negative pitch. Each point of the surface area will be wiped by the other screw. The screw flanks are developed in the crosscut from a cycloidal family and both screws roll off one another in a fashion resembling that of gear wheels. In the wiping points, the relative velocity depends on the distance of these points from the rotating axis and is in the area of $-2\pi n(R_a - R_1) \leq V_{rel} \leq +2\pi n(R_a - R_1)$. In contrast, with counterrotating arrangements, the screws in the wedge area roll off one another in a rolling motion with locally different gliding velocities. This gliding velocity is small compared to the circumferential speed and is typically $+2n(R_a - R_1)$. Corotating screws do not roll off one another at any point, but instead have a translation movement, where one crest edge wipes a screw flank tangentially with equally high relative velocity. The relative velocity is high compared to the circumferential speed and is $V_{rel} = 2\pi n(R_a + r_1)$. In the wedge area the screws have equal pitch and equal pitch direction are intermeshing.

The open area in the wedge part between the two screws is different. In counterrotating designs, there are closed C-shaped chambers with a very small volume such that no material transfer from one screw to the other takes place. In this type of design the distributive mixing efficient is greatly reduced.

In contrast, corotating screws form V-shaped wedge areas that have four to five times more volume. This enables material to be transferred from one screw to the other, resulting in a renewal of material layers and surfaces. The net result is a higher degree of mixing.

Self-cleaning screws are often employed to prevent material from adhering to the screw root. Material that adheres to the screw root can degrade, because of a broad residence time distribution, and eventually fall off and be carried out with the product, showing up as contamination.

Self-cleaning action is achieved in both counterrotating and corotating screws, through an opposite roll-off motion or wiping motion. This self-wiping is achieved in different ways in each system, each having different degrees of effectiveness. With counterrotating screws, the roll-off process between the screw crest and screw root and between the screw flanks simulates the action of a calender. The necessary shear velocity required to wipe the boundary layers is proportionately lower because of the low relative velocity. Also, the material is drawn into the roller gap and is squeezed onto the surface.

In corotating screws, one crest edge wipes the flanks of the other screw with a tangentially oriented, constant relative velocity. There is a higher relative velocity in this arrangement, and hence there is a sufficiently high shear velocity available to wipe the boundary layers. The calender effect does not occur; however, a more efficient and uniform self-cleaning action is achieved.

The rolling motion of the screws in the counterrotating arrangement causes a calender effect which results in pressure between the roll nip surfaces. This

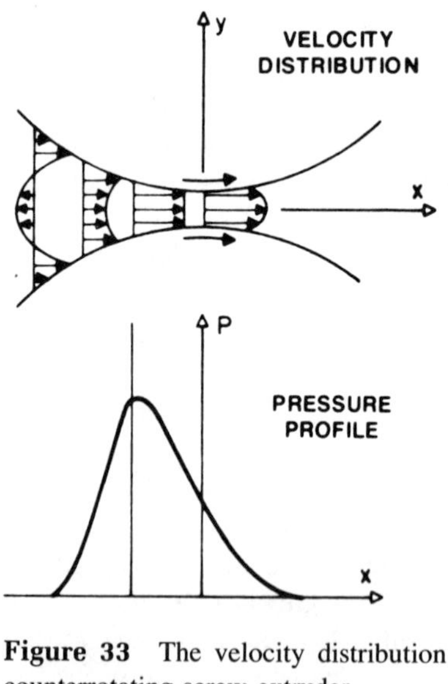

Figure 33 The velocity distribution and pressure buildup in the calendar gap of a counterrotating screw extruder.

pressure acts to push the shafts apart. Figure 33 illustrates the pressure and velocity distribution in the roller gap. The resultant parting forces on the two shafts can cause wear of metal surfaces to occur.

Wearing of the screw surface will increase proportionally to the relative velocity of the shafts. To minimize this effect intermeshing, self-wiping, counterrotating machines should be operated at relatively low speeds.

Since corotating screws have no calendering effect between the crest and the flank of the screws, there is considerably less wear. Therefore, corotating machines can be operated at much higher screw speeds with greater throughputs. A typical corotating, self-wiping, twin-screw extruder with 170-mm-diameter screws can process 9000–11,000 lb/hr of polypropylene at a screw speed of 300 rpm.

Some attention must be given to the design of the feed intake zone of twin screws. Figure 34 schematically shows the cross section in the feed zone of a counterrotating screw machine. To maximize feed intake the screws turn outward on top and inward at the bottom. If the screws turned inward at the top, the material would have to be pulled in by the calender gap and there would be little free volume available for the material to be drawn in. The polymer is conveyed to the lower wedge where it is partially compressed and conveyed as a unit volume toward the downstream melting zone. To maximize the intake free volume, multiflighted deep-cut screw elements are sometimes used. The conveying principle of the counterrotating screws is based on closed chamber unit volumes, which is desirable for feeding solids such as plastic resin pellets. The entire free volume can be utilized, assuming optimum pitch angle. It should be noted, however, that because of the calendering effect, solids are forced between the screws and crushed (refer to Figure 34B). This action causes the screw

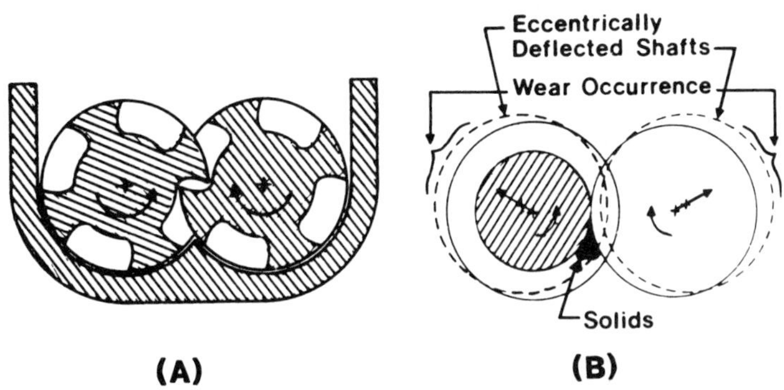

Figure 34 (A) Feed intake zone of counterrotating screw system. (B) Solids crushing in the wedge region of counterrotating screws.

elements to be forced apart in an angle at the top, where only the barrel can provide support, which can result in severe wear. The wearing action can be minimized if the gap between the screws is relatively large and if the screw speed is low. However, the conveying volume is proportional to the screw speed, and hence a low screw speed means lower throughput. Larger flank and root clearance measures can reduce the throughput.

In contrast, corotating screw elements convey the material from one screw toward the lower wedge (depending on wedge resistance); the material is compressed and then picked up by the other screw and conveyed further. This is illustrated in Figure 35. The wedge area provides a twist restraint on the resin, which reduces the tendency of the material to rotate in the channel and increases conveying capacity in the axial direction. The twist restraint in the wedge area varies with the number of flights per unit length. The larger the wedge angle, the larger the twist restraint and the greater the cross-section reduction in the wedge area during transfer of the material from one screw to the other. Screw bushings with wide crests can cause significant changes in direction of material flow, leading to larger twist restraints. High conveying angles can be achieved with fully intermeshing, self-wiping corotating screws. The conveying angle is the angle between the vertical line to the screw axis and the true conveying direction of the product. With single-screw machines the conveying angle can be calculated by analyzing the machine design parameters and the physical properties of the solids. Such relationships do not yet exist for corotating twin-screw extruders since the function of the wedge area has not been mathematically formulated. Therefore, experimental measurements of the conveying angle must be made clear. Figure 36 shows the conveying angle of a fully intermeshing, self-wiping corotating twin screw compared with a single-screw system in the processing of high-density polyethylene powder. Note that each time the pitch approaches $H = D_a$ a maximum is reached. Furthermore, the slope of the pitch in the corotating screws is lower than that of single screws. The conveying angle of

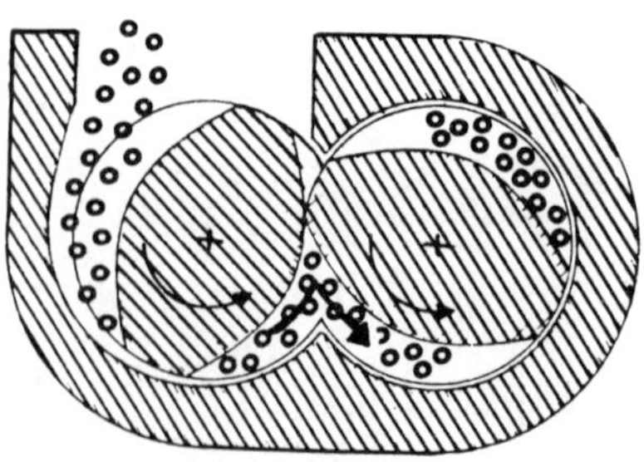

Figure 35 Feed intake zone of a corotating screw system.

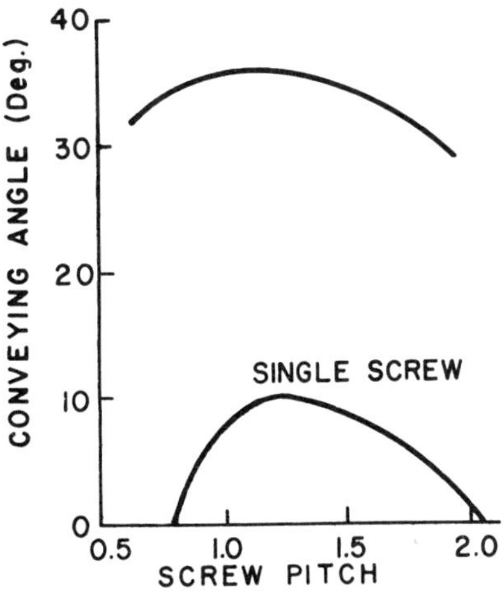

Figure 36 Comparison of the conveying angles of a twin screw and single-screw machine as a function of screw pitch *H*. Data are for HDPE reported by Eise et al. (1981).

fully intermeshing, self-wiping corotating screws is generally two to three times greater than that of a single-screw machine, as noted by Eise et al. (1981).

We now direct attention to the melt zone of the machine, which can be divided into two sections:

1. Compacting of the fluffy products in order to minimize air pockets. The air in most cases escapes through the feed opening. Simultaneously, a melt film is formed on the barrel walls and the resin solids fuse because of heat and pressure.
2. Generation of melt through the shearing section of the product and mixing of the unmolten particles with the already formed melt. As a result, a large heat exchange surface is generated between the solid particles and the melt.

With fully intermeshing, self-wiping counterrotating screws, the compaction of the material is achieved by several methods:

1. By continuously reducing the screw channel
2. Using various melt zones with different screw pitches and number of flights
3. By continuously increasing the crest widths
4. By continuously decreasing outer and/or inner diameters

In machines having a tight clearance between screws, a compression over several screw flights cannot be obtained nor will all of the air completely escape out of the individual chambers toward the feed. It is for this reason that counterrotating screws are constructed with loose clearances. This means that the advantages of the closed chambers offered by the counterrotating principle are ignored. The melting process is similar to that in single-screw machines, with the following actions occurring:

1. Development of a melt film on the barrel wall,
2. Scraping off the melt film by means of a pushing side of the flank,
3. Collection of melt in front of the pushing flank.

The rolling motion of the counterrotating screw produces the following additional effects:

4. In the wedge intake a melt whirlpool is created which draws in solid particles and softens them.
5. The plasticized mass is now taken in by the calender gap and collected at the trailing flank.

Because of the calendering effect in counterrotating screws, within one rotation the chamber volumes are axially displaced by distance H. One screw can roll off the other one time; therefore, only a fraction of the chamber volume passes the calender gap. In spite of the high axial velocity, a relatively low shear action of the calender gap is achieved, which is dependent on the low relative velocity $-2\pi n(R_z - R_i) \leq V_{\text{rel}} \leq +2\pi n(R_z - R_i)$ and the width of the calender gap. If the machine has tight clearances a high shear velocity results but only small amounts of the product are drawn into the gap. High clearance draws in a greater amount of the chamber volume, however, with a low shear effect. Increasing the screw speed results in displacement of the melting zone toward the discharge end, which can result in an inhomogeneous product. Additional mixing devices (e.g., interrupted flights) result in improved mixing and uniform heating of the product. In general, however, the plasticizing process is only marginally improved. Unmolten particles can often be observed in the discharge during such operations.

In corotating screws (fully intermeshing, self-wiping) the compression of the intaken solids is produced through restriction which is caused by the reversed screw (conveying in opposite pitch direction). These reversed screws are referred to as left-handed screws. Figure 37 shows several screw arrangements for the plasticating zone. In the left-handed screw restriction, pressure buildup is generated by the upstream right-handed screws. Plasticizing results from the back-up length generated by the left-handed screws. Melting of the resin is similar to that of a single screw, that is, generation of melt film on the barrel wall and sintering of the product; scraping off the film from the barrel wall by the pushing flank; and collection of a melt pool in front of the pushing flank.

Through proper specification of pitch angle and screw speed it is possible to shift the zero shear strain point of the axial flow components outside of the screw

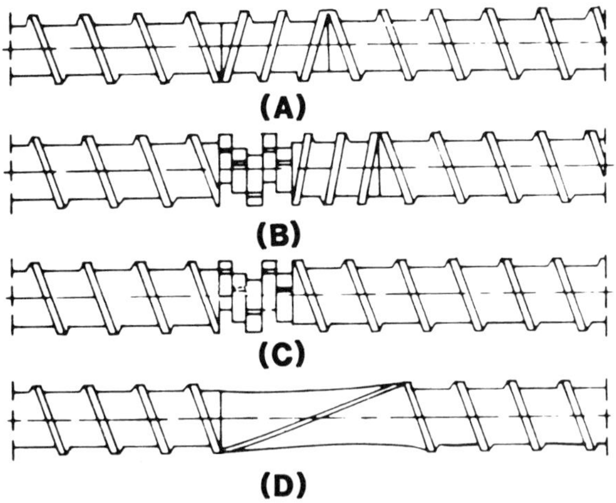

Figure 37 Various screw arrangements in the plasticating zone of a twin-screw, co-rotating machine.

channel. This can result in greater shear stress imposed on the suspended solids particles than is possible with counterrotating screws. In the wedge area the material is picked up by the other screw and does not undergo a calendering effect. The stresses are applied between the flanks by constant relative velocity, $V_{\text{rel}} = 2\pi n(R_a + R_i)$.

With the help of special mixing and shear elements such as kneading discs or self-handed elements with large pitch, plasticating can be performed. These elements improve the self-wiping action. Kneading discs are designed to have the same cross section as the corresponding screws. Discs with various widths and staggered angles can be arranged along the shaft. A kneading block works as follows. When the barrel is rotated relative to the disc, a wedge-shaped flow develops. Figure 37 shows the velocity distribution and pressure profile in such a wedge when partially filled. Figure 38 shows the condition when the kneading discs are completely filled and are under pressure. During superimposing pressure, the convergent as well as the divergent wedge can generate pressure. At the maximum pressure (i.e., when $(dP/d\rho = 0)$), there is pure drag flow and at this point, ρ^* of the wedge, the shear stress is constant. The entire amount of material is drawn into the wedge. Through variation of radial clearances, speed, and number of flights, all of which influence the shape of the wedge, leak flow and shear effect can be varied in the shear gap. This type of wedge effect does not take place in counterrotating screws because of the slope of the flank. The finite width of the disc causes a melt flow in axial direction on both sides. The location

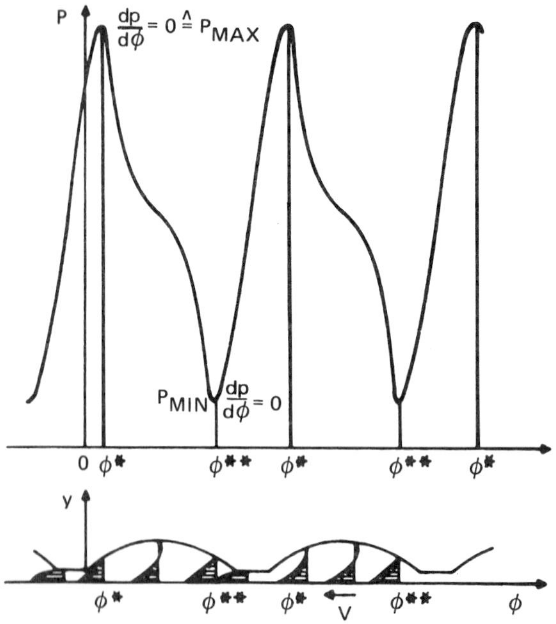

Figure 38 The pressure profile and velocity distribution with melt under pressure along a kneading disc.

of the pressure maximum is shifted only slightly; however, the pressure intensity drops considerably. By varying the disc widths the pressure profile in the cross section can also be varied. By staggering in several directions, the material flows axially on both sides into adjacent channels, resulting in a high degree of mixing from channel to channel.

Counterrotating screws essentially have plug flow caused by their low clearance. To increase the longitudinal mixing, the flanks and the root clearances must be enlarged. A system open in length and cross directions is purposely created to produce better homogenization of material.

In contrast the corotating screw conveys material by means of frictional force. As such, there exists a flow in the partially filled melting zone, where each particle is subjected to the minimum shear velocity of $\dot{\gamma} = 2\pi Ra/n$. In this situation there is no shear-free dead zone. Varying the screw pitch and speed provides different degrees of fill in the screw channel.

Counterrotating screw systems offer a positive conveying capability. Melt conveying against pressure in closed chambers does have disadvantages. The chamber volume is intermittently released, which leads to pressure and throughput fluctuations. To minimize these fluctuations, passage and conveying capabilities of the screw geometry must be in phase with each other. Overfeeding or underfeeding of the screws increases the magnitude of these fluctuations. Another way of minimizing the fluctuations is to purposely open the clearances to approach an open lengthwise and open cross-direction system.

In corotating screws, the screw crests of corotating screws are connected together to form the free wedge volume. Even in a fully intermeshing corotating mechanism, an open system exists. Partial positive conveying is obtained because of the influence of the wedge resistance. Downstream restrictions such as dies cause the melt to accumulate along the screw channel upstream, usually over several turns of the screw. Through proper screw configuration and appropriate processing parameters, the development of dead shear zones in the screw channel can be prevented. Optimal processing conditions can be calculated and adjusted which results in uniform shear stress distribution and minimal backup length.

As noted earlier, the effects of velocity distribution, stress distribution, and strain imposed in the material are distinct in each process. In addition to basic geometric variations possible in corotating and counterrotating mechanisms, there is an additional degree of process flexibility possible with corotating mechanisms such as screw speed and throughput. Fully intermeshing counterrotating screws form lengthwise and cross-wise closed C-shaped chambers, resulting in a plug flow similar to that observed in a screw pump. In contrast, fully intermeshing corotating screws do not form closed chambers, but rather length-wise open channels and crosswise closed chambers when mechanical clearances are disregarded. Polymer conveyance in this case depends on drag flow. Forced conveyance can be defined as an angle of conveyance based on (opposing) screw intermesh, where material is conveyed forward by each screw rotation for at least the screw crest width. Because of the presence of an uninterrupted open channel, material exchange can take place lengthwise along the machine, resulting in a mixing system based on the drag flow within the lengthwise open screw channel.

Because of geometric differences, there are basic differences in velocity and stress distribution of down-channel flow for intermeshing corotating and counterrotating twin-screw mechanisms, which have a significant effect on the plasticizing and mixing capabilities of twin-screw systems.

To gain better perspective, consider a coordinate system superimposed on the rotating screw surface so that the screw surface is stationary relative to the observer, while the barrel rotates at a constant velocity.

Positioning the coordinate system on the screw surface of twin-screw intermeshing mechanisms would not be meaningful since the eight-shaped barrel opening and corresponding screw would rotate in relation to the stationary screw where the coordinate is placed. Therefore, the coordinate system should be placed on the inner barrel surface parallel to the screw axis. It is moved along in the axial direction of the screw with the velocity $V_1 = Hn$ where H designates screw pitch and n screw rpm. With a constant screw speed and the same screw geometry along the machine length, the boundary conditions become constant. Following Eise et al. (1981), the coordinate system is such that the X-axis is normal to the screw channel, the Y-axis perpendicular to the barrel surface, pointing toward the rotating axis, and the Z-axis parallel to the screw channel as shown in Figure 39.

From a force balance, where Y_{0z} is the integration constant which denotes the location of zero stress point of down-channel flow. The counterrotating screw arrangement forms closed C-shaped chambers; therfore, the flow Q in the Z-direction for the selected coordinate system is

$$\frac{Q}{W} = \int_0^h V_z\,dy = 0 \tag{75}$$

where W = channel width.

Assuming a Newtonian fluid, equations (74) and (75) can be integrated and solved for the velocity and stress distributions:

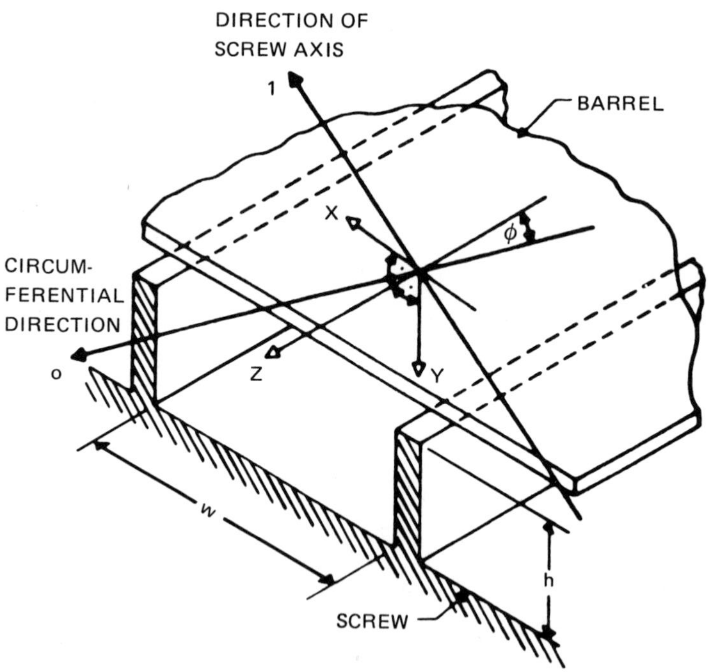

Figure 39 Coordinate system for modeling a twin-screw machine.

$$V_z/V_0 = \frac{1}{\cos\phi}\left\{3\left[\left(\frac{y}{h}\right)^2 - \frac{y}{h}\right](\sin^2\phi + 1) + \frac{y}{h}\cos^2\phi + \sin^2\phi)\right\} \quad (76)$$

$$\frac{\tau_{yz}}{n\dfrac{V_0}{h}} = -\frac{1}{\cos\phi}\left[3\left(2\frac{y}{h} - 1\right)(\sin^2\phi + 1)\cos^2\phi\right] \quad (77)$$

The velocity and stress distributions for a conventional screw pitch ($\phi = 17.66°$) are shown in Figure 40A for the counterrotating system. At $y/h = 0.36$ there is a zero stress point. The same zero stress point exists for each screw geometry.

In corotating screws open channels are formed. The material therefore can pass from one screw to the other along the Z-direction. The flow Q can never reach zero in a selected moving coordinate system when mechanical clearances are neglected. Pressure drop can develop in the Z-direction, which varies, depending on operating conditions, between zero (pure drag flow) and maximum value. The highest pressure is developed when the discharge end is closed. Although of no practical significance, such an analysis establishes the maximum pressure buildup capabilities of the machine. In this limiting case the observer witnesses a through flow which corresponds to a theoretical plug flow, for which the throughput is

$$Q = W\int_0^h V_z\,dy \quad (78)$$

Upon integration, the maximum pressure gradient is obtained:

$$\left.\frac{\partial P}{\partial z}\right|_{max} = \frac{6\pi Dn\eta\cos\phi}{h^2} \quad (79)$$

For this maximum pressure gradient, the following velocity and stress distributions are derived:

$$\frac{V_z}{V_0} = \cos\phi\left[\tan^2\phi + 4\frac{y}{h} - 3\left(\frac{y}{h}\right)^2\right] \quad (80)$$

$$\frac{\tau_{yz}}{n\dfrac{V_0}{h}} = -2\cos\phi\left(2 - 3\frac{y}{h}\right) \quad (81)$$

The distributions are shown plotted in Figure 40B, indicating that the zero shear stress point exists in the screw channel at $y_{0z}/h = 2/3$. In the open discharge

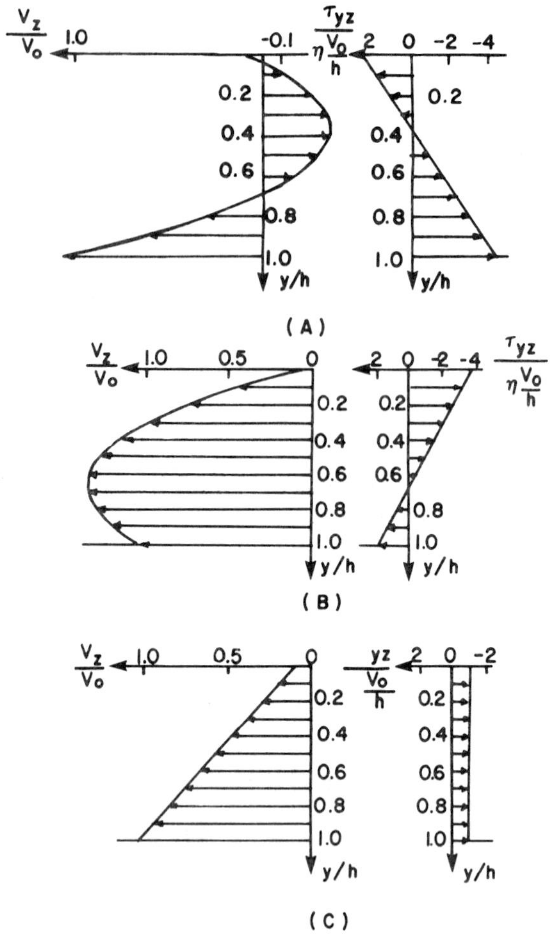

Figure 40 (A) Velocity and shear stress distributions of down-channel flow for counter-rotating screws. (B) Distributions of down-channel flow for corotating systems with closed discharge. (C) Distributions for corotating screws with drag flow.

case, pure drag flow ($\partial P/\partial z = 0$) exists and the velocity and stress distributions can be determined by solving Eq. (78):

$$\frac{V_z}{V_0} = \frac{y}{h}\cos\phi + \frac{\sin^2\phi}{\cos\phi} \tag{82}$$

$$\frac{\tau_{yz}}{n(V_0/h)} = -\cos\phi \tag{83}$$

Velocity and stress distributions in partially filled channels are given in Figure 40C. As shown, there is a constant stress component for the down channel flow in the screw channel. The relationship between pressure gradient, velocity, and location of zero stress point can be obtained from the following equations:

$$\frac{V_z}{V_0} = \cos \phi \, \frac{\left[\frac{1}{2}\frac{(y)^2}{h} - \frac{y_{0z}}{h}\frac{y}{h}\right]}{\left(\frac{1}{2} - \frac{y_{0z}}{h}\right)} + \tan^2 \phi \qquad (84)$$

and

$$\frac{T_{yz}}{n\frac{y_0}{h}} = \cos \phi \frac{\left(\frac{y}{h} - \frac{y_{0z}}{h}\right)}{\left(\frac{1}{2} - \frac{y_{0z}}{h}\right)} \qquad (85)$$

In the preceding analyses the effects of screw clearances and the wedge area were not considered. Also, it was assumed that a Newtonian fluid was used. With viscous materials, parabolas of a higher order for velocity distribution are observed. The shear rate at the wall is increased accordingly as the areas with lower shear stresses get broader. Significant qualitative differences are not obvious; however, the quantitative differences can be significant.

The velocity distribution and the shear stress distribution of the down channel and cross flow for the counterrotating screws are given by Eise et al. (1981). With both flow components, the apex of the velocity distribution, which position is identical to the zero shear stress point, is located in the screw channel. For the down-channel flow, it is $y_{0z}/h = 0.36$, and for the cross flow it is $y_{0x}/h = 0.66$. The velocity distribution in fully intermeshing, counterrotating screws which convey materials in C-shaped chambers can only be influenced by means of the screw pitch. The boundary layers, particularly at the screw root, are subjected to the greatest stresses and, since adherence to the wall is a prerequisite, they are also subjected to the greatest amount of shear.

The velocity distribution of corotating screws can be influenced by changing the throughput or by changing the screw speed as well as by changing screw pitch. Eise et al. (1981) have shown that, with the same boundary conditions, different velocity distributions and, therefore, different shear stress distributions result. Equation (84) states that the velocity distribution is dependent not only on the pitch angle ϕ, but also on the position of the zero shear stress point y_{0z}. This, however, is a function of the pressure gradient, the viscosity, the screw speed and the screw geometry as shown by the following relation:

$$\frac{y_{0z}}{h} = \frac{\pi D n \eta}{\cos \phi h^2(\partial p/\partial z)} (\sin^2 \phi - 1) + \tfrac{1}{2} \qquad (86)$$

The dependency between y_{0z}/h and the flow Q can be seen from equations (78) and (84). The velocity distribution therefore can be altered considerably by preselecting the throughput and/or screw speed.

The relationships between the intensity, uniformity, and location of maximum and minimum shear stresses define the basic differences between corotating and counterrotating screws with identical screw geometries. These differences can be summarized as follows:

The shear stress τ_b at the barrel wall is greater with counterrotating screws than with corotating screws.

The shear stress τ_s at the screw root is greater with counterrotating screws than with corotating screws.

The maximum shear stress τ_{max} in the counterrotating screws (not considering the wedge area and mechanical clearances) is identical to the shear stress at the screw root, τ_s. In corotating screws it is identical to the shear stress at the barrel τ_b.

In counterrotating screws the minimum shear stress is at $y/h = 0.38$; however, in corotating screws it is between $0.66 \leqslant y/h \leqslant 0.9$.

In counterrotating screws, the change of the shear stress $\partial\tau/\partial z$ in the channel depth is considerably larger than with corotating screws, assuming that in actual production closed discharge does not occur. This means that the shear stress distribution of counterrotating screws is considerably broader than in corotating screws.

The shear stress distribution is only dependent on the screw pitch H in counterrotating screws. In contrast, in corotating screws not only does the shear stress distribution change with the pitch angle ϕ, but it is also dependent on the throughput and the screw speed.

Because of the first two differences between the designs, as well as the increased tendency toward mechanical wear in counterrotating screws, the operating speeds of counterrotating screws must be relatively low. When considering throughput, the superior conveyance of the counterrotating screw is offset by the higher operational speeds of the corotating screw.

In counterrotating screws, the material at the screw root is drawn into the wedge area where it is exposed to high strain. In counterrotating screws the minimal shear stress is inside the chamber. On the other hand, by properly selecting operating conditions in corotating screws, the shear stress is nearer the screw root.

In corotating screws the shear stress distribution under actual operating conditions is in the area of $1 < y_{0z}/h < \infty$ and is considerably more uniform. In comparison to counterrotating screws, the layers at the screw root are subjected to less shear strain. These layers are wiped by the crest of the oncoming screw into the wedge area and are then recombined near the barrel wall. The material layers from the area of the barrel wall are forced in the wedge area into the bottom of the channel of the oncoming screw. In corotating screws, because of the absence of the low shear stress zone and the layer transfer rearrangement in the wedge area, the shear strain on the product is more uniform. This is a favorable condition for uniform plasticizing and homogeneous mixing. Because the external heating during the plasticizing process plays a secondary role, higher screw speeds and large screw diameters can be employed.

COUNTERROTATING, TANGENTIAL TWIN-SCREW EXTRUDERS

The counterrotating, intermeshing-type extruder is most commonly used for the processing of rigid PVC for pipe or siding. This type of extruder provides positive conveying characteristics suitable for feeding rigid PVC powder dry blend, and when combined with low speed screw (typically 50 rpm or less) provides a low shear environment suitable for the processing of materials with poor thermal stability.

The corotating, intermeshing system does not have positive conveying characteristics like the counterrotating intermeshing system. As noted, conveying takes place by drag flow. As the channel is axially open, this system, like the counterrotating, tangential type can be applied to process engineering (e.g., venting).

The counterrotating, tangential type was the first twin-screw extruder to be manufactured in the United States. The counterrotating, tangential screw arrangement has been used in a broad range of process applications throughout the polymer processing industry over the last 33 years. Nichols (1982) compares the twin-screw extruder with single-screw extruders.

As noted above, tangentially opposed screws are not positive conveying devices. However, unlike single-screw extruders, which depend upon frictional forces for solids conveying, in the twin-screw extruder the lower apex applies a "keying" effect analogous to grooved feed bushings in single-screw extruders. This effect continues through all of the functional zones and can be readily observed in the partially filled vent zones as a rolling bank of polymer on the lower apex. The outstanding feed acceptance of these extruders allows extremely uniform performance while starve feeding, which is the feeding method most commonly employed with all twin-screw extruders. While starve feeding has been applied successfully to single-screw extruders, its use has been limited because of feed acceptance problems. The highly advantageous feed acceptance of counterrotating, tangential extruders also provides for the possibility of sequential feeding. A benefit of sequential feeding is that separate ingredients of a compound can be handled with dedicated feeding equipment, thus eliminating the need for preblending. Other advantages of sequential feeding include higher total throughput rates, lower product temperature, and improved product mixing.

Melting in single-screw extruders was described earlier. In the counterrotating, tangential twin-screw extruder, the melting mechanism varies considerably from the single-screw model in that the keying effect of the lower apex and the interaction of the staggered screw flights tend to promote a considerable amount of exchange between the channels of the two screws. Thus, the polymer melt and solid granules become mixed together instead of having a clear separation of melt pool and solid bed.

In single-screw systems, screw designs to accelerate melting for high efficiency have developed along two opposing concepts. In the first case the melting rate is controlled by providing a barrier between the solid bed and the melt pool to assure that the solid bed does not break up prematurely and become encapsulated in the melt. An opposing concept is the wave screw. In this design concept a conventional feed and melting zone are employed until the point at which about 50% of melting is completed, and then the melt and solids are mixed together. This is accomplished by varying the metering channel depth in a sinusoidal pattern, alternating between very shallow, high shear zones and rather deep, low shear zones. The effect of this design is to promote the distributive mixing of the now thoroughly broken up solid bed with the melt pool, thus using the residual melt pool heat to complete the melting of the remaining solids, resulting in a very low average extrudate temperature. This mechanism is similar to that which is postulated for the counterrotating, tangential extruder except that the mixing effect of twin-screw extruders is generally conceded to be significantly superior to that which can be obtained in a single-screw design.

A comparison of the characteristic curves of the single-screw and the counterrotating, tangential extruders is given in Figure 41.

The single-screw characteristics at zero pressure or open discharge are de-

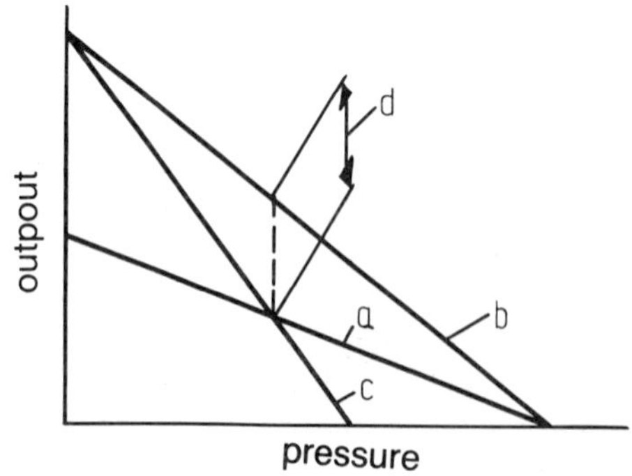

Figure 41 Comparison of screw characteristics of a single screw (a) to two single screws (b) and a twin screw (c).

fined by the drag flow equation, and the slope of this screw characteristic over a range of pressure is established by subtraction of the pressure flow characteristic for various pressures. The screw characteristic curve for two single screws is the algebraic sum of the result for one single screw. The screw characteristic for the counterrotating, tangential twin-screw machine has a steeper slope than the characteristic of either one single screw or two single screws. The distance between the intersection of the twin-screw characteristic and the single-screw characteristic lines and a point at equal pressure on the characteristic line of two single screws can be defined as the longitudinal mixing flow through the open area at the apexes, as shown in Figure 41. This flow component is a pressure flow term, and hence as pressure flow increases, pumping capacity is correspondingly reduced. Several components combine to produce the total mixing capability, namely, the rotation of the material in the screw channel as would normally be experienced in a single-screw extruder, interchange of material between the two screws, and the longitudinal mixing flow component. These mechanisms combine to homogenize materials going through the apparatus.

In general, all extruders are assumed to be nominally "once through equipment." This implies that preblending, or bulk mixing, is essential for good mixing results, and, in fact, for single-screw extruders where feed acceptance is often poor, preblending is a necessity. In counterrotating, tangential twin-screw extruders however, since feed acceptance is unusually good, as long as sufficient accuracy is provided in the associated feeding devices, uniform product quality can be obtained by feeding separate streams either at the same axial location or sequentially in various axial locations.

Two types of mixing that are important are distribution and dispersion. Distributive mixing is used for any operation employed to increase the randomness of the spatial distribution of the ultimate particles without reducing their size. In simple mixing the ultimate particles of the system pass from a less probable to a more probable arrangement. The term *dispersive mixing* applies to those mixing processes that reduce the size of the ultimate particles as well as randomize their positions. Dispersive mixing depends on exposing agglomerates to ultimate particles to sufficiently high shear stresses to cause them to break down. The level of shear stress required for deagglomeration to occur is related to the size of the ultimate particles and to the nature of the bonds holding the agglomerate together. The shear stress in the molten polymer is proportional to the product of viscosity and shear rate. In general, the shear stress in normal melt pumping channels is not sufficiently high to provide adequate dispersive mixing. In counterrotating, tangential extruders, it is often beneficial to introduce all of the components requiring dispersion in the initial feed so that they will be exposed to the high shear stresses existing in the melting zone where the viscosity is very high. Consequently, the polymer melt is already prepared for distributive mixing in a lower viscosity zone downstream. Alternatively, high shear stress mixing devices, such as reverse flight compounders or very restrictive cylindrical compounders may also be used in the fully developed melt region, but this may lead to undesirable temperature rise in the polymer melt causing degradation. Distribution is somewhat limited in single-screw extruders, and various auxiliary devices must be used to improve distributive mixing in these machines.

One approach is the use of a static mixer, which provides a high degree of radial mixing as a result of the placement of a series of fixed blades which redirect the polymer flow path. This type of device is usually employed between the end of the extruder screw and the extrusion die.

In many applications there is the additional requirement of removing undesired volatiles from molten polymers. This has resulted in the development of various types of vented extruders. The screw section under the vent cannot be completely full, or else a path is not provided for the escape of the volatiles, and further, if any positive pressure exists in this region, the polymer will flow out of the vent opening. The relationships between the pumping section upstream of a vent section (which pumps against zero pressure) and the downstream pump (which must overcome the die resistance) are well defined if certain simplifying assumptions are made. The key assumptions involved by the designer are:

The flights in the vent region between the two pumping sections are deeper than the pumping sections, and therefore are only partially filled.

The upstream pumping (or melt metering) section is assumed to be full and pumping against zero discharge pressure, and is pumping at a rate consistent with the drag flow calculation. Figure 42 illustrates a typical screw design (single) along with the pressure profile for a two-stage venting system. The assumption that the first-stage metering section is the rate-controlling section of the entire screw leads to the following relationships: since the first metering section does not pump against a restriction, its output rate is defined by the drag flow relationship:

$$Q_{\text{net}} = Q_{d1} \tag{87}$$

Since the second metering section must overcome the resistance of the die assembly, its rate must be defined by the algebraic sum of the drag and pressure flow as noted earlier:

$$Q_{\text{net}} = Q_{d2} - Q_{p2} \tag{88}$$

Combining the preceding expressions provides the mass balance for the screw

$$Q_{\text{net}} = Q_{d1} = Q_{d2} - Q_{p2} \tag{89}$$

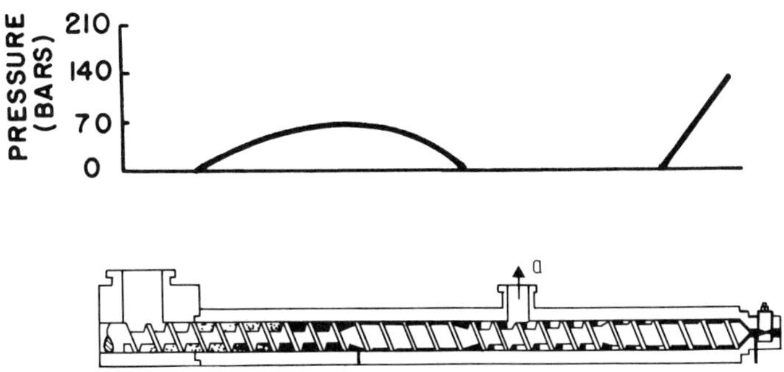

Figure 42 Screw design and pressure profile for a two-stage vented screw.

There are two imbalances in this expression—vent flow and surging. Vent flow Q_{vent} is

$$Q_{d1} > Q_{d2} - Q_{p2} \tag{90a}$$

or

$$Q_{d1} = Q_{d2} - Q_{p2} + Q_{vent} \tag{90b}$$

Surging is given by the following inequality:

$$Q_{d1} < Q_{d2} - Q_{p2} \tag{91}$$

At steady-state conditions, statement (91) cannot exist. It does, however, indicate a real condition in which the pumping capacity of the second metering section is such that until it "fills up," it pumps inefficiently, and when full, it pumps with such great efficiency that it empties before the first-stage pump can replenish it. This in fact is the surging due to a poorly designed two-stage screw.

In the design of a vented extruder screw, several criteria are needed in addition to the requirements of rate, melt quality, and average melt temperature. These criteria are devolatilizing capability, pressure stability, and pressure-generating capability. Several design factors affect pressure-generating capability. Inspection of the terms in the drag and pressure flow equations, neglecting leakage flow, help to define these.

The drag flow equation is

$$Q_d = F_d \frac{\pi D^2 h[1 - (ne/t)] \sin \phi \cos \phi}{2} N \tag{92}$$

The pressure flow equation is

$$Q_p = F_p \frac{nDh^3[1 - (ne/t)] \sin \phi \cos \phi}{2} N \tag{93}$$

where the geometric shape factors F_d and F_p depend on the ratio channel depth h and channel width W. This dependency is shown in Figure 43.

The preceding relationships show that drag flow is directly proportional to the channel depth and the screw speed and is not related to the polymer viscosity, nor is it affected by the metering section length. In contrast, pressure flow is proportional to the third power of the metering section depth, is independent of screw speed, is a function of polymer viscosity, and is inversely proportional to metering section length. This would suggest that the second pump depth is the most convenient variable to modify to correct a vent flow problem due to an imbalance as described by Eq. (90a). In some cases, this is a correct action to take, but often it can aggravate the problem. Simplifying equations (92) and (93) and substituting appropriately in equation (89), we obtain:

$$K_1 h_1 = K_1 h_2 - (K_2 h_2^3 \, \Delta P) \tag{94}$$

where K_1 represents all of the terms in the drag flow equation (92) except the channel depth (h_1 or h_2) and K_2 represents all of the terms in the pressure flow equation (93) except the channel depth (h^3), and head pressure (ΔP).

By rearranging Eq. (94), the following is obtained:

$$\Delta P = C \frac{h_2 - h_1}{h_2^3} \tag{95}$$

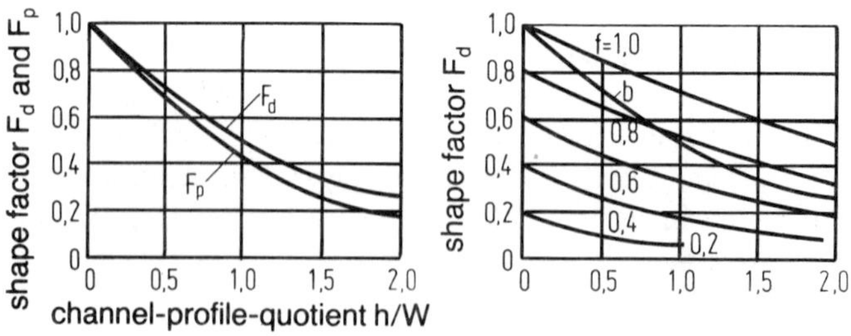

Figure 43 Shape factors for drag (F_d) and pressure flow (F_p) depend on channel profile (h/W).

This relationship can be expressed in a dimensionless form and is shown graphically in Figure 44 as pressure capability versus pump ratio for the two metering zones. Note that the pressure-developing capability reaches a maximum (for Newtonian fluids) at 1.5:1 pump ratio. With non-Newtonian polymers, the maximum pressure development can occur at approximately 1.75:1 pump ratio. Hence, the second-stage pump cannot be greatly deepened.

Nichols (1982) reports that the practical limits of pump ratio are a minimum of 1.5:1 and a maximum of 2.0:1. Working toward the lower limit tends to enhance stability but limits pressure capability, while working toward the upper limit may help maximize pressure capability but may also result in somewhat less stability, depending on how the extruder is operated. The use of a fixed orifice valve with degradable polymers or an adjustable valve with stable polymers is recommended.

Since there is a limitation on the second pump depth, the only two variables which can significantly affect pumping (pressure-generating) capability are viscosity and the length of the second metering section. The extruder manufacturer does not have the ability to specify the polymer viscosity, and therefore the only variable left to work with is the second pump length.

Attention is now given to devolatilization. Devolatilization rate is generally thought to be limited by diffusion of the solvent molecule through a thick film of polymer. From this viewpoint, the polymer fill can be considered as an infinite slab; hence the approach to equilibrium is limited by end-to-end mixing (back mixing) and that surface generation is combined with surface exposure time.

A counterrotating, tangential extruder has unique devolatilizing capability and is found in widespread use in some of the most demanding large-scale applications at rates of 5000–7500 kg/hr dry product, starting with solvent levels as high as 50%. Single-screw extruders, by comparison, are rarely used for volatile levels above 1%.

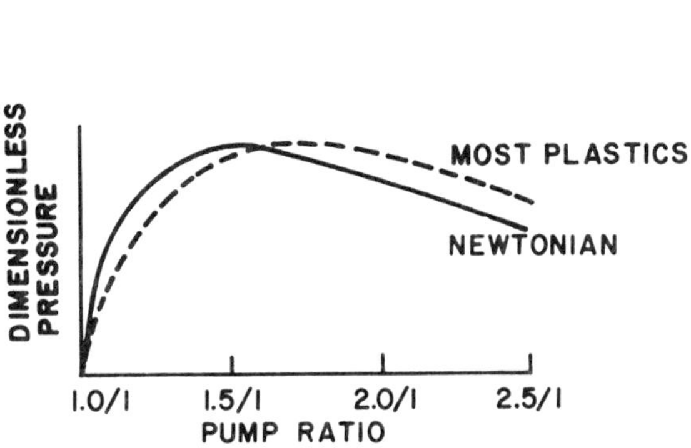

Figure 44 Plot of dimensionless pressure capability versus pump ratio.

With counterrotating extruders, the additional pressure flow through the open apex area (the longitudinal mixing flow component) is detrimental to high pressure generation. Therefore, counterrotating, tangential extruders often utilize single-screw discharge pump sections. These can be designed as an integral part of the main extruder screw operating at the process screw speed, or they can be incorporated as cross-head extruders of different diameter and speed. To calculate the final discharge pump depth in a multistage system, one need only determine the apparent metering capability of the first pump (in twin screw extruders, this is often determined by a feed rate setting) and then the single-screw equations apply directly to design the discharge pump. To calculate pumping in the twin-screw sections, based on an isothermal Newtonian model using drag flow and pressure flow correction factors to the basic single-screw equations, use the following equation:

$$q = \tfrac{1}{2}WHV_{bz}F_{DTW} - \frac{WH^3}{12\eta}\left(\frac{\Delta P_T}{\Delta Z_T}\right)F_{PTW} \tag{96}$$

where

$$F_{DTW} = \frac{4f}{1 + 3f} \tag{97}$$

$$F_{PTW} = \frac{4}{1 + 3f} \tag{98}$$

Parameter f is the fraction of helical length occupied by zone 1 and typically has a value of 0.9. The ratio $q_S/q_D \simeq 0.5$, and twin screw extruders will typically deliver about 1.73 times the flow rate of a single-screw machine.

Because of good devolatilizing capabilities, typical applications of the counterrotating, tangential twin-screw extruder are desolventizing during polymer manufacture from solvent solutions, latex coagulation during polymer manufacture from latex emulsions, and compounding of plastics. This includes, among other things, coloration, fiberglass addition, and preparation of hot melt adhesives.

NOTATION

$A_{1,2}$, $B_{1,2}$	physical and geometrical extrusion constants
a	slope of channel
A_r	Arrhenius shift factor
Br	dimensionless Brinkman number
c_s	specific heat
D	hopper diameter
D_b	inside barrel diameter
d	die diameter
E	activation energy
e	width of flight

F frictional force per unit area
F_d, F_p extruder geometric function and drag flow/pressure flow shape factors, respectively
f fraction of helical length occupied by zone 1 (Z_1/Z_T)
G mass flow rate
g Bagely correction factor
H_s mechanical heat generation rate
H' heat of fusion
h channel depth
IV intrinsic velocity
K geometric extrusion parameter
k rheological coefficient
k_m melt thermal conductivity
k' swell index coefficient
L length
L_s channel length filled with polymer
MI melt index
M_n number average molecular weight
M_v viscosity average molecular weight
M_w weight average molecular weight
M_Z, M_{Z+1} third and fourth moments of molecular-weight distribution
m_0 flow consistency index
N screw speed
N_i number molecular weight of ith species
n power law exponent
n' swell power law exponent
P pressure
ΔP_t pressure rise per turn in twin extruder
p number of flights in parallel
Q'_s theoretical volumetric throughput (i.e., no friction)
Q^* screw discharge coefficient
q heat flow
R universal gas constant
R_s, R_b radii of screw root and inside barrel, respectively
T absolute temperature
T_0 reference temperature
t temperature
t_c residence time
U velocity
u mean flow velocity
V_{sy} bed velocity in y-direction
W average channel width
X average solids bed width
Y distance
Z total melting length
ΔZ_t helical length per turn ($\Delta Z + \Delta Z_2$)
z axial distance

Greek Symbols

α solvent property constant in Mark–Houwink viscosity correlation
α_s thermal diffusivity of bed
Γ nominal wall shear rate
$\dot{\gamma}$ shear rate
δ melt layer thickness
η viscosity
η_a apparent viscosity

$\dot{\eta}$	inherent viscosity
θ	angle of plug advance relative to screw axis
ν	speed of screw rotation
ξ	non-self-conveying friction force per unit surface area
ρ	density
τ	wall shear stress
ϕ	helix angle of flight
ϕ'	solids coefficient of friction
ψ	friction coefficient or force
$\dot{\psi}$	transformation variable

REFERENCES

Baldwin, F. P., and G. Ver Strate, *Rubber Chem. Technol., 45,* 709–881 (1972).

Bridgwater, J., and A. M. Scott, Flow of solids in bunkers, in *Handbook of Fluids in Motion,* N. P. Cheremisinoff and R. Gupta, eds., Ann Arbor Science Publishers, Ann Arbor, MI, 1983.

Carley, J. F., *Mod. Plast., 60*(2):77 (1971).

Chung, C. I., W. J. Hennessey, and M. H. Tusim, *Polym. Eng. Sci., 17*:9 (1977).

Dekker, J., *Kunststoffe, 66*:130 (1976).

Eise, K., H. Werner, H. Herrmann, and U. Burkhardt, An analysis of twin-screw extruder mechanisms, in *Advances in Plastics Technology,* Vol. 1, No. 2, Van Nostrand Reinhold, New York, April, 1981.

Flory, R., *Principles of Polymer Chemistry,* Cornell Publishing, Ithaca, New York, 1952.

Johnson, P. S., *Rubber Chem. Technol., 56,* 3 (July–August 1983).

Kacir, L., and Z. Tadmor, *Polym. Eng. Sci., 12*:387 (1972).

Keller, R. C., paper presented at Rubber Division Meeting, ACS, Los Angeles, CA, April 24, 1985.

Klein, I., and D. I. Marshall, *SPE J., 21*:1376 (1965).

Klein, I., D. I. Marshall, and C. A. Friehe, *SPE J., 21*:1299 (1965).

Kresge, E. N., C. Cozewith, paper presented at Rubber Division Meeting, ACS, 1984.

Krevelen, D., van, *Properties of Polymers,* Elsevier, New York, 1972.

Marshall, D. I., I. Klein, and R. H. Uhl, *SPE J., 21*:1192 (1965).

Munstedt, H., *Kunststoffe, 68,* 92–98 (1978).

Nichols, R. J., *Kunststoffe German Plastics, Vol. 9,* Carls Hanser Verlag, Munich, 1982, pp. 510–605.

Scholte, Th., in *Developments in Polymer Characterization,* Vol. 4, J. Dawkins, ed., Applied Science Publishers, 1983.

Sezna, J. A., *Elastomerics, 16*:(8) (Aug. 1984).

Slade, P. E., *Polymer Molecular Weights, Part I,* Marcel Dekker, New York, 1975.

Squires, P. H., *Soc. Plast. Eng. Trans., 4*(1):7 (1964).

Tadmor, Z., and E. Broyer, *Polym. Eng. Sci., 12*:378 (1972).

Tadmor, Z., and I. Klein, *Engineering Principles of Plasticating Extrusion,* Van Nostrand Reinhold, New York, 1970.

Tadmor, Z., and I. Klein, *Principles of Plasticating Extruders,* Van Nostrand Reinhold, New York 1969.

Tadmor, Z., D. I. Marshall, and I. Klein, *Polym. Eng. Sci., 6*:185 (1966).
Throne, J. L., *Plastics Process Engineering,* Marcel Dekker, New York, 1979.
Ver Strate, G., Ethylene propylene elastomers, in *Encyclopedia of Polymer Science and Engineering,* McGraw-Hill, New York (in press).
Zavadsky, E., and J. Karnis, *Rheol. Acta, 24*:556–565 (1985).
Zavadsky, E., J. Karnis, and V. Pechoc, *Rheol. Acta, 21,* 470–474 (1982).
Zavadsky, E., J. Karnis, and V. Pechoc, Conf. Chemocomp., Antwerp, Belgium, 1982.
Zavadsky, E., *Chem. Vlakna, 28*:160 (1978).

7
Laboratory Techniques for Processability Testing

INTRODUCTION

The term *processability* refers to the relative ease with which neat or compounded polymers can be handled in production operations and equipment. The most frequently applied production operations are mixing, milling, extrusion, molding, and curing. The end-user manufacturer is concerned with the processing quality or processability of the polymer-based article and is sensitive to factors that may affect the material's behavior and variability during the various manufacturing stages. In the rubber industry, terms such as *plasticity* and *recovery* are used heavily as relative ratings of material processability. Plasticity is generally associated with the relative ease of deformation (i.e., the inverse of effective viscosity). In other words, a highly plastic elastomer or rubber is one that deforms or flows readily. Devices used to test for plasticity, such as the Mooney viscometer, attempt to measure resistance to deformation. Unfortunately, terminology is not always consistent, even within a particular industry. For example, the parallel plate compression test described in Chapter 2, although widely used in the rubber industry, provides a measure of material stiffness and refers to the measured value as plasticity. The ASTM standard for the parallel plate compression test refers to the measurement (i.e., the measured compressed thickness) as the plasticity number. Additional contradictory terminology stems from the fact that in unvulcanized rubbers, permanent deformation is an attractive feature. That is, the material should not be elastic. Some definitions often used by technical sales personnel include definitions of plasticity in this feature (e.g., material susceptibility to and retentivity of deformation). Laboratory test methods aimed at measuring the property of elastic recovery have evolved; one such example is the Goodrich plastometer. It is important to note that although such tests are a rheological characterization of the polymer, they in fact do not provide a true measure of viscosity or resistance to flow, but instead provide a measurement of a combination of viscous and elastic properties.

The specific type of laboratory test depends on the manufacturing operation or operations that will process the elastomer. For example, if the interest lies with the mixing characteristics of a polymer, then the laboratory processability test might be directed at assessing the ease with which the compounding ingredients can be incorporated and dispersed in the polymer. If, on the other hand, extrusion is of interest, tests might be directed at evaluating the ease with which the compound can be fluxed, transported by the screw, and extruded, and/or the dimensional stability of the extrudate. In applications of injection molding, interest may be in the relative ease of flow of the material, the absence of excessive heat buildup in the runners and gates, the speed of vulcanization and degree of mold shrinkage. Operations such as calendering, milling, and various other processes can each have their own individual requirements.

The major obstacle in material testing is the ability to scale up laboratory characterization data to full-scale processing equipment. This in fact is a formidable task that is further complicated by economic constraints of manufacturing. Manufacturing operations are manpower intensive and costly. These factors often mandate that processability tests be made simple, fast, and inexpensive, especially when they are implemented in plant operations for product quality control purposes. Unfortunately, the simpler the test, very often the less correlation it has with processability in full-scale equipment. Another serious problem that plagues even the more sophisticated processability testing labs is the scale of equipment, since geometric differences may exist between small-scale apparatuses and commercial equipment. The classical approach to scale-up makes use of the principles of dimensional analysis. The method of similitude requires two fundamental criteria in order to scale up properties or equipment: these are geometric similitude and dynamic similitude. Very often, lab-scale versions of such equipment as mixers or extruders have significantly different geometries than their commercial counterparts. In addition, to minimize experimental efforts and cost, testing is often limited to a very narrow range of conditions. When the laws of similitude are in fact ignored, efforts in laboratory-scale processability testing can be wasted and projections as to a product's performance in a commercial operation can be misleading.

The importance of processability testing is that it provides a basis for projecting polymer behavior in commercial operations. If adequate correlations can be developed between full-scale processing operations and the lab, then a basis for product quality testing can be established. In most commercial operations there are several stages of processing, and hence laboratory testing must often be implemented in steps to address each of them in order to regulate product performance. The principal areas evaluated in processability testing are raw material characterization, control of mixing, characterization after mixing, regulation during article shaping, and properties prior to and during vulcanization. The importance of each of these areas are discussed below.

This chapter describes various laboratory techniques for testing, characterizing, and projecting polymer performance in industrial-scale processing equipment. Although the discussions largely address techniques employed in and examples drawn from the handling of elastomers, the general concepts are readily extended to all types of polymers. It should be noted that rheological characterization plays a large role in processability testing; however, this subject has already been treated in Chapters 2 and 6. The latter chapter in particular illustrates the application of a capillary rheometer in assessing the extrusion performance of elastomers. Additional comments concerning rheological characterization are made here.

PRINCIPLES OF RHEOLOGY AND PROCESSING

As noted above, processability testing must be performed in separate stages in order to be effective in regulating an entire processing operation. The first obvious step is to characterize the neat polymer properties that are most relevant to the behavior in the ensuing processing operations such as mixing, molding, extrusion, and so on. As an initial step, the material's viscosity and elasticity, as well as its response to temperature and shear, should be examined in a mixing cycle similar to that of its intended operation. Failure characteristics in terms of such properties as elongation at break and gum green strength provide insight into the material's behavior when subjected to shearing forces on a mill or internal mixer. In conjunction with this, characterization of the molecular weight and molecular-weight distribution (MWD), as well as detection of the presence of long-chain branching, should be conducted using gel permeation chromatography (GPC). The use of GPC and high-speed solution viscosity testing on polymer samples represent invaluable tools for controlling the polymerization process so that polymers can be tailor-made for specific processing operations. Also, stress relaxation, which affects properties such as die swell, depends on the MWD. The most pronounced effects of MWD can be observed in the early stages of processing. Batch-to-batch variations in MWD can be one of the most significant causes of problems in mixing; hence, measurements related to MWD are important. One cannot realistically implement processability testing without a characterization of molecular weight and MWD.

Neat polymer characterization should be followed by studying the parameters which affect the mixing process. Conventional approaches to controlling mixing have used the criteria of time and temperature to establish when individual batches should be dumped. With stricter product specifications emerging in new consumer products, more precise and reproducible control techniques of mixing cycles are being imposed. One approach to controlling mixing cycles that requires further study is that of considering the energy input at different stages in the cycle. It should be noted that processability testing is usually conducted at the termination of mixing. One of the most important tests performed involves measuring/rating filler dispersion (refer to discussions in Chapter 5). The degree of mixedness is important to end product properties such as tensile strength, elongation or shrinkage, stress relaxation, heat aging characteristics, and behavior of the mix during the forming operation. Both mix viscosity and elasticity should be measured at this stage.

During the forming or shaping operation, temperature and shear rate must be controlled. Both these parameters affect the flow through the apparatus and during extrusion. The types of tests usually conducted relate to machine capacity, die swell, and the appearance and reproducibility of the extrudate.

Tests are also usually conducted both before and during the vulcanization process. Prior to vulcanization, measurements of the curing rate and/or scorch sensitivity can be made. During vulcanization the heat history of the polymer is a critical parameter to monitor. In the ideal case, thermal history should be homogeneous throughout the product so that the degree of cure and absolute cure state is the same throughout.

In any of the processing operations, polymer flow through the apparatus depends on the viscoelastic properties of shear and elongational flow. In addition, flow or deformation is affected by the degree of adhesion or friction between the surfaces of the processing approaches and the polymer itself. This latter factor is often overlooked in processability testing because of differences between full-scale equipment and laboratory prototypes. In some applications, however, it can be an important factor which may account for inability to scale up processing behavior.

As illustrated by the extrusion example in Chapter 6, polymer processability is related to the polymer type, molecular weight, and molecular-weight distribution. In addition, it also can be a function of the degree and type of branching, the physical form of the polymer (e.g., granules, powder, slats, strips), the nature of the downstream processes, and the temperature and shear history of the process. Another consideration alluded to in Chapter 5 is that the mixing process and final physical properties of the material also depend on compounding ingredients. In particular, parameters such as filler type, filler concentration, degree of polymer–filler interaction, extent of bound rubber, addition of oils and various process aids, as well as prevulcanization or scorch can also alter processability and final product properties.

As noted very early in this chapter, what is often ignored in processability testing for the sake of convenience is the condition of hydrodynamic similarity. Figure 1 shows typical shear rate ranges for different commercial processing operations. With most conventional processability testers (e.g., the Mooney viscometer), testing is done at shear rates far below actual conditions (typically 2.5 s^{-1}). Although shear rate itself should not be considered the governing criterion for dynamic similitude (a better basis would be a modified Reynolds number or Euler number), it does illustrate that interpretation and extrapolation of laboratory-scale processing evaluations is risky and is almost entirely empirical as practiced by industry today. Nonetheless, testing does provide some guidance, and since a wide range of processability tests have evolved through the years, some latitude exists in selecting a technique or method which at least approaches one or more of the processing operations.

In conjunction with this, one must realize that the viscoelastic properties of polymers and their mixes are not constants. They, in fact, will change with time in both a reversible manner and an irreversible manner. Samples stored for some period of time may have to be milled and/or premolded into a test piece, and even this can be dangerous as it can alter the processability of the specimen. Sample preparation therefore also enters as a variable, and it is important to establish a standard procedure early on so that data generation can at least be performed on a normalized basis. The ASTM for rubber testing notes that the preparation of specimens for processability measurements should be controlled

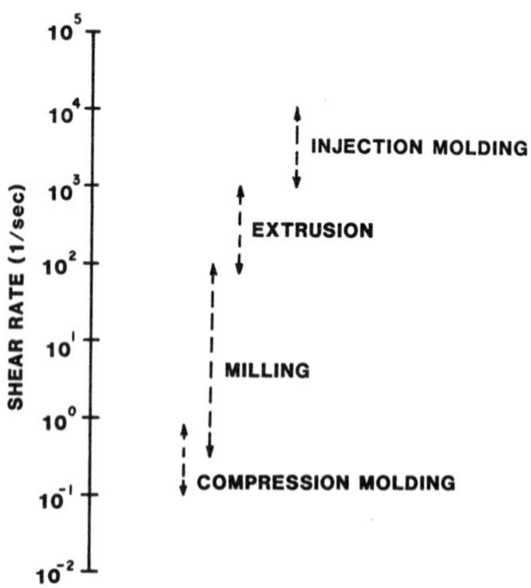

Figure 1 Typical shear rate ranges for commercial processing equipment.

carefully if reproducible results are to be obtained without modifying the properties to be measured. This is especially the case when performing tests on neat elastomers. ASTM D1646 provides recommended procedures for sample specimen preparation of rubbers. Some of these procedures involve milling practices. In some cases it may be necessary for a degree of nonpermanent structure to be broken down either by pretreatment or during the test itself.

As noted as early as Chapter 1, processability primarily depends on the viscoelastic properties of the polymer. It is important, therefore, to understand the characterization terminology used to distinguish viscous and viscoelastic properties. By way of review, one can establish relationships of fundamental rheological properties to processing behavior through the use of processability testing instruments such as rotational, oscillatory, and capillary viscometers. The practical problem of interest is to establish relationships between the fundamental viscoelastic properties and the theories of polymer processing (i.e., milling, extrusion, calendering, building, and preforming operations). White (1969, 1977) provides a review of the fundamental viscoelastic property relationships to polymer processing theory.

Viscoelastic properties of an elastomer at small deformations can be explained by a modulus such as the shear modulus:

$$G^* = G' + iG'' \tag{1}$$

where

G' = in-phase, elastic, or storage modulus
G'' = out-of-phase or loss modulus

Parameters G' and G'' are both temperature and frequency dependent. A parameter known as the frequency factor relates measurements over a temperature range; that is,

$$\ln \alpha_T = \frac{C_1(T - T_0)}{C_2 + T - T_0} \tag{2}$$

where T_0 is a reference temperature. In rubber applications the reference temperature is usually chosen to be about 50°C above the glass transition temperature of the polymer, in which case the constants C_1 and C_2 correspond to the values 8.86 and 101.6°C, respectively. Equation (2) relates the material's response at one shear rate and temperature to a different shear rate through an appropriate change in temperature. The limitation of Eq. (2) is that it applies only to small deformations. In contrast, the response of viscoelastic materials at large deformations is highly nonlinear. Nakajima and Collins (1974, 1976) review large deformation theory. Most textbooks refer to the classical model of viscoelasticity, which uses the spring and dashpot explanation. The Voight and Maxwell models most often alluded to describe only the extreme or very limited behavior of elastomeric materials. More sophisticated models are basically extensions of the simple Maxwell spring model and are based on two or more parallel Maxwell elements such that the dashpots are simulated by power law viscosities [see Turner et al. (1979), for example].

In Chapter 1 Newton's law of viscosity was used as the starting point for describing viscoelastic response:

$$\eta = \frac{\tau}{\dot{\gamma}} \tag{3}$$

where

η = dynamic viscosity
τ = shear stress
$\dot{\gamma}$ = shear rate

The facts of life are, however, that polymer melts are non-Newtonian, and hence a plot of τ versus $\dot{\gamma}$ results in a curve, not a straight line, even on logarithmic coordinates. In practice, however, we overlook this and attempt to approximate response at least over a limited range of strain rates by using the empirical power law relationship:

$$\tau = k\dot{\gamma}^n \tag{4a}$$

$$\eta_a = \frac{\tau}{\dot{\gamma}} = k(\dot{\gamma})^{n-1} \tag{4b}$$

It is important to note that the flow curves of different elastomers may cross over; that is, each material can actually have a different n-value at a given shear rate. As such, a single-point viscosity measurement for the purposes of relative comparisons or rankings of polymer behavior may be meaningless. Again, molecular-weight distribution can play a dominant role in establishing the pseudoplasticity of an elastomer. For example, in general, very narrow MWD polymers tend to be more Newtonian in their flow response curves (i.e., $n \sim 1$) over a wide range of shear stress. Most rubbers that are broader in MWD, showing the most probable distribution (i.e., $M_w/M_n > 2$–3) in gum state display an n-value of about 0.3. Another factor to consider is that many materials, particularly mixtures of different MWD polymers, can in fact show substantial Newtonian response at low shear stresses, but become distinctly non-Newtonian at higher stresses. An example of this is illustrated in Figure 2.

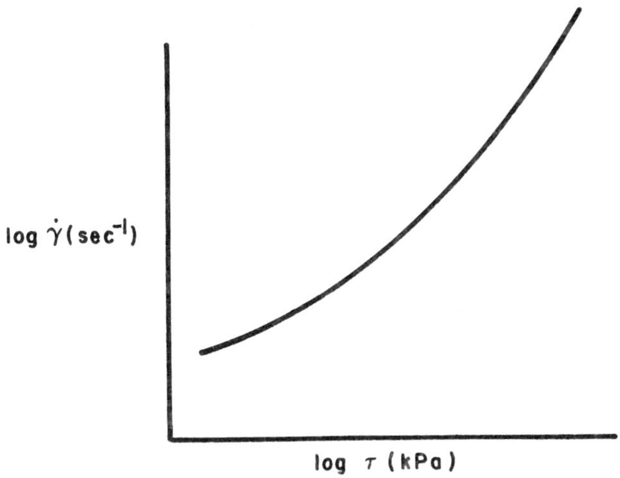

Figure 2 Flow curves for typical EPDMs.

There are a variety of instruments used in rheological characterization, already given an overview in Chapter 2. Among the more useful instruments is the capillary rheometer because it can cover a reasonably wide range of shear rate conditions. Also, the characteristics of capillary flow approach, at least conceptually, what occurs in an extruder and certainly through an extruder die. Because the problem of capillary flow has already been treated in an earlier chapter, only some additional comments are given here. The analysis presented in Chapter 2 is based on the assumptions of steady-state flow, isothermal conditions, and no slip at the capillary wall, from which a simple force balance results in an expression for the shear stress at the capillary wall. In most commercially available instruments such as the MPT (Monsanto Processability Tester) the polymer melt extrudes into the atmosphere, and hence the pressure drop is taken as the difference between the pressure in the reservoir and atmospheric pressure. An important drawback of this instrument is that the measurement ignores end effects due to viscous and elastic deformations at the entrance to and exit from the capillary. The obvious approach to overcoming this shortcoming is to use very large L/R ratio capillaries so that end effects are minimized; however, there are practical limitations to this approach. For elastomeric materials, L/R values can exceed 80 to minimize end effects. In other applications, such as in the characterization of black liquors in the paper industry, the author has found $L/R > 2000$ may be necessary in order to reduce end effects to negligible contributions. The specific L/R range depends on the viscoelasticity of the material and no generalizations can be made. A screening study to determine the magnitude of end effects is needed prior to implementing the capillary rheometer as a tool for polymer property characterization studies. A second approach is to apply computational corrections to τ. Large entrance losses occur with elastic non-Newtonian materials. As outlined in earlier chapters, in measuring rheological parameters of polymeric fluids with a capillary instrument, two corrections are usually needed, namely, the Bagley and slip correction calculations. The Bagley correction attempts to account for the net effect on the observed pressure drop due to viscous and elastic deformations of the polymer melt at the ends of the capillary. It does this by forcing the capillary measurements to behave as if the capillary's effective length l_{eff} were greater than its true length. In the calculations the increased effective length, l_{eff}, can be expressed in terms of multiples of the radius and the true shear stress at the wall can then be computed based on this additional length. Recall that l_{eff} is determined by measuring the pressure drop in a series of capillaries of varying L/R, then plotting ΔP versus L/R, extrapolating to $\Delta P = 0$ at each of the different shear rates. The intercept values on the L/R axis equal $-l_{\text{eff}}$ for each shear rate. In theory, the Bagley correction term should be independent of the capillary length, but in reality it does vary somewhat with L/R because of the elasticity of polymers. The degree of variability should again be established in a screening study. In addition, the Rabinowitsch equation (refer to Chapter 2) must also be applied as part of this correction to account for the "true" shear rate at the wall. This problem has been illustrated in Chapter 2 for a power-law model fluid.

The second form of correction addresses the condition of slip (also described in Chapter 2). Slip is most prevalent with unfilled rubbers and most often observed at low temperatures. There is, however, a large void in the literature concerning conditions under which slippage occurs for filled elastomers. The effect of slip on processability in mixers, calenders, mills, and commercial-scale mixers has not been reported in the open literature at present and clearly is of more than academic interest.

In addition to capillary flow, the problem of elongational or extensional flow is of importance to processability testing. In practice, extensional flow occurs at the die entrance of an extruder and in the approach to a nip of a two-roll mill. Cotten and Thiel (1978) have noted that much of the pressure drop observed at the head of an extruder may be related to extensional flow, particularly with short dies, which are often favored in rubber applications. Returning to some basics presented in Chapter 1, the strain rate of a polymer is

$$\dot{\epsilon} = \left(\frac{l}{l_t}\right)\frac{\mathrm{dl_t}}{dt} \tag{5}$$

where l_t = length of a given element at time t.

Stress–strain testing is a standard characterization tool used in elastomer applications, but there is an important principle to be kept in mind. In normal testing, l_t is in fact not a constant over the experiment. With a purely elastic material, the stresses that develop are related to the deformation and do not depend on the rate at which this deformation is applied. In contrast, a Newtonian material develops stress that is proportional to the rate of deformation. Hence, a measurement of the extensional viscosity of a polymeric material can only be made under the condition of constant deformation rate during a stress–strain experiment. If the strain rate is constant, then

$$\dot{\epsilon} = \frac{[\ln(l_t/l_0)]}{t} \tag{6}$$

where l_0 = initial length of test specimen. The total strain is thus

$$\epsilon = \ln\left(\frac{l_t}{l_0}\right) \tag{7}$$

The extensional viscosity is normally defined as the ratio of stress to extensional rate; that is,

$$\eta_\epsilon = \frac{\tau}{\dot{\epsilon}} \tag{8}$$

Furthermore, from Eq. (5) we can deduce the following exponential relationship:

$$l_t = l_0 \exp(\dot{\epsilon} t) \tag{9}$$

Equation (9) states that for the experiment to be performed at a constant extensional rate, the separation of any two molecules within the bulk elastomer must grow exponentially. In conventional tensile testing machines this condition is probably never reached. With filled rubbers (e.g., carbon black filled elastomers) specimens never actually reach steady-state viscosity. Instead, the stress continues to grow with increasing strain up to a rupture point. Rupturing occurs because of rapid orientation of the molecular chains during extensional flow.

The most widely used instruments used in elastomers processability and rheological characterization studies are those based on rotary shear. These include the Couette viscometer, cone and plate viscometers, the Weissenberg Rheogoniometer, the Mooney viscometer, and similar instruments. These devices have already been discussed elsewhere in this book. A few comments on the Mooney viscometer, however, are made here because of its relative industry dependence to rubber applications.

The Mooney viscometer is a shearing disc viscometer in which a flat serrated disc rotates in the elastomer sample in a flat grooved or serrated cavity.

This creates an element of Couette flow along with rotary shear opposing flat surfaces. However, the flow geometry is more complex and it is likely there are transition flow patterns in the intermediate regions of the rubber sample. As noted earlier, there are two standard rotor sizes—the larger one, used most frequently, and a smaller one used for highly viscous, stiffer rubbers. The most commonly reported measurement is designated ML 1+4 (100°C), meaning the Mooney viscosity measured at 100°C after 1 min preheat and 4 min rotor rotation. Typical Mooney viscosity curves for different rubber samples are shown in Figure 3. There are several inherent problems with the Mooney viscometer, most of which are associated with improper maintenance. Rotor disc wear, negligence in cleaning, poor temperature and rotor speed control, and entrapment of air pockets in rubber samples can affect the accuracy and reproducibility of measurements. As with any heavily used instrument, particularly if applied in frequent quality control testing, it is advisable to run periodic calibration checks and to control chart the measurements as illustrated in Figure 4. Such practices will signal the need for dead weight calibration and/or replacement of rotors.

Another potential problem with the Mooney viscometer is slippage. Instabilities in Mooney viscometer curves for EPDMs have been reported by Markert (1976) at short times and at rubber temperatures below 100°C. This was probably

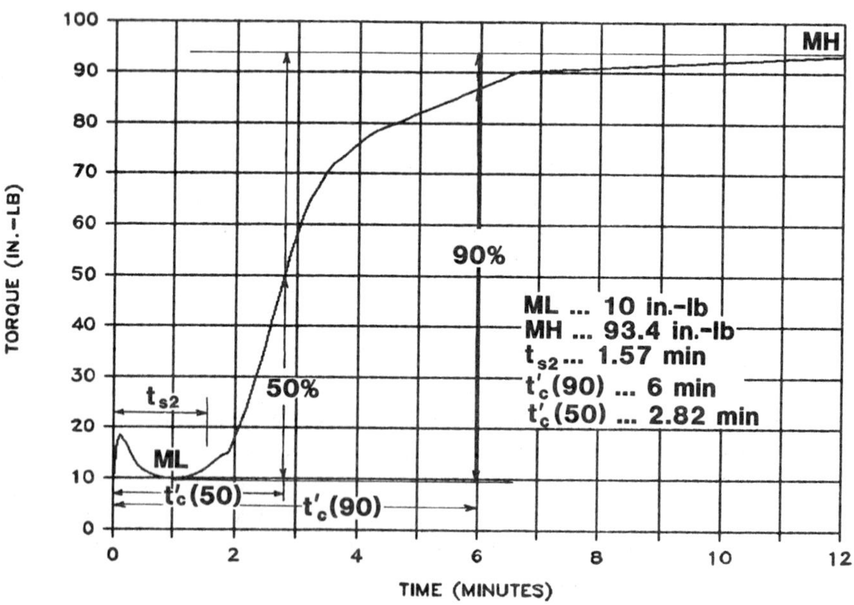

Figure 3 Typical Mooney viscosity curves for rubber samples.

due to shear breakdown of the polymer when a critical stress was exceeded. Similar observations have been made by the author for very stiff (very high Mooney) polymers, an example of which is shown in Figure 5. The instability may be attributed to wall-slip effects, or the apparent wall-slip effects may be due to structure breakdown close to the wall. One approach to overcoming this problem is to blend the stiff rubber sample with a low Mooney viscosity sample on a mill. Care must be taken not to break down the high Mooney component, and so blending should be performed at moderate to low shear rates and temperatures. The Mooney viscosity of the blend can then be measured, and since its overall viscosity is lower, the slippage is reduced. Figure 6 shows the Mooney viscosity response curves for several blend ratios of two polymers. Note that at the 50/50 blend, the torque curve shows no signs of instability and the measurement is reproducible to within plus or minus 0.5 Mooney points. From a measurement of the low Mooney viscosity rubber, the true Mooney level of the stiffer component can be estimated from the log-law blending rule:

$$a \ln(\mathrm{ML}_1) + (1 - a) \ln(\mathrm{ML}_2) = \ln(\mathrm{ML}_B) \tag{10}$$

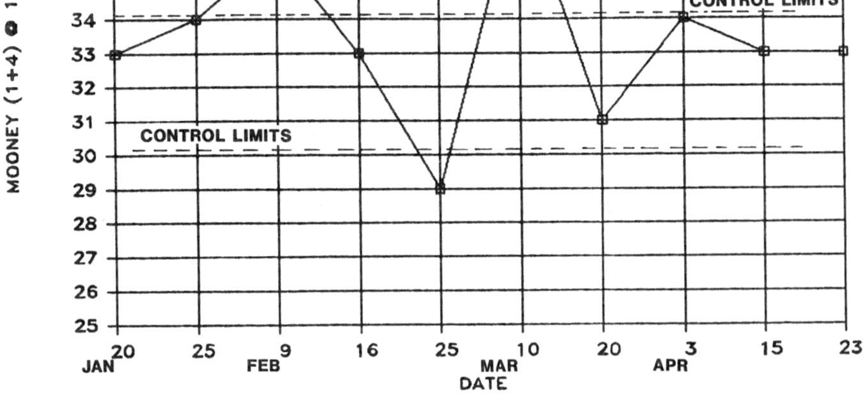

Figure 4 Control chart to cross-check calibration of a Mooney viscometer.

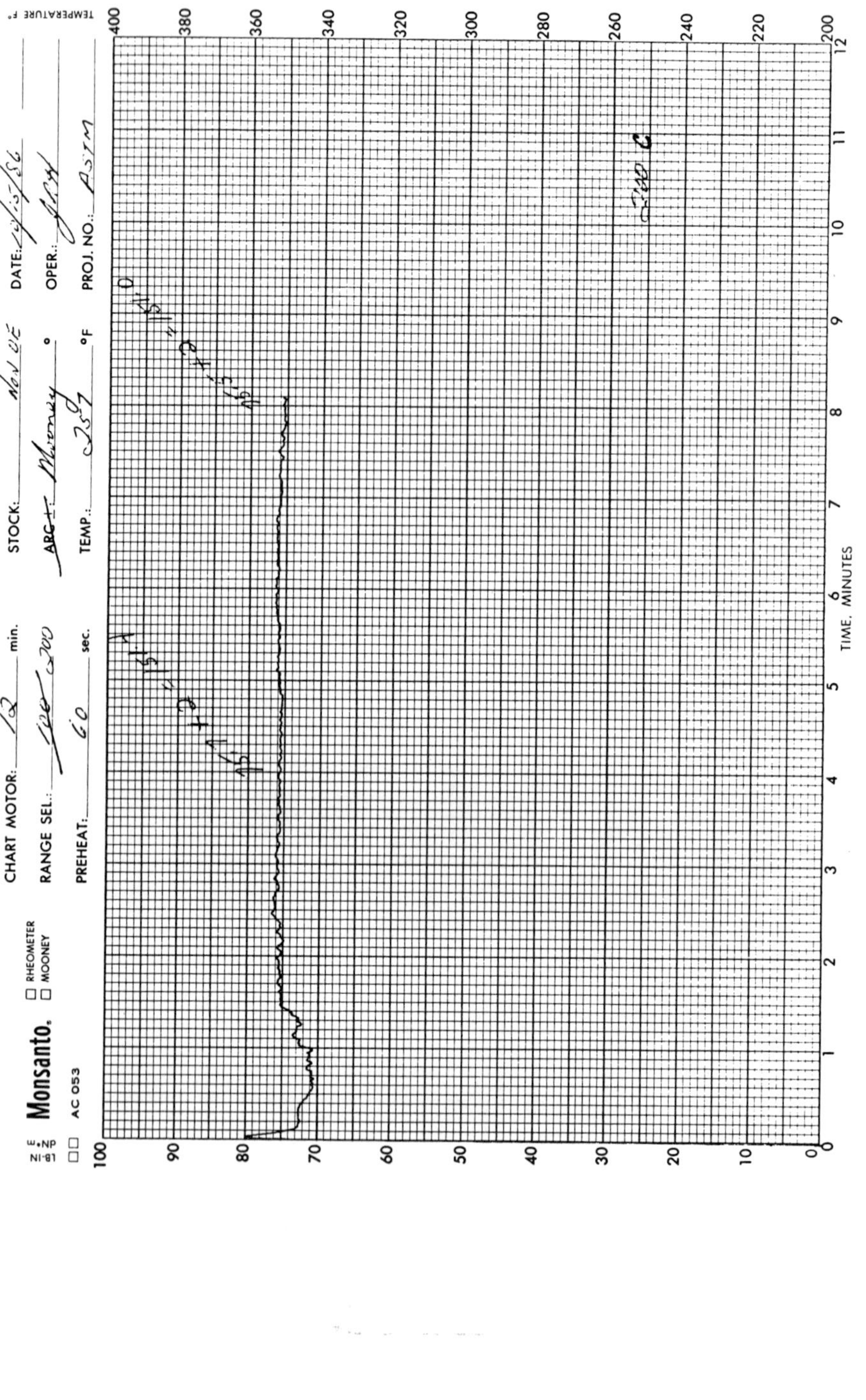

Figure 5 Example of shear instability (slippage) for a Mooney viscosity curve.

where

ML_1, ML_2 = Mooney viscosity for blend components 1 and 2, respectively
ML_B = Mooney viscosity of the blend
a = fraction of blend for component 1

Hence, knowing ML_1 (from a measurement of the low Mooney sample), ML_B, and a, the Mooney viscosity of the stiffer component (ML_2) can be solved from Eq. (10). A final note on the Mooney viscometer is that the instrument is most often not utilized to its full capabilities in industrial laboratory environments. Measurements are normally restricted to a fixed, very low rotor speed; however, the instrument is capable of performing tests over a relatively wide range of shear rates. The commercially available instrument can be run at shear rates close to 100 s^{-1}. Depending on the polymer sample, however, heat generation may become a problem, along with the potential for regimes of slip to be encountered. It does, however, offer the potential to be used in a similar fashion as a capillary rheometer.

Another commonly used type of instrument is based on the principle of and eccentric rotating disc, of which the most well known is the Rheometrics Mechanical Spectrometer. The instrument can conduct experiments over a variety of oscillatory and continuous shear modes and test conditions. The device is highly versatile in that it can be used to characterize a wide range of properties and materials ranging from solutions to solids. The instrument is equipped with a variety of adaptors of different geometries, such as cone and plate, parallel plate, biconical, concentric cylinder, an eccentric rotating disc. The last adaption is known as the ERD system (also called the orthogonal rheometer). Rubber is positioned between two parallel, but not coaxial discs which rotate at the same speed and in the same direction. This action subjects the rubber specimen to sinusoidal oscillations. By measuring the forces on the anvils along the principal axes, one can calculate the real and imaginary components of the shear modulus, which forms the basis for determining the normal stress effects. Macosko and Davis (1974) and Willey et al. (1974) provide detailed discussions on this instrument.

DIFFERENTIAL SCANNING CALORIMETRY AND PROCESSABILITY TESTING

A differential scanning calorimeter (DSC) is an instrument used for the measurement and characterization of the thermal properties of materials. Figure 7 shows a block diagram of a typical DSC. The measurement principle involves bringing the average temperature of the sample holder to a desired initial temperature and

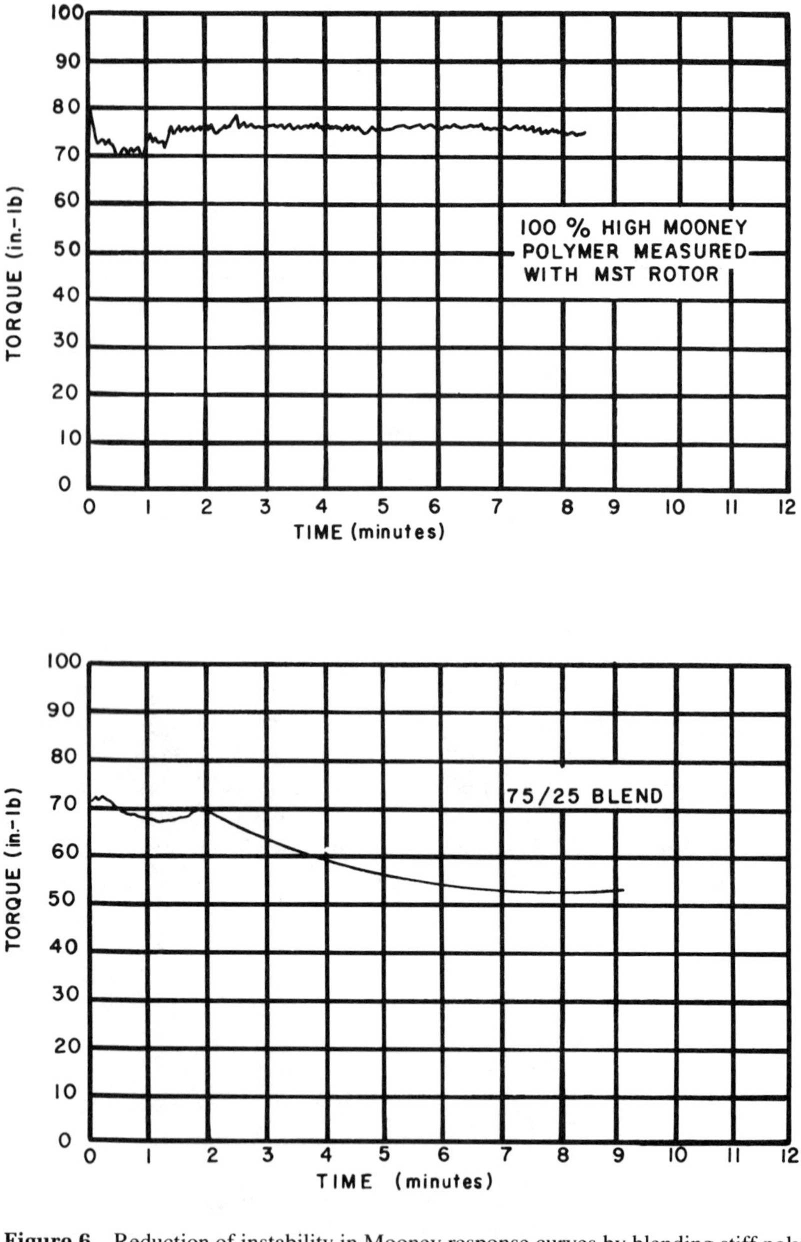

Figure 6 Reduction of instability in Mooney response curves by blending stiff polymer with a lower-viscosity standard rubber.

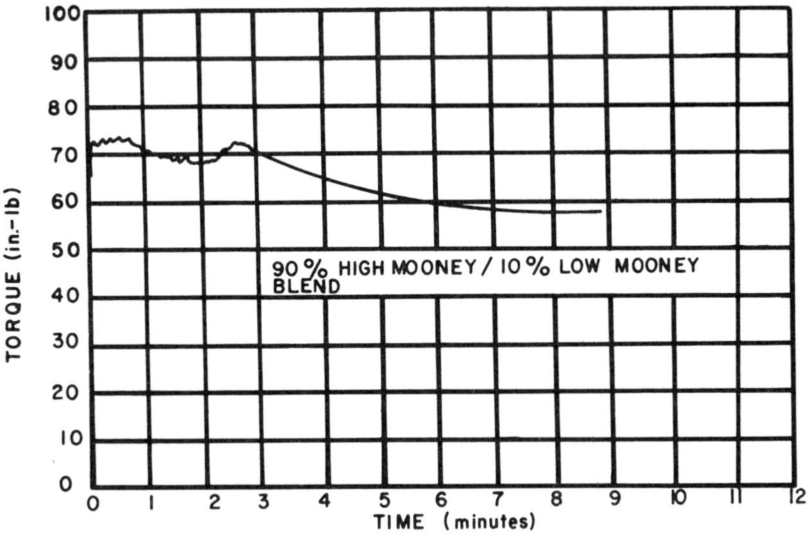
90 % HIGH MOONEY / 10 % LOW MOONEY BLEND
TORQUE (in.-lb)
TIME (minutes)

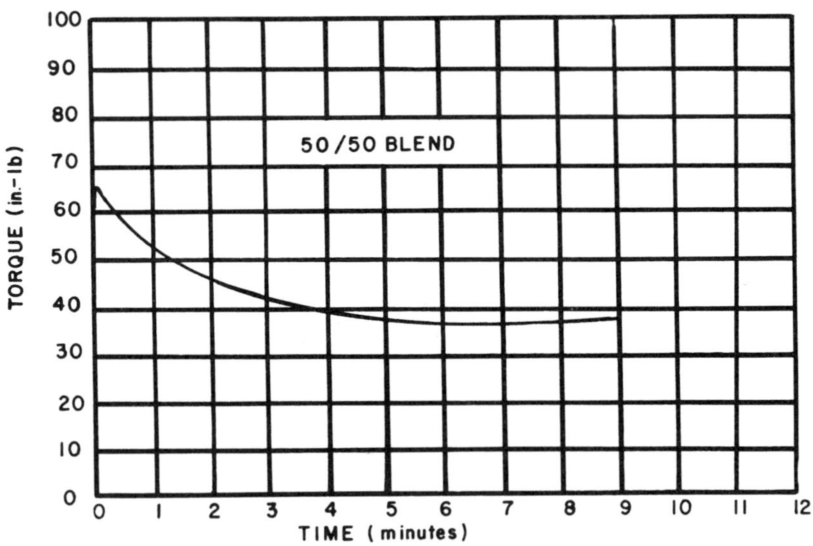
50/50 BLEND
TORQUE (in.-lb)
TIME (minutes)

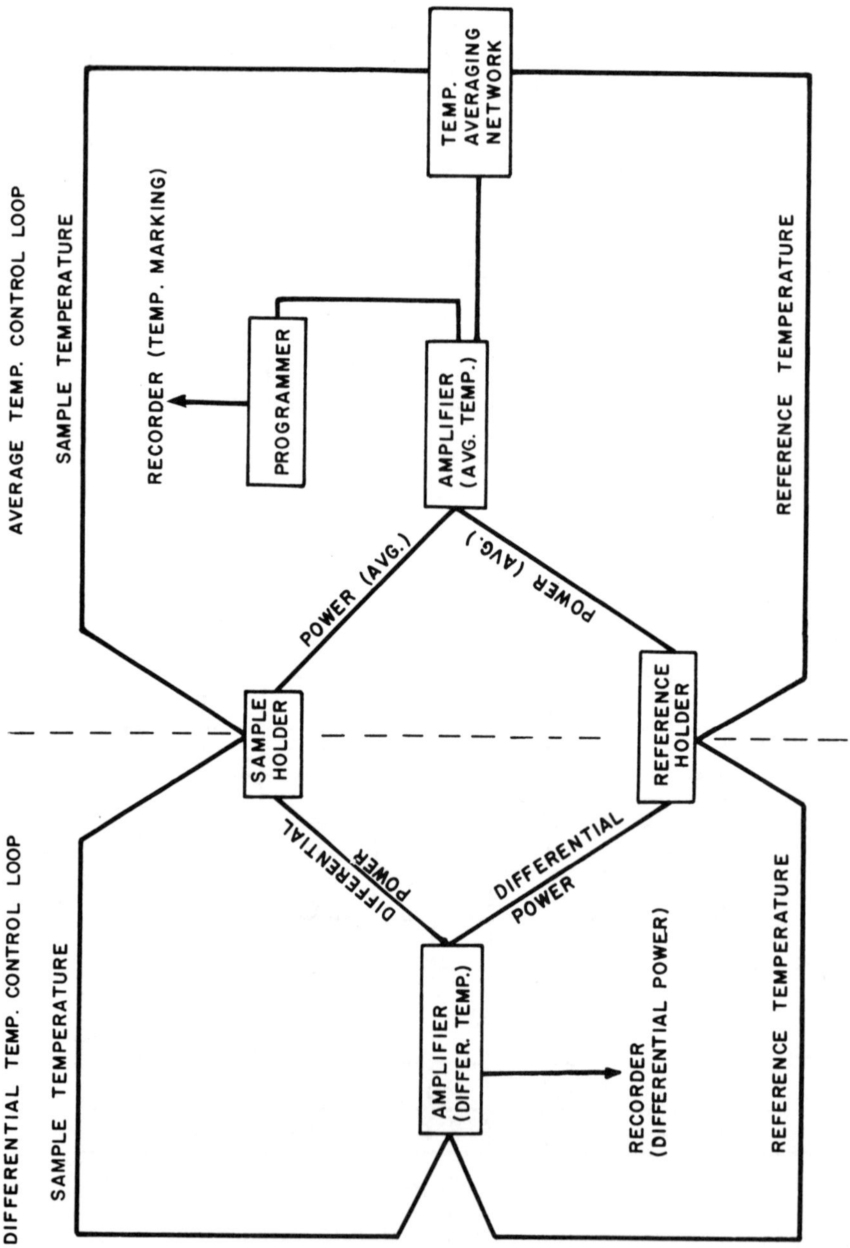

Figure 7 Block diagram of typical DSC instrument.

then, by means of a microprocessor, a desired change in average temperature of the sample holder to a final temperature is achieved. When a transition such as melting, boiling, dehydration, or crystallization occurs in the sample, an endothermic or exothermic reaction takes place. The change in power needed to maintain the sample holder at the same temperature as a reference holder (i.e., its programmed temperature) during the transition is recorded as a peak. Usually the chart abscissa denotes the transition temperature and the peak area indicates the total energy transfer to or from the sample.

The sample holders are comprised of two platinum alloy cups which are mounted in the cavities of a solid aluminum block, and which contain a heater and sensor. For measurements above ambient conditions the instrument is usually capable of circulating a cooling fluid through the aluminum block to maintain it at a constant temperature, independent of the temperature range of the measurement. Since this is a direct calorimetric measurement, each sample holder must have its own built-in heater and a temperature sensor. A high-gain, closed-loop electronic arrangement provides differential electrical power to the heaters to compensate for temperature fluctuations. This means that for practical purposes, the holder temperature is unaffected by the sample behavior. The differential power required to maintain the balanced condition can be recorded directly in millicalories per second on an appropriate recorder. This measurement is equivalent to the rate of energy absorption or evolution of the sample.

Figure 8 illustrates typical calorimetric data derived from DSC measurements. Endothermic transitions and increases in heat capacity are represented by upscale departures from the ordinate baseline, whereas exothermic reactions are

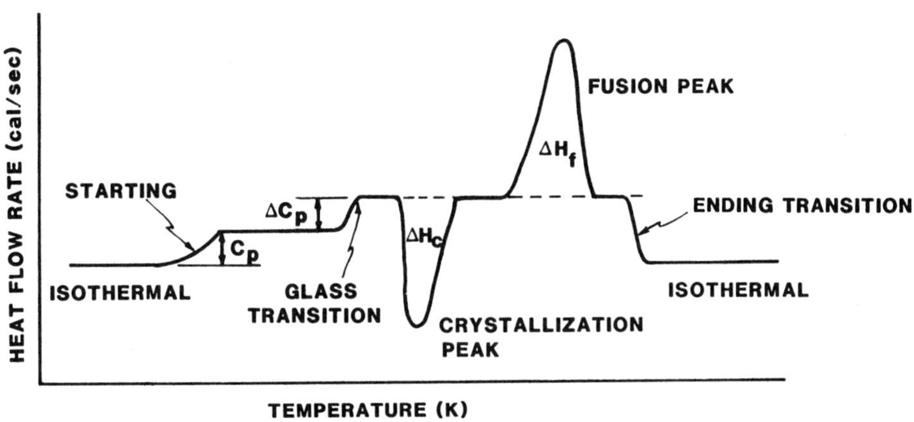

Figure 8 Thermal analysis data.

represented by downside departures from the baseline. The abscissa is usually calibrated in degrees Kelvin.

A DSC instrument will measure the rate of energy absorption or evolution by the sample in units of millicalories per second. In some experiments this rate can be measured directly (e.g., in kinetic reaction studies). In measurements of heat capacity (or specific heat) the ordinate displacement is measured directly since $C_p = dH/dT$, which is related to the heat of absorption dH/dt:

$$\frac{dH}{dt} = \frac{dH}{dT} \times \frac{dT}{dt} \tag{11}$$

where dT/dt is the temperature scanning rate.

The total energy associated with a transition or reaction is obtained by integrating over the peak with respect to time:

$$\Delta H = \int \frac{dH}{dt}\, dt \tag{12}$$

Both solid and liquid samples can be analyzed. Solid samples can be in film, powder, crystal, or granular form. Quantitative accuracy is usually unaffected by sample shape; however, the qualitative appearance of the thermogram may be affected by the sample configuration. For maximum peak sharpness and resolution, a configuration which provides good contact between the sample surface and pan is desirable. Thin films or discs of sample or fine granules spread in a thin layer on the pan bottom give good instrument performance. Polymer films can be sampled by cutting out sections of the film with a cork borer.

A sharp, first-order transition, such as the melting point of a pure material, displays an isothermal behavior. Since the temperature of the sample does not increase during the transition, the temperature of the transition is that tempera-

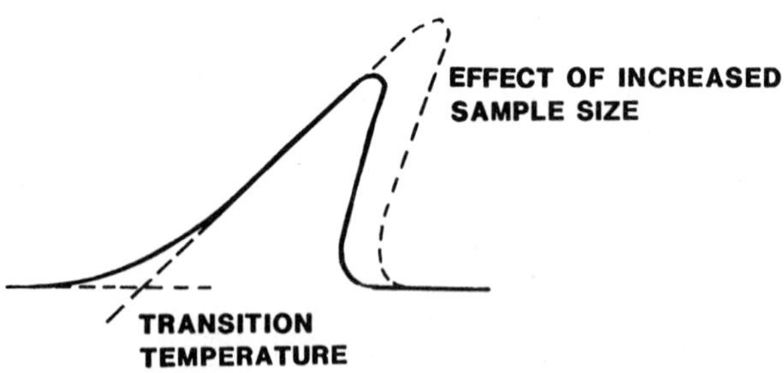

Figure 9 Determination of transition temperature.

ture at which the transition is first observed (refer to Figure 9B). Note that this temperature is not influenced by variations in sample size, whereas the peak temperature can be.

Crystalline or semicrystalline materials usually undergo melting. Possible exceptions occur if the material sublimed or decomposed prior to reaching the melting point. Melting is the most commonly encountered thermal behavior and is observed as a sharp endothermic peak on a DSC thermogram. The area under the melting peak is a direct measure of the heat of fusion for the sample. The sharpness of the melting peak is affected by such factors as purity and crystallite perfection.

For nonpolymeric materials, an impurity will depress its melting point and significantly broaden its melting range. Figure 10 illustrates typical melting peaks for different purity samples of benzoic acid. These types of curves enable one to distinguish samples of different purities. In contrast, polymers inherently melt over broad temperature ranges. Figure 11 shows the thermogram for PTFE, illustrating that the material has a relatively sharp melting peak but melts over a 150°C temperature range. The broad melting characteristic is attributed to crystallite size or imperfections through the polymer, and is not evidence of impurity. The smaller, less perfectly formed crystals tend to melt at lower temperatures because of surface free energy effects. The melting characteristics of a polymer are influenced by a number of factors that affect the degree of perfection of crystallinity, such as thermal history. Gray (1970) has shown that the percent crystallinity of the polymer can be determined from the measured heat of fusion.

An important aspect to consider is the presence of multiple melting peaks.

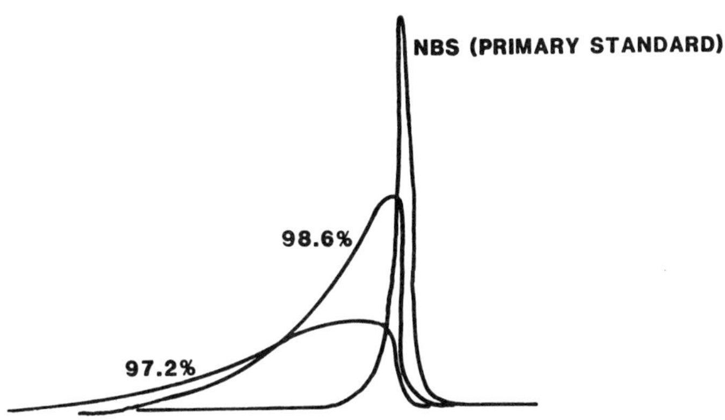

Figure 10 Effect of purity on melting peak shapes of benzoic acid.

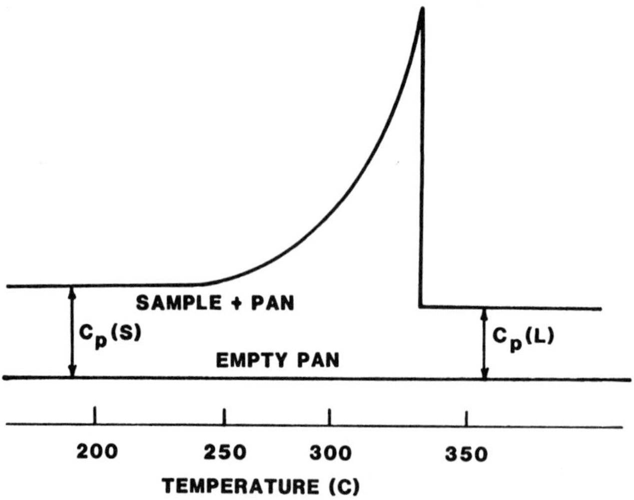

Figure 11 Melting peak for PTFE.

This can be attributed to several factors. First, crystallization is very sensitive to the thermal history that the polymer sample is subjected to during processing. If the material is maintained at different temperatures, crystal growth will be promoted in those regions. As a result, on heating, a minimum can be detected at the annealing temperature, since those crystals formed on annealing will melt above the annealing temperature and those crystals formed on cooling from the annealing temperature will melt below that temperature. An example of processing pretreatment is illustrated in Figure 12. The three EPDMs are chemically and structurally identical, but they were processed and annealed under different conditions. Thermograms of this type offer the possibility of controlling processing conditions. Also, the influence of thermal history can be distinguished from structural differences by introducing a constant thermal history to the samples under study, such as cooling at 20°C/min from above the melt. If the rescans between samples are then identical, the initial differences may be attributed to thermal history. On the other hand, if the rescans continue to show differences, then the samples are likely to be structurally different.

Multiple melting peaks can be observed in blends and mixtures. When two polymers are mixed together, they may retain their individual thermal characteristics. Figure 13 illustrates the thermogram of a blend of a high- and low-density polyethylene. Branching tends to impede crystallite perfection; hence the low-density constituent melts at the lower temperature, being almost completely resolved from melting of the crystallites formed from the linear molecules. Scans

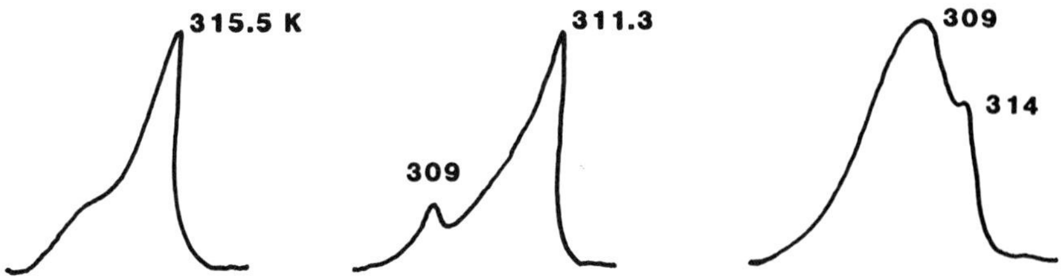

Figure 12 Effect of pretreatment on an EPDM's melting behavior.

such as waxes are comprised of mixtures of chemically similar and structurally different molecules. Melting thermograms of these materials can show multiple peaks as well. Wax thermograms in general are highly characteristic and can be used as a means of identification.

Some comments concerning copolymers are warranted. There are basically two types, random and block. Random copolymers are different from the homopolymers, whereas a block copolymer usually retains the characteristics of the homopolymers. A block copolymer thermogram resembles that of a physical blend.

In terms of characterizing polymers and evaluating their processability, one of the more important properties to define is crystallinity. The term *crystalliza-*

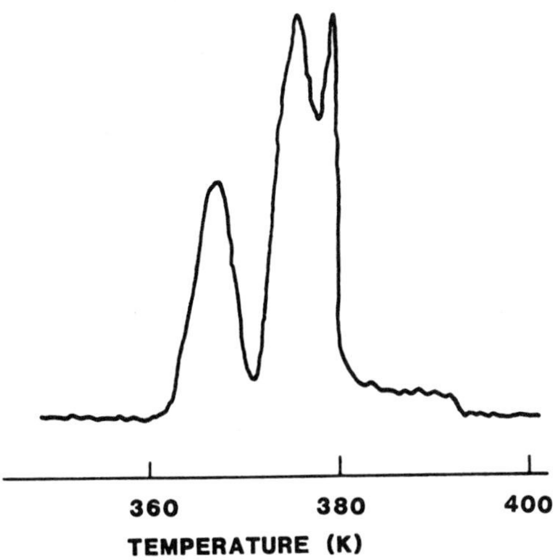

Figure 13 Example of multiple melting peaks from a physical blend of a branched and linear polymer.

tion includes four general types with respect to DSC measurements, namely, program cooled crystallization, cold-crystallization, recrystallization, and isothermal.

Program cooled crystallization refers to materials that melt on heating and then crystallize on cooling as witnessed by a sharp exotherm on the DSC trace.

Since all materials will supercool, the rate of crystallization may be faster than the rate of melting. In this case, the crystallization exotherm may be taller than the melting endotherm. If complete crystallization has occurred, the peak areas (i.e., energies) will, however, be the same. A problem in crystallization studies is, however, that a number of materials do not crystallize completely, or even partially, on cooling. For example, organic constituents that have been crystallized from solution often tend not to crystallize from the melt. With polymers, the process of crystallization is not instantaneous but, instead, involves the processes of nucleation and propagation. As a result, the extent of crystallization depends also on the rate of cooling.

In cold-crystallization, polymers crystallize very slowly so that they can be rapidly cooled to a completely amorphous glassy state. When they are heated through the gassy region, molecular mobility increases and the substance crystallizes. This behavior is referred to as cold-crystallization since it occurs at temperatures far below the melting point. An example of this is shown in Figure 8 for a copolymer sample that has been shock-cooled (solid line) from the melt. The cold-crystallization exotherm is at 296 K.

As noted earlier, polymers tend to melt over a broad temperature range because of the surface-free energy of the smaller, less perfectly formed crystals. Once these crystals are melted, however, the free energy of the liquid can be higher than that of the more perfectly formed crystals which are stable in the temperature range. Consequently, it is within thermodynamic principles for the liquid to crystallize as shown in Figure 14. Controlled cooling of the copolymer resulted in

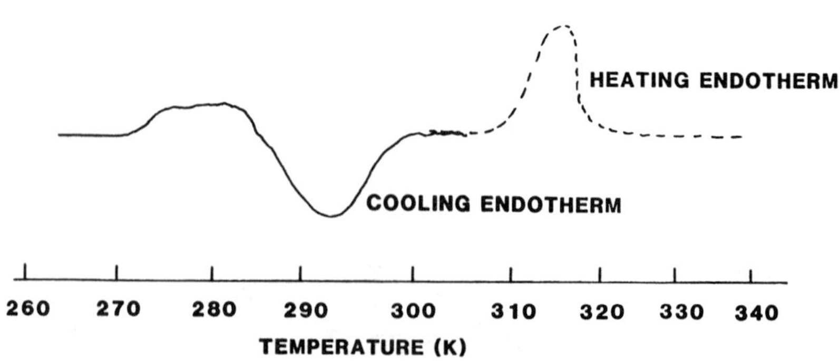

Figure 14 Thermograms of a copolymer.

partial crystallization, as witnessed by the absence of the cold-crystallization exotherm observed for the amorphous sample. On the other hand, the morphologies of the crystals formed give rise to recrystallization, as evidenced by the exothermic peak at about 315 K, during the melting cycle of the experiment. These crystallization effects on heating influence determination of the true crystallinity of the sample. Simply considering the area under the fusion peak as a measure of crystallinity is likely to result in large errors.

A final note on crystallization concerns tests that are conducted under isothermal conditions. It is important to realize that isothermal operation can affect the degree of crystallization as well as the size distribution of the melting curve. It is important to standardize measurements to account for the degree of supercooling on the rate of crystallization.

Another property of polymeric materials of importance is the glass transition. At the glass temperature (T_g) the material changes from a brittle glassy state to a plastic or liquidlike state. The glass transition is thought of as a second-order transition since it involves a discontinuous change in a secondary thermodynamic quantity such as the specific heat. Note that a DSC thermogram is essentially a continuous plot of specific heat as a function of temperature; hence the glass transition will be observed as a discontinuity or step change in the ordinate of the trace. The magnitude of the step change will be proportional to the amorphorous content of the polymer. The T_g is influenced by several factors such as molecular weight, the degree of cure, plasticizer, polymerization, and so on. Perkin-Elmer Corp. (1973) discusses the effects of various parameters on T_g. Table 1 gives typical values of T_g and melting temperatures for various polymers.

Table 1 Glass Transition and Crystalline Melting Points of Different Polymers

Polymer	$T_G(K)$	$T_m(K)$
Silicone rubber	150	215
Polybutadiene	173	253
Butyl rubber (polyisobutylene)	205	
Natural rubber (polyisoprene)	203	245
Polyvinylidene fluoride	234	483
Polypropylene	253	449
Polyvinyl acetate	303	
Nylon G	320	498
PVC	355	453
Polystyrene	373	503

The DSC is used routinely in polymer characterization studies. Specific thermodynamic properties measured include specific heats, heats of fusion, and quantitative and qualitative analysis of polymer blends and copolymers, the technique can be used for distinguishing folded chain from extended chain morphologies, for evaluating the effects of comonomers or chain substitutes on morphology and general thermal behavior, for crystallinity determinations, and for determining thermal stability and crystallization rates. In addition, the DSC technique can be used for evaluating the effects and effectiveness of additives such as plasticizers, antioxidants, and accelerators, as well as catalyst and nucleating effects on the polymer's thermodynamic properties. Although DSC methods are not generally thought of in terms of processability testing, the technique enables one to establish useful correlations between DSC derived data and finished product quality. Many polymer processing operations are critically dependent on the control of thermal treatments; examples are curing heat-setting, annealing, shock-cooling, and hot-drawing. Each of these techniques can be studied and controlled using DSC techniques. A simple example of this was given in the last chapter, where heat of fusion data generated by the DSC were correlated with compound green strength and extrusion rates. Another example is given in Figure 15. Figure 15 shows the thermograms of two polymers made with the same catalyst system (solution polymerization) but at greatly different reactor mixing conditions. Note that both polymers show the same heats of fusion, and therefore have the same level of crystallinity. In addition, they both show near-identical melting ranges and peaks, but upon cold-crystallization, polymer B (well mixed in reactor) crystallizes more than 20° below polymer A. We could conclude that polymer B is a more homogeneous polymer in terms of its ethylene distribution. In an injection molding operation we may expect poly-

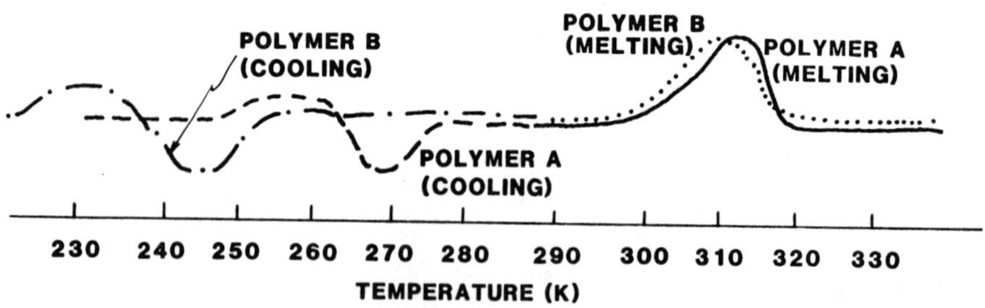

Figure 15 Cold-crystallization detected in DSC. Differences between polymers can be related to mold shrinkage.

mer A to display the problem of mold shrinkage. In simple mixing/milling tests both polymers were compounded identically, rolled into sheets, and allowed to set up for 8 hours. It was observed that polymer A showed nearly twice as much shrinkage as polymer B. Hence, we may conclude that DSC can be employed routinely for processability testing in a manner that involves empirically relating thermodynamic properties to product processing performance.

It is important to try to relate thermodynamic observations to the molecular properties of the polymer in order to establish processability correlations. For elastomers, the key molecular requirements are as follows:

Weakly interacting chains
- above the glass transition temperature
- noncrystalline unstrained

Presence of cross-links
- both chemical and physical

Presence of a small number of chain ends after cross-linking
- high-molecular-weight ends
- reactive-chain ends

Presence of a two-phase system
- display crystallization on extension
- display polymeric domains
- allow incorporation of fillers and reinforcing agents.

Figure 16 summarizes the general categories of stress–strain properties of polymers.

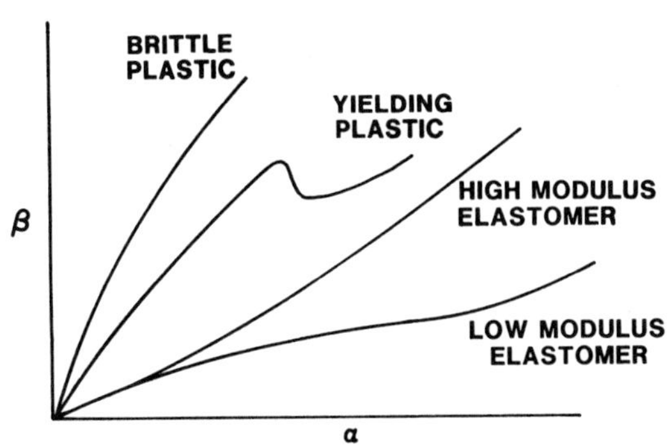

Figure 16 Stress–strain properties of polymers.

IMPORTANCE OF MOLECULAR PROPERTIES AND VULCANIZATE PROPERTIES

Figure 16 illustrates the types of stress–strain relationships that characterize various polymer materials. From dynamic mechanical spectroscopy, the amount of energy dissipated from a block of material subjected to stress is

$$E'' = E' \tan \delta \tag{13}$$

where

E'' = loss modulus (i.e., energy dissipated)
E' = storage modulus (i.e., energy stored elastically)
δ = phase angle between the applied stress and resulting strain

Phase angle δ is analogous to the power factor in basic electricity. The relationship between loss modulus and storage modulus to temperature is shown in Figure 17.

In the processing of polymers the material is subjected to stresses which invoke physical extension or strain on the molecular structure. The simple dynamic shear experiment shown in Figure 18 shows that when a force F is applied to a block, stress is induced, causing deformation or strain to occur. On a molecular scale this means an extension or rather unentanglement of the long polymer molecules. This deformation results in an increase in interatomic dis-

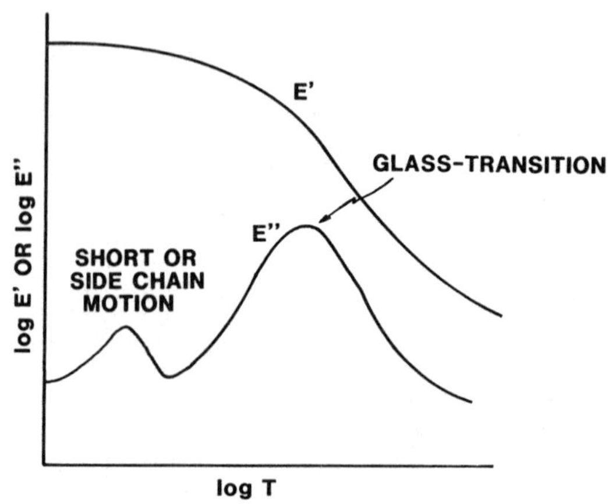

Figure 17 Relationships for storage and loss moduli.

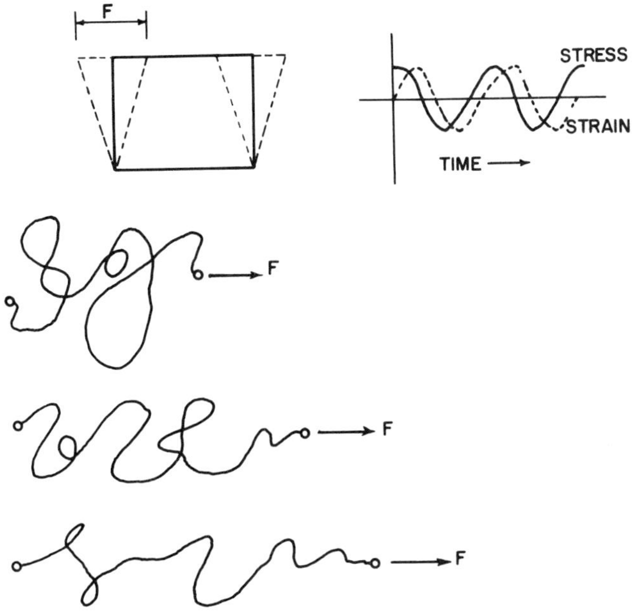

Figure 18 Applications of force to polymer molecules.

tance, and there is a transformation in volume and shape of the material. From thermodynamic principles the change in free energy upon extension is

$$\Delta F = \Delta H - T\,\Delta S \tag{14}$$

Enthalpy, ΔH, can be thought of in terms of the familiar spring model described in numerous textbooks and alluded to in Chapter 1. For elastomeric materials it is representative of the increase in intra-atomic distances. The entropy term denotes the loss in the number of possible chain configurations during the deformation.

Time and temperature dependence of the stress–strain properties of elastomers plays an important role in their processability. With elastomers it is possible to identify a failure envelope which characterizes the physical and rheological nature of the material (see Figure 19). From the viewpoint of engineering applications the tensile strength of elastomers is of great significance in determining end uses. The tensile strength of elastomeric materials depends on a number of factors, the most critical ones being the extension rate and temperature the material is subjected to, the type of elastomer (i.e., single-phase noncrystallizable, crystallizable, polymers which contain domains such as glass domains or ionic domains), the type of cross-linking, and, of course, the molecular network.

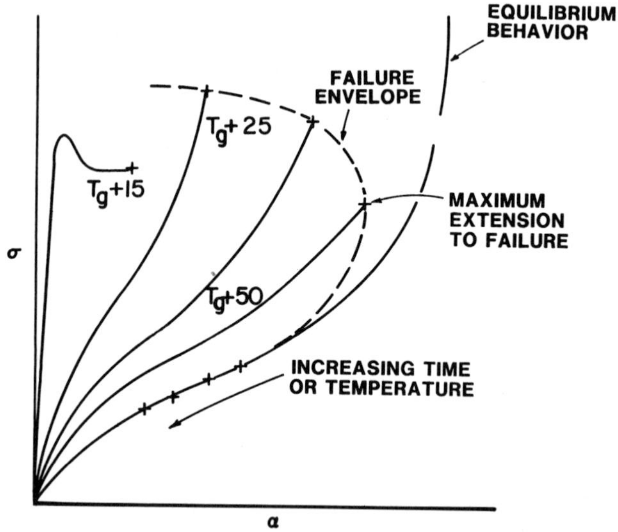

Figure 19 Time and temperature dependence of stress–strain properties.

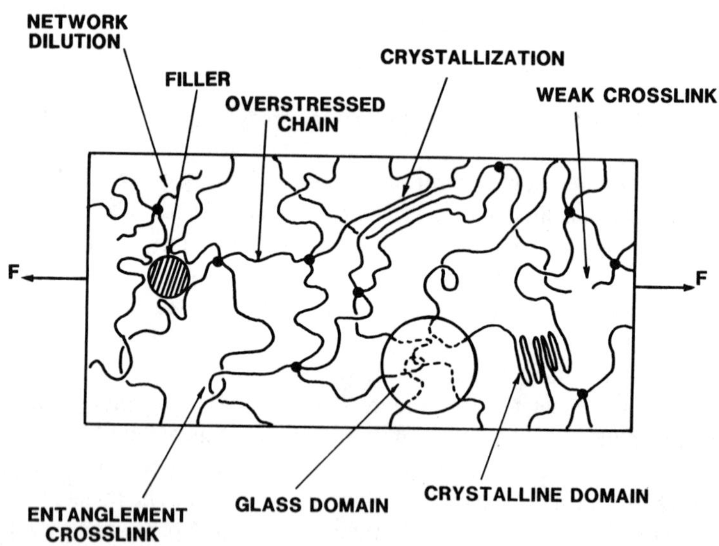

Figure 20 Tensile strength mechanisms.

Elastomers gain their strength properties largely by the incorporation of filler material; however, there are more complex contributions. The various sources of strength include the material's viscoelastic properties, molecular network perfection, the orientation of chains, the degree of energy dissipation by fillers, the ability of fillers to deflect and stop the propagation of microcracks, various polymeric domains which serve as fillers, the deformation of domains, and various interactions among the above. Figure 20 summarizes the various mechanisms contributing to tensile strength.

It is worthwhile reviewing some basic principles of vulcanization for elastomers before discussing processability. Figure 21 presents a simplified picture of the molecular structure of an unvulcanized and sulfur-cured vulcanized rubber network. The sulfur molecule in this case forms a cross-link network by bonding between the rubber molecules. The properties of the vulcanizate will depend on the cross-link density. The properties of engineering interest are tear strength, fatigue life, material toughness, stiffness, tensile strength, set properties, hysteresis, and others (refer to Figure 21). There are a battery of testing procedures to assess each of these properties, but in conventional rubber applications the starting place is usually the cure response of the vulcanizate. To assess this a device called an oscillating disc rheometer is used, a simplified diagram of which is shown in Figure 22. Typical torque response curves are shown in Figure 23.

Important parameters related to the vulcanization process are the time elapsed before it begins, the rate at which it occurs, and the extent to which it occurs.

The multitude of rubber products used throughout the world could be manufactured without the step of vulcanization. The process of vulcanization involves the cross-linking of the chainlike rubber molecules, which are transformed into an elastic network. There are a variety of cross-linking agents, the most predominant of which are sulfur, organic peroxides, and various polyfunctional coupling agents. In conventional cured rubber applications, sulfur is predominantly used because of its high vulcanizate properties, the wide latitude offered in compounding, and its low cost. A large advantage of sulfur package curing systems is that they provide delayed action, which is needed in the processing, shaping, and forming stages of the rubber article prior to its becoming an intractable vulcanized network. Accelerators coupled with sulfur can, in addition, be used to increase the rates of cross-link formation after the delay period.

Elastomeric materials are characterized by their ability to forcibly retract to their approximate original configuration after being subjected to deformation. The forces responsible for retraction can be increased and the degree of permanent deformation remaining after being removed from the deforming force can be reduced by vulcanization. In other words, vulcanization contributes to enhancing the property of elasticity and therefore is responsible for reducing plasticity. A very simple viewpoint of an elastomer is that on a molecular scale it is comprised of a network of linear polymer chains that are joined together by cross-links. It is

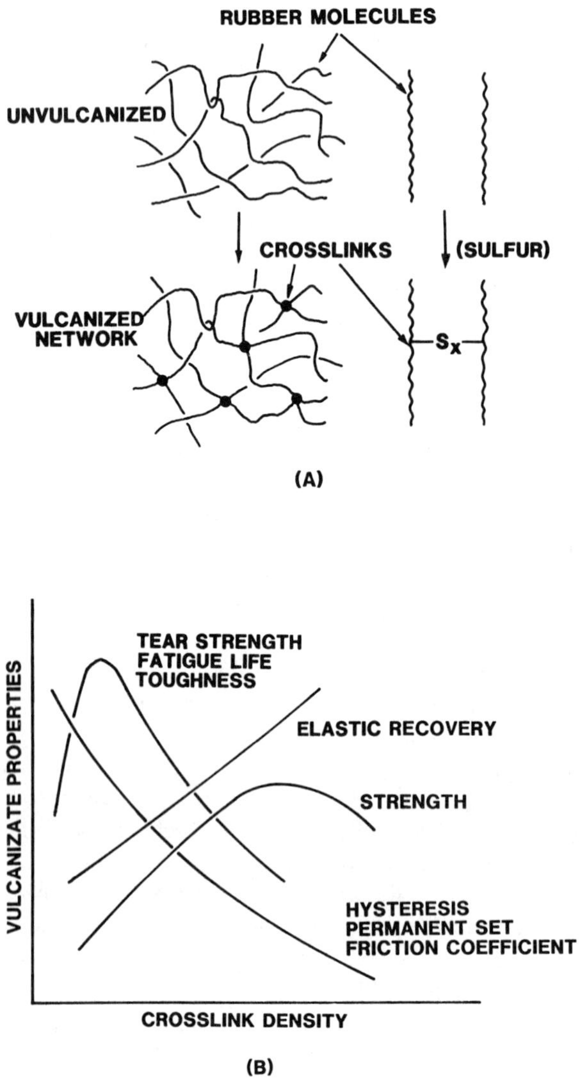

Figure 21 (A) Effect of vulcanization on rubber matrix. (B) Graphical illustration of how vulcanization properties are affected.

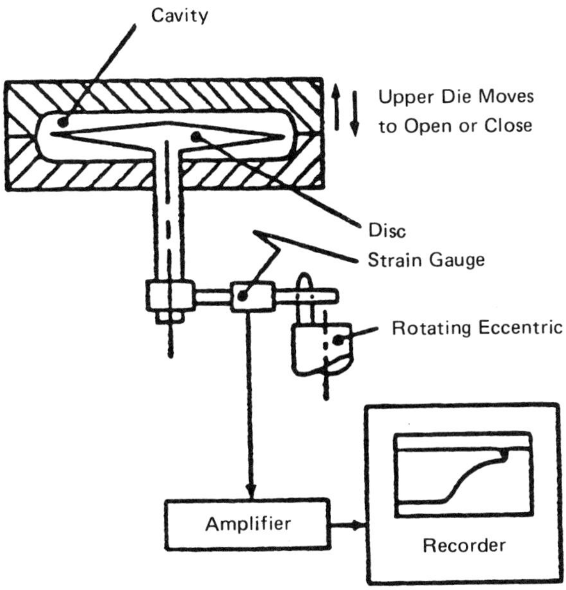

Figure 22 Oscillating disc rheometer.

the relative number of network-supporting polymer chains per unit volume of elastomer that establishes the magnitude of retractive force capable of resisting deformation. The number of supporting chains can be increased by raising the number of junctures. The process of vulcanization is a chemical means of producing network junctures. It does this by inserting cross-links between polymer chains. In practice, sulfur vulcanization is performed by heating the elastomer and vulcanizing agents together in a mold.

As noted earlier, properties such as fatigue life, tear strength, and toughness can be characterized by the energy needed to break. A measure of the deformation energy is hysteresis, which is a rate-dependent property. Hysteresis is practically defined as the ratio of the viscous component to the elastic component of deformation resistance.

A practical processing problem is to prevent or minimize the risk of premature vulcanization. This is referred to as *scorch*. For an elastomeric material to be of engineering and consumer interest, it must have sufficient scorch resistance; that is, it must be capable of having sufficient time delay in vulcanization to enable processing (mixing, milling, forming, molding, etc.) of the final article. In other words, the process of cross-link formation must be controllable within the limits of the processing requirements of the application. One key

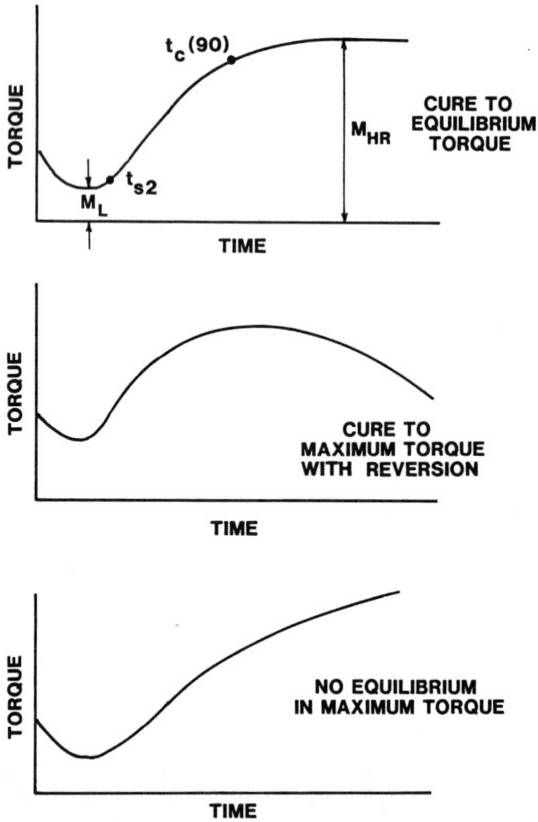

Figure 23 Vulcanization properties characterized on an oscillating disc curemeter.

requirement then in selecting a polymer–compound system for a particular end-use application is that processing time and scorch resistance be compatible.

Scorch resistance can be studied on the laboratory scale and should be measured at the intended processing temperature. It is measured as the time required for the onset of cross-linking. By using a Mooney viscometer one can detect this time from a sharp increase in viscosity. A rheometer (or curemeter) such as that shown in Figure 16 can supplement this evaluation by providing a measure of the extent of cure.

The curemeter shown in Figure 22 is based on the principle of oscillating discs: a sample of rubber is cured in a cavity, while a disc, embedded in the sample, is in oscillatory rotation covering an arclike motion. The resistance to the disc's oscillation is measured as a function of time, producing the curemeter response curve shown in Figure 23. This type of curemeter (rheometer) chart characterizes both the induction time and the rate of cure (i.e., the vulcanization rate). Mooney viscosity measurements over time provide the processing safety guideline in terms of scorch resistance. This latter measurement is usually reported as time in units of minutes for a prespecified change in Mooney units. In the rubber industry, scorch times are typically reported as minutes for a five-point rise in Mooney viscosity, or over a three-point rise. The oscillating disc rheometer is then used to characterize the vulcanization kinetics at a temperature typifying full-scale processing.

APPLICATION OF TORQUE RHEOMETRY

Mixing and extrusion constitute two of the most widely used processing operations for elastomeric products. It is critical therefore that representative laboratory testing procedures be used to characterize and project the polymer's performance prior to manufacturing finished articles. Small-scale internal mixers (midget and B-Banbury size) mixers capable of neat rubber capacities ranging from 50 or more grams of rubber to kilogram charges are usually standard equipment in a rubber processing laboratory. As a minimum test, the work input to the machine may be used with some success to assess the state of mixedness during a mix cycle and to determine when to dump the batch. A laboratory-scale Banbury or similar mixer can serve as a processability tester to obtain data on the changes in resistance to deformation during mixing, the times to specific properties of the power trace being used as indexes of processability of the neat polymer, or the degree of dispersibility of carbon black, mineral fillers, or other additives in a given polymer.

More advanced processing laboratories are equipped with instrumented miniature mixers called torque rheometers, which make it possible to measure the torque on the rotors. This can provide a basis for developing quantitative indexes of processability as a function of the time of mixing. Companies such as Haake-Buchler and Brabender also provide an adaptable screw mixer section that is a miniature extruder. The torque rheometer is equipped with dual temperature and pressure sensors as well as a torque meter, and hence provides a measure of the work accomplished in the process material during mixing or extrusion under controlled conditions of temperature, shear, and time. Hence, the device is capable of detecting changes in state, viscosity, or chemical reactions on a real-time basis. It is important to note that the torque rheometer's data can be interpreted in terms of shear force (dyne/cm^2), shear rate (s^{-1}) and viscosity. The data can be interpreted in terms of work and power. Torque data measure power requirements by adding shaft rotation rate to the torque balance expression.

Conventional polymer processing is comprised of several types: polymerization (i.e., resin cooking), internal mixers (master blending), roll mill/calendering (i.e., preparation for processing and mixing), grinding/cutting (preparation for extrusion), extrusion (melting and pumping), pelletizing (preparation for final processing), final process of extrusion, injection molding, blow molding, etc. In practical terms, each of these operations is handled in a batch mode, with intermediate product conveyed from one step to another. At each staged operation a specific shear stress and shear rate is required and is often quite different from the shear rate and shear stress of the next process step. This means that the material could be transferred faster or slower to a successive processing step than the following stage requires for feeding. In addition, the polymer will undergo continuous shear stress and varying temperature stress which can in fact alter the chemical structure. From the standpoint of processability testing, one must provide a wide latitude of testing in order to keep track of the polymer's viscosity during all process phases.

Torque is related to the concept of shear stress, and the speed of the rheometer is related to shear rate. The relation of shear stress to shear rate is the apparent viscosity. For this reason, torque rheometers can be considered a hybrid device between the viscometer and a scaled-down processability tester. A basic torque rheometer is comprised of a drive, a torque (viscosity) measuring system, a sensor, a temperature control unit, and a data recorder. The fundamental work in the area of torque rheometry has been based on using a mixer as a sensor on the basic measuring instrument. As noted earlier, the various processing steps such as mixing, extrusion, calendering, and molding subject the polymer to different shear rates. Thus any polymer can behave differently in these various processes. In fact, any two materials may be perfectly suitable in one process but not suitable in another. Figure 24 provides an example of the influence of shear rate on the stress of two rheologically different polymers. Since apparent viscosity is

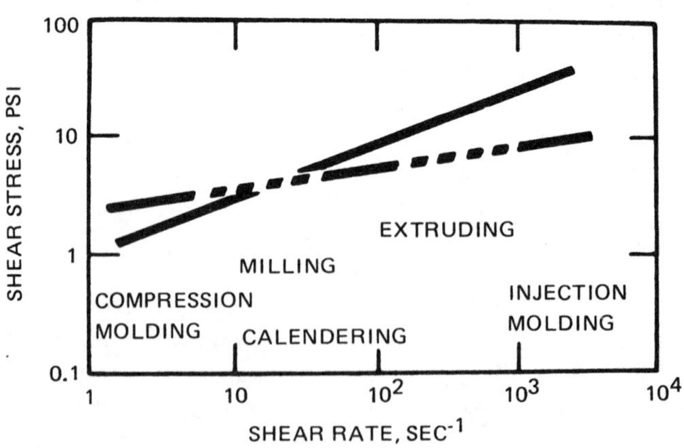

Figure 24 Example of flow curves for two polymers.

defined as the ratio of shear stress to shear rate, it is obvious from Figure 24 that no apparent viscosity varies as a function of shear rate but that the polymers have different viscosities relative to each other. In processing operations involving conditions of low shear rate, these two polymers might be used interchangeably; however, there would be significant differences in operations involving higher shear rates. It is important then that the materials be tested in the same range of shear as anticipated in the actual processing to form a basis for selection.

Plastics processing operations usually subject the polymer to very high shear rates and temperatures for a period of time. With increased utilization of plastics and elastomers, production rates have increased, meaning that the capacities of processing equipment have increased, exposing new polymers to even higher shear rate conditions. As noted in earlier discussions, many of these processing operations require close tolerances on extruded and molded articles. Both the high production rates and close manufacturing tolerances require that the feedstock polymer have exact and duplicate properties from production to production. Prediction of material processability cannot be based on only a knowledge of the polymer's solution viscosity or average molecular weight as determined in low shear rate tests. Extrapolation of viscosity data obtained under low shear rate conditions such as a melt index can lead to erroneous conclusions. Two resins or elastomers can have greatly different flow responses as a function of shear rate, and hence have significantly different apparent viscosities in regions of high shear rate. This discussion becomes clearer by diagnosing a typical flow response curve for a polymer. Figure 25A illustrates the generalized flow proper-

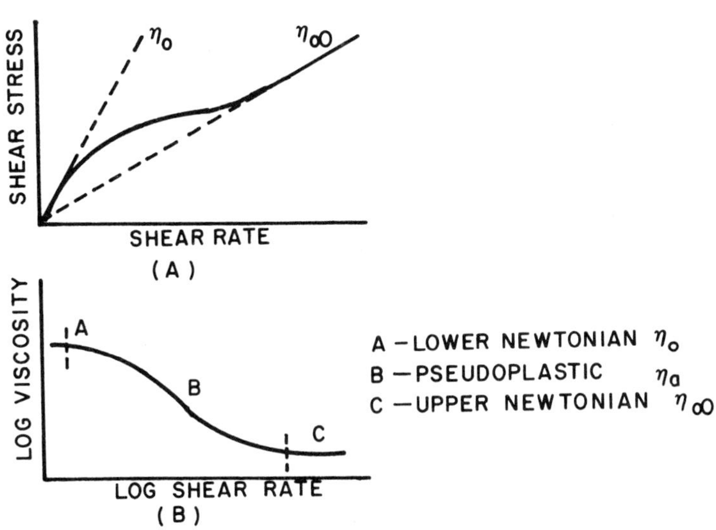

Figure 25 Generalized flow curves for polymers.

ties in terms of a plot of shear stress versus shear rate on linear coordinates. Figure 25B shows the same plot on logarithmic coordinates. In examining these plots we note the following. First, at very low shear rates, the response of the polymer is Newtonian in the vicinity of the so-called zero shear viscosity. In the intermediate region, the polymer is pseudoplastic over a wide range (several decades) of shear rate. Finally, at very high shear rates, the polymer is again observed to approach that of Newtonian behavior.

These characteristics can be attributed to the molecular structure of the polymer. Under low shear rate conditions, the thermal motion of the polymer chain segments dominates and overcomes any tendency toward molecular alignment. In this case the molecules are random and exist in a highly entangled state. In this state there is the greatest resistance to deformation.

When the material is subjected to higher shear rates, molecules begin to untangle and align. In this state the resistance to polymer chain slippage of molecules past one another is reduced. At very high shear rates, the polymer molecules are nearly totally untangled and aligned, and eventually reach a state of minimum flow resistance.

The pseudoplastic region of the plot in Figure 25B is approximated by the power-law model. Figure 26 shows a plot of viscosity versus shear rate for several polymers. Different rheometers discussed in Chapter 2 cover the various shear rates in polymer characterization. Note that the low-density polyethylene

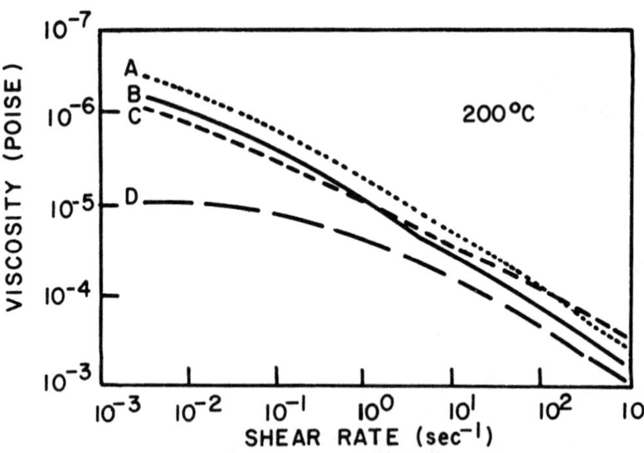

Figure 26 Viscosity versus shear rate curves for several polymers: Curve A: HDPE (MI = 0.2, $\overline{M_w}$ = 1.7 × 10^5); Curve B: HDPE (MI = 0.2, $\overline{M_w}$ = 1.4 × 10^4); Curve C: HDPE (MI = 0.08, $\overline{M_w}$ = 2.0 × 10^4); Curve D: LDPE (MI = 3.5, $\overline{M_w}$ = 2.0 × 10^4).

(LDPE) in Figure 26 shows Newtonian flow behavior in the low shear rate region. The values for MI in Figure 26 represent melt index, which is a low shear rate test. Since MI is obtained at low shear it has a higher correlation with molecular weight and zero shear viscosity. On the other hand, it should not be expected to correlate with polymer flow response in the shear rate region's normally imposed by commercial processing operations.

Rheological measurements depend on the test specimen, shear rate, temperature, pressure, and the specimen's prior history. Each of these parameters must be carefully controlled during testing in order to obtain reproducible data.

Conventional rheological testing considers the polymer only in the melt state as a function of shear rate and temperature. From a processor's viewpoint, however, there are many potential difficulties caused by the effects of work done to the polymer prior to, and in transition to, the melt state. Thermosetting plastics also present processing problems related to the rate of cure and its effect on polymer rheology.

Extrusion can be considered a typical operation. The important considerations in this operation were summarized in Chapter 6. When dealing with plastics, the resin powder in the feed hopper must be free-flowing, to provide uniform feed; the material then must be conveyed to and compacted in the cool feed zone. The lubricity of the compound must be controlled so the initiation of fusion is neither too fast (high torque) nor too slow (poor melting). The rate of fusion, the apparent viscosity at the hot end of the extruder, and the high temperature–high shear stability in the forming zone must all be considered. Furthermore, the viscoelastic character of the compound also influences extrusion throughout the entire operation.

The basis for processability testing and projection of results from small-scale testing can be established in the following manner. When polymer samples known to provide good processing performance in large-scale equipment are compared with samples known to cause processing problems, the operating conditions of shear rate and temperature can be optimized in the lab-scale apparatus to provide a correlation with a given factory process.

With a torque rheometer, work energy is measured, which in turn reflects the flow behavior (or viscosity) of the polymer sample under shear and thermal stress. This in turn is a function of the rotor speed (i.e., process speed) and temperature during the various viscous stages the polymer undergoes. It is important then that the torque rheometer measure not only torque but also temperature and rpm. Also, in comparing several stocks, no measurable torque differences may be apparent; however, significant temperature differences may exist. A torque rheometer, then, is capable of providing real-time data on torque and melt temperature under specific set temperature and process speed conditions. Information on the amount of work needed to bring a polymer from its solid state to a homogeneous melt as well as the time required for processing can be obtained. In addition, the instrument can be used much like a Mooney viscometer to ascertain scorch time, but at shear conditions more closely approximating the factory operation. In plastics handling, it is possible to ascertain under what conditions the polymer will chemically break down or establish the elasticity of the resin. On this last point, polymers can be ''fingerprinted'' in terms of the bandwidth or the torque response curve. As an example of a processability–product quality test, one could establish an ''ideal'' curve based on an acceptable, well-processing polymer under specified conditions in actual process equipment. A bandwidth can be established within which the polymer's torque curve must lie in order to be acceptable for the process under consideration. Deviation outside the bandwidth would indicate that the polymer was unacceptable or that process conditions had to be altered. Although this is strictly an empirical approach, with experience it can form the basis of an acceptable method of product quality control.

The torque rheometer can be compiled with a mixer as described earlier to provide a pilot device for mixer–processability studies. In this fashion a mix can be prepared under different conditions and the optimum operating range defined and scaled up to production levels. A torque rheometer–mixer arrangement enables different samples to be homogenized to a uniform viscosity, or it can be used to stage identical power inputs to provide repetitive samples in a uniform manner at a standard viscosity and temperature. One advantage of carefully controlled mixing is that it can also provide the basis for specimen preparation for other physical testing such as rubber cure time, tensile strength, and modulus testing.

Another use for the mixer arrangement is in the study of compatibilizers. For example, one can study different ratios of polymer blends with compatibilizers. Properties in this case are highly dependent on particle size and dispersion qualities. The effect of mixing different polymer systems in various ratios with different amounts of ''couples'' can be carried out. Such a test can be conducted by subjecting the formulation to varying mixing times and speeds. The objective is to maximize physical properties and minimize the amount of compatibilizer, therefore reducing cost.

Figure 27 provides an example where SBR polymers may require different black incorporation times and have different average torque levels. Note that the torque response curves show different amounts of mixing work (i.e., work being the area under the torque curve). It is also important to note that the torque level also provides an indication of the degree of reinforcement. The time required for reinforcing or extruding fillers to be incorporated into a polymer is a compounding parameter that can be readily measured.

Chemical reactions such as rubber scorching, polymer cross-linking, gel formation, and thermosetting of resins can be readily studied in a torque–rheometer–mixer arrangement. By proper selection of the mixer configuration

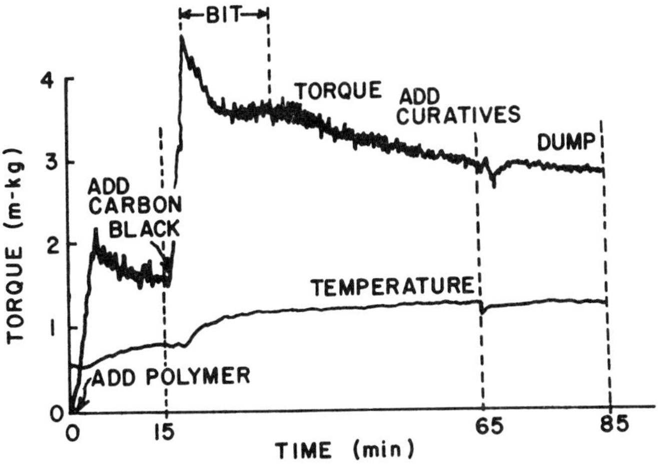

Figure 27 Effects of carbon black and curatives in rubber as measured in a torque rheometer.

and rotor speed, the shear input can be prespecified. Changes in torque level can be monitored as the reaction takes place. In other applications accurate temperature control may be the most critical parameter. Figure 28 provides an example of the effect of compounding ingredients in rubber on the reaction.

An example of a torque response for a typical thermoset is shown in Figure 29. In this example the measurable variables are the loading peak, the point of maximum flow (i.e., minimum viscosity), the point of maximum cure (the second peak), the time to maximum flow, and the time to maximum cure.

An example of rubber scorch evaluated in a mixer–torque rheometer is shown in Figure 30. The curves shown are for two rubber compounds which differ in their accelerator concentration. The specimen having the lower amount of accelerator shows a greater scorch safety but also has a slower rate of cure. A torque rheometer can also be applied in studying the cross-linking characteristics of polyethylene and in testing the thermal gelling of plastisols.

Mixer–sensor torque rheometry studies can separate out phenomena that take place successively and/or concurrently during processing. For example, extrusion involves several stages (solid flow, fusion–melting rates, viscosity levels, viscosity changes). Each of these regimes can be readily identified on a torque curve equipped with a mixer sensor. A torque curve derived from an extruder sensor arrangement provides data on the overall combined effect.

The important features of a typical mixer–sensor torque rheometer rheogram are summarized in Figure 31. Phase I of the rheogram denotes a change of state

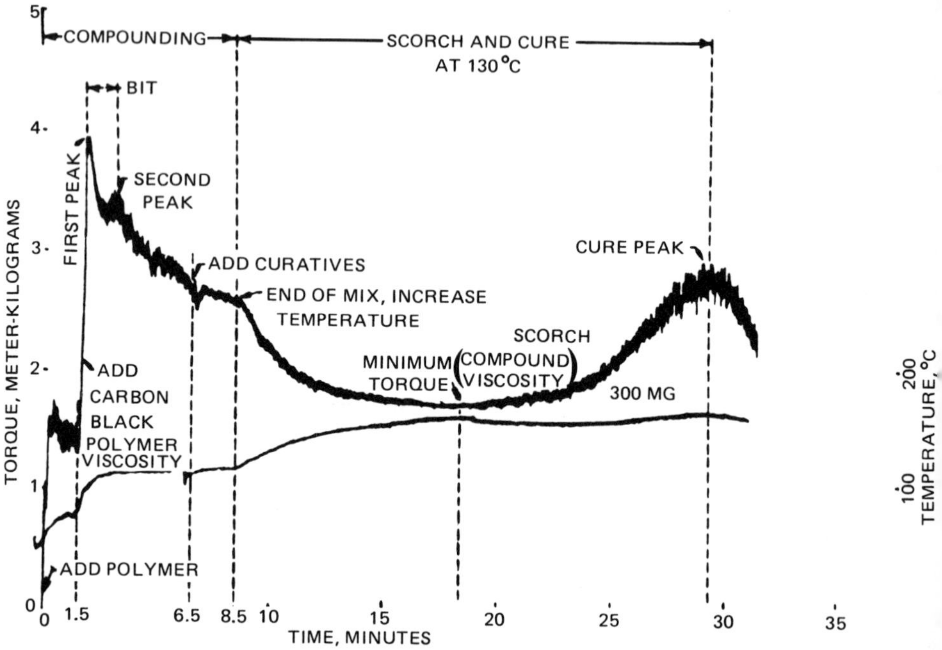

Figure 28 Rubber testing in a torque-recording internal mixer.

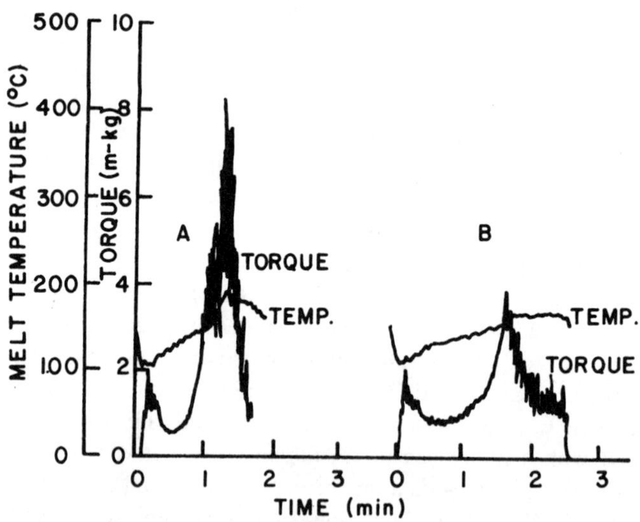

Figure 29 Study of the cure of two thermosetting plastics.

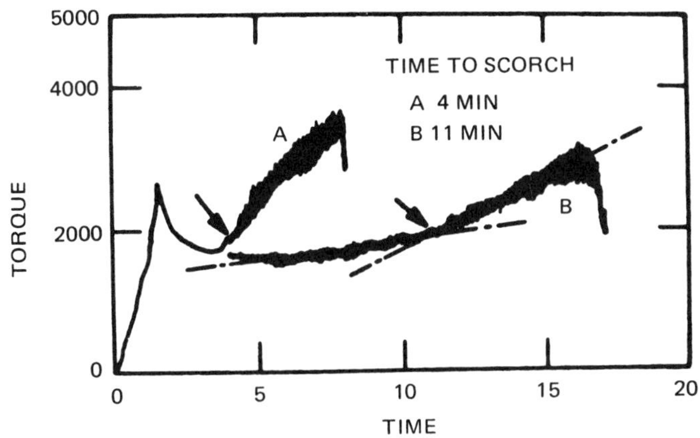

Figure 30 Study of rubber scorch and cure in a mixer–sensor torque rheometer.

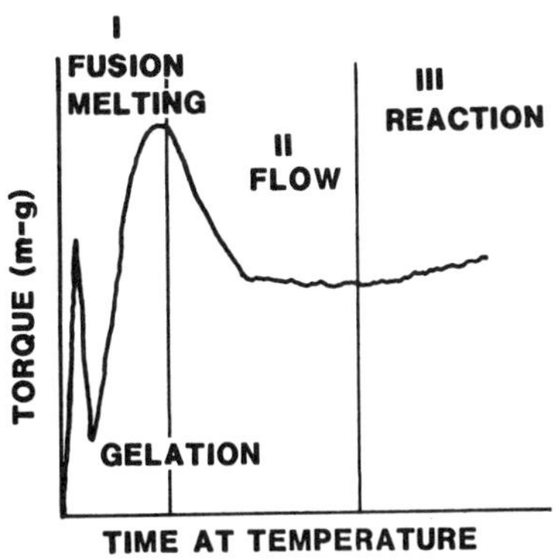

Figure 31 Important features of a torque rheogram.

for the material (e.g., the melting of a thermoplastic resin), or structural change occurring in a liquid such as the gellation of a plastisol. Phase II typifies the beginning of viscous flow, and the effects of temperature and shear can be observed in this region. Phase III denotes the region where possible reactions take place, such as the curing of an elastomer. The reaction region would also show undesirable phenomena such as the degradation of polymers, or scorch, in which case a torque increase would be observed.

It is also possible to correlate torque measurements with some finished product properties. For example, with thermoplastics, one can mix the formulation until a stable viscosity level is achieved. After the rotors have been left off for some brief time period to allow the compound to rest, the drive can be turned on again. The initial shock torque that is measured will often be higher than the last recorded torque prior to stopping the drive. This observation exemplifies the recovery of viscosity or structure of a thixotropic material during testing. The ratio of shock torque to former torque of a molding compound can be found to correlate with the property of impact resistance of some blow-molded articles.

Another example of the use of a torque–rheometer–mixer system for processability–product quality testing is given by the rheograms of Figure 32. In this example two production lots of the same EPDM of an extrusion grade masterbatch were tested in a Banbury-style mixer attachment (system used was a Rheocord System 40, Haake-Buchler). A customer reported that the two materials processed differently; however, both gum and compound Mooney viscosity measurements were the same for the two samples. A comparison of the torque response curves for the two compound samples is shown in Figure 32. The polymer that reportedly processed well displayed a unique torque pattern at an average level. The bandwidth decreased slightly during the first 5 min of mixing; however, the minimum torque level was constant throughout the experiment. The poor processing sample required a much longer mixing time before the same steady bandwidth was achieved. A higher average torque was observed for the poor-processing compound, which indicated that a greater amount of energy was needed to process the material. The wider bandwidth also indicated that the sample was more nervy and had greater elasticity. The transition from a rough elastic state to a smoother, more plastic property is probably indicative of improved carbon black–mineral filler dispersion.

Another processing experiment readily conducted in a torque–mixer rheometer is a fusion test. The *time of fusion* of a polymer refers to the time period from the loading of solid polymer into a mixer cavity to the point at which it melts into a plasticlike mass. This time span depends on the base resin and its compounding ingredients (e.g., plasticizers, peptizers, stabilizers) as well as the polymer itself. The test is most often associated with the characterizing of PVC resins but applies to virtually all types of polymers. The test essentially is aimed at measuring the fusion time, which can be observed by the peak torque required at the time of fusion as shown in Figure 33. Fusion or melting characteristics of a polymer can be related to processing conditions (e.g., the compression zone in a plastics extruder). The amount of energy or work involved in fusion and the time to fusion establish both the degree of fusion and the point where it occurs in a continuous processing operation such as extrusion. The torque response curve shown in Figure 33, derived by mixing a polymer at a constant rotor speed, can be related to an extrusion operation in the following manner. If the material is a plastic resin, for example, the high shear melt viscosity of the polymer is the most important property in the compression and metering sections. The rheo-

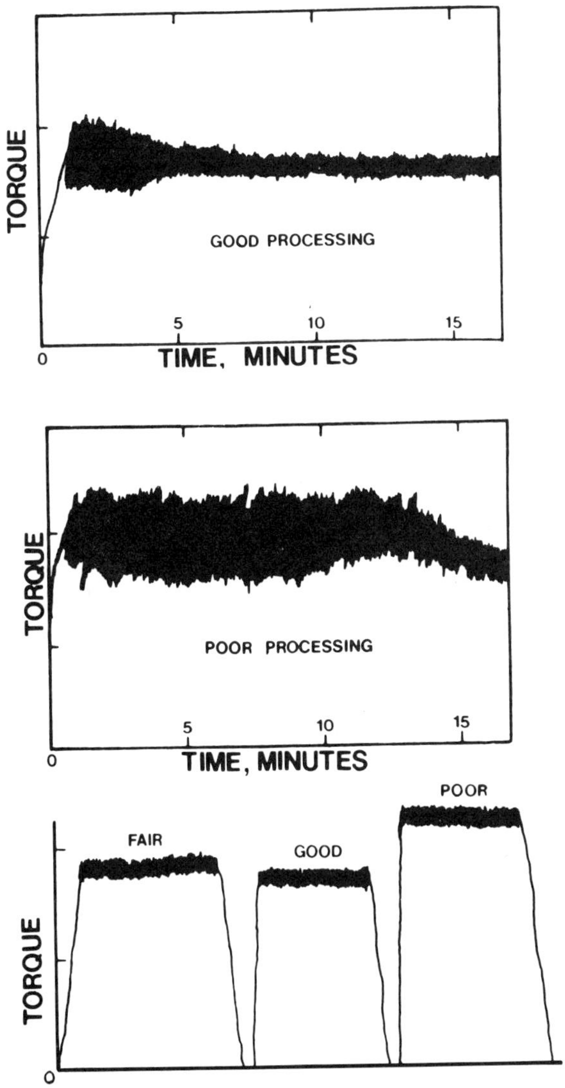

Figure 32 Example of use of torque rheometry to distinguish between processing characteristics of polymer samples.

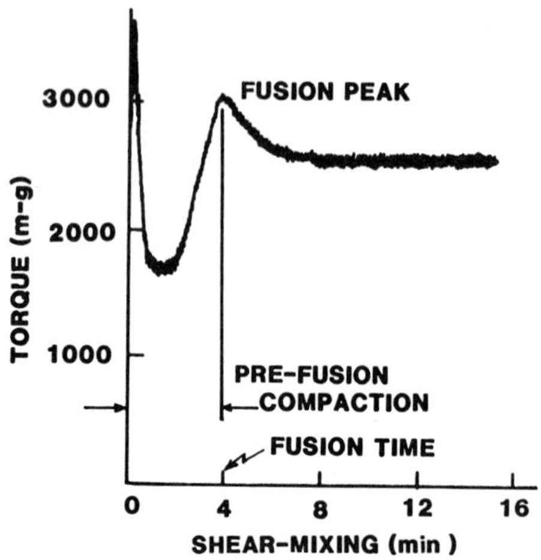

Figure 33 Fusion peak and time as measured in a torque–mixer rheometer.

gram of Figure 33 shows the material charging and fusion peaks to be relatively close together, after which the material decreases in torque to some constant viscosity level. In the extruder, similar behavior takes place; that is, the solid polymer is subjected to a shearing force, which brings it to a point of fusion, after which it achieves a constant viscosity and is transported through the metering section of the extruder.

In many rubber and plastics processing problems it is also of interest to characterize mixing behavior beyond the fusion peak. For example, in rigid PVC formulations the addition of lubricants is often made. The purpose of a lubricant is to promote proper flow of the plastic melt. The effects of lubricants on melt viscosity and material stability can be simulated over a range of temperatures and shear rates which simulate the feed end of an extruder. Figure 34 illustrates a

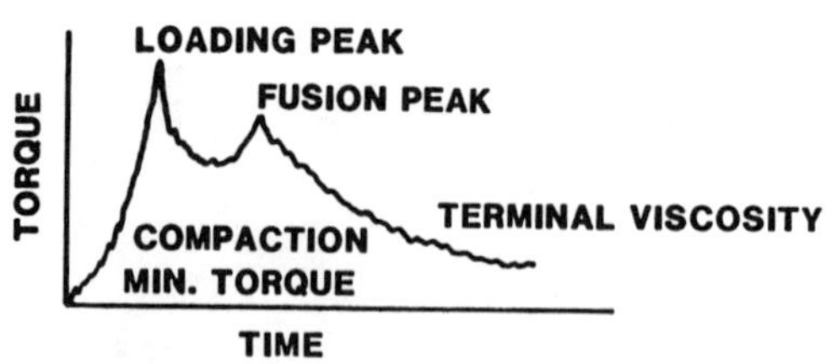

Figure 34 Use of mixer–torque rheometer in characterizing the viscosity of an elastomer.

typical rheogram, noting the important features. A resin or polymer having insufficient lubricity is typically characterized by a high loading peak and a short fusion time. In contrast, a compound with excessive lubrication will experience slippage and usually a slow temperature rise during the mix as characterized by a resulting long fusion time. This can create a serious problem in the extrusion of plastics, as the resin could pass through the extruder without being properly fluxed. However, proper polymer lubrication and optimization of the fusion peak depends in part on the equipment geometry and flow conditions. Scale-up of processing observation can be risky in this area, and it is a more prudent policy to implement a series of tests on different compounds to develop relative processability ratings. As noted, a lubricant is used to promote flow. In the extrusion of rigid plastic materials (e.g., PVC pipe) lubrication reduces frictional heat generation as well as affects polymer–metal and polymer–polymer interfacial action. Lubricants also affect the fusion rate; however, this is a strong function of the particular stabilizer ingredient used in the compound.

All of the torque curves described are essentially viscosity-sensitive signals that can be related to shear stress via conventional viscometry. The speed setting on a torque rheometer is related to shear rate. This means that if several polymers are evaluated independently at some identical rotor speed and temperature, any variations in torque among the specimens is indicative of viscosity differences. A simple experiment to characterize the viscosity of elastomer samples involves running the torque rheometer at a constant speed to achieve a stable torque response, as shown in Figure 34A. By running a succession of different speeds, polymer samples can be evaluated over a range of shear rates, as is accomplished in a conventional capillary rheometer (see Figure 34B). A problem however is that the geometry of a torque rheometer is complex, and hence it is extremely difficult to interpret rheograms in terms of fundamental rheological parameters. This means that from the standpoint of processability testing, the instrument is most useful in providing a relative index of processability and scale-up to full-size processing machines should be done cautiously. It is possible to relate rotor torque and speed through a power law relation, as is done with shear stress and shear strain; that is,

$$\mathrm{T} = C(n)KS^{n} \tag{15}$$

where

T = torque
n = power law index
$C(n)$ = function which depends on n and machine geometry
K = constant in the power-law shear stress–shear rate relationship
S = rotor speed

The power law index n obtained in a torque rheometer will agree with the value derived from data obtained from a capillary rheometer. This means that from capillary-rheometer-derived K and n values, one can calibrate a torque rheometer, in which case $C(n)$ can be deduced as an explicit function of the machine's geometry and the value of n.

To summarize, the rheogram derived from a mixer–torque rheometer provides processing information in three general areas. First, it can detect the times at which peaks and troughs occur in torque–time or power–time curves, which in turn are representative of the interactions between polymer rheology and various processing–mixing steps. The second area is that the total work performed by the mixer on the elastomer or polymer can be measured. This is important for two reasons. Mixing energy is an identifiable index of processing in itself and can be compared with different polymer samples when experiments are conducted over identical mixing times, rotor speeds, and temperatures. These data in turn can form a basis for scaling up or projecting processing behavior to full-scale equipment. Also, in some processes it may be more desirable from the standpoint of dispersion and polyblending to dump a masterbatch on the basis of totalized mixing power or torque, rather than using the conventional criteria of mixing to a prespecified time and temperature. This is an option that needs to be explored for the particular polymers and processing operations under consideration. A third area of information derives from the fact that the mix cycle history in the initial stages and even at the end of first-stage mixing, at least with rubber, is often characterized by the magnitude of short-period variations in power or torque, as shown by the width of the torque response band on a rheogram (refer back to Figure 32). This can form a basis for fingerprinting a polymer's rheological and processing characteristics and could provide the basis for product quality testing in some instances.

The torque rheometer can be adapted with a screw mixer section to perform extrusion experiments. The use of Garvey dies, slit dies, or capillary dies can provide information on extrudate dimensions and stability, appearance ratings and swell. The standard rating for extrusion through a Garvey die is shown in Figure 35. Systems such as the Haake-Buchler-Rheocord System 40 or Brabender Plasticorder have drives that can adapt several types of mixer cavity–rotor configurations and screw mixer chambers. Figure 36 illustrates some of the capabilities of the Rheocord System 40 in extrusion experiments on several EPDM samples. The extruder retrofitted with dual temperature and pressure sensors along the axis of the barrel can simultaneously record pressure–temperature–torque responses as a function of experiment time. Through the use of a microprocessor and a control monitor, the operator can preprogram a set of extrusion conditions. For example, the extrusion characteristics of a polymer compound can be studied by automating the experiment so that the drive automatically changes to different screw speeds or temperatures as step or ramp functions (see Figure 37). Thus, a large data base on

GRADING NUMBER	DEFINITION
A	SWELLING & POROSITY
B	SHARPNESS & CONTINUITY OF 30° EDGE
C	SMOOTHNESS OF THE SURFACES
D	SHARPNESS & CONTINUITY OF THE CORNERS

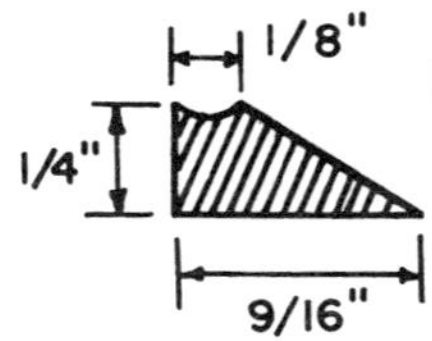

DIE PROFILE (ORIGINAL PROFILES GENERATED ON 1/2 ROYLE TUBER AT TEMP. OF 180°F)

EXTRUSION RATING BASED ON FOUR SEPARATE GRADINGS, EACH RANGING FROM 1 = VERY POOR TO 4 = EXCELLENT. THE PERFECT RATING IS 16, AND 12 IS SATISFACTORY FOR MOST PRODUCTION EXTRUSIONS PROVIDED THE 30° EDGE RATES AT LEAST 3.

Figure 35 Garvey die rating system. (Test is described in *Industrial and Engineering Chemistry*, *34* (1942), and is a method for evaluating the extrudability of rubber stocks. The extrusion strip has two 90° corners, a 150° edge, and a 30° edge.)

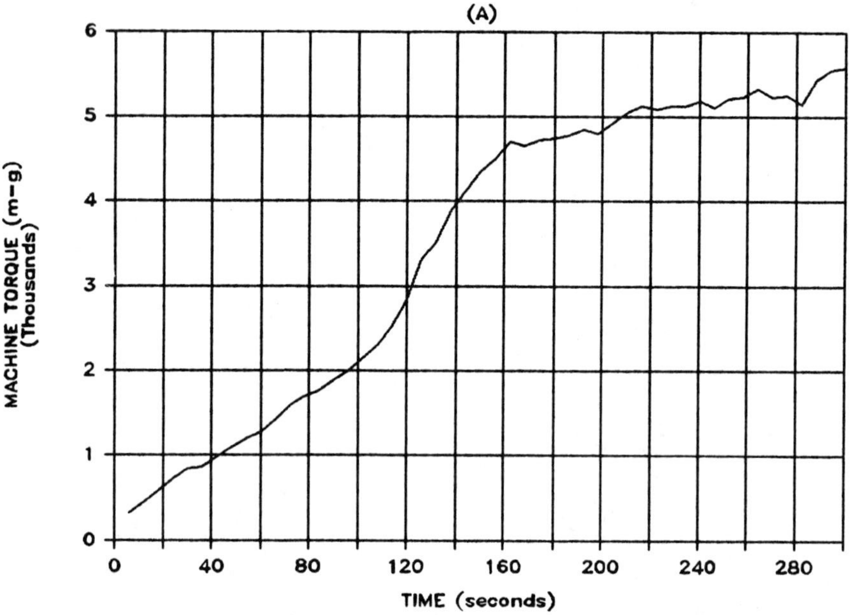

Figure 36 Torque/temperature/pressure response at several locations along extruder can be recorded simultaneously for an extrusion test in the Rheocord System 40.

the extrudability of different compounds or polymer candidates can be generated in a rather short time by running continuous extrusion tests preprogrammed on the instrument. The investigator can then postanalyze cross-correlation plots of torque versus screw speed or torque versus head–upstream pressure, and so on, to obtain relative process ratings between samples. The references cited at the end of this chapter provide more detailed readings on the theory and uses of torque rheometry in polymer processing studies.

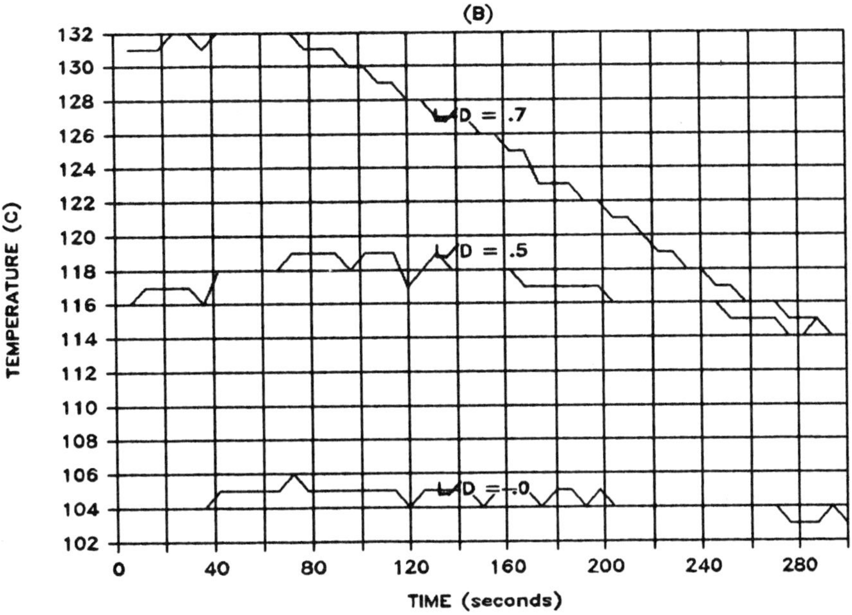
(B)
TEMPERATURE (C)
132
130
128
126
124
122
120
118
116
114
112
110
108
106
104
102
L/D = .7
L/D = .5
L/D = 1.0
0
40
80
120
160
200
240
280
TIME (seconds)

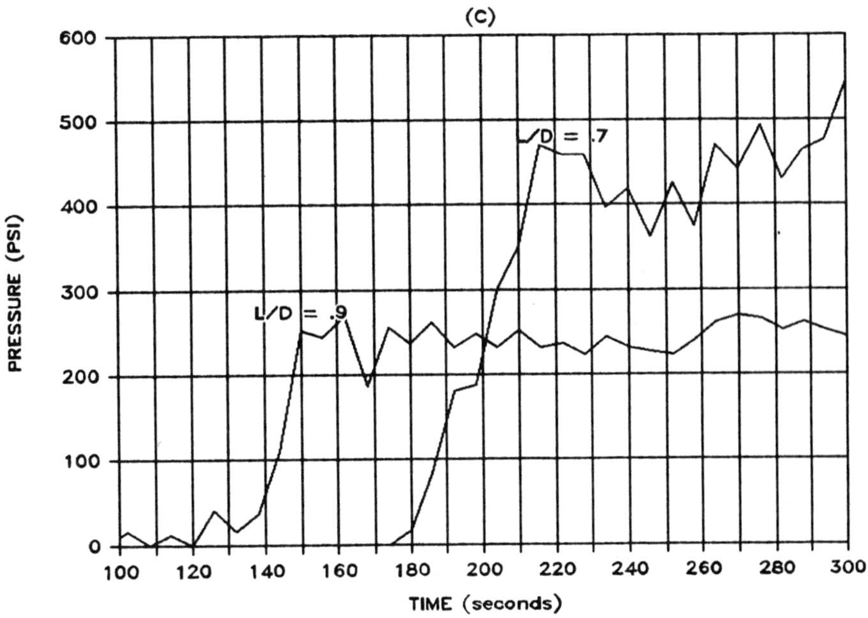
(C)
PRESSURE (PSI)
600
500
400
300
200
100
0
L/D = .7
L/D = .9
100
120
140
160
180
200
220
240
260
280
300
TIME (seconds)

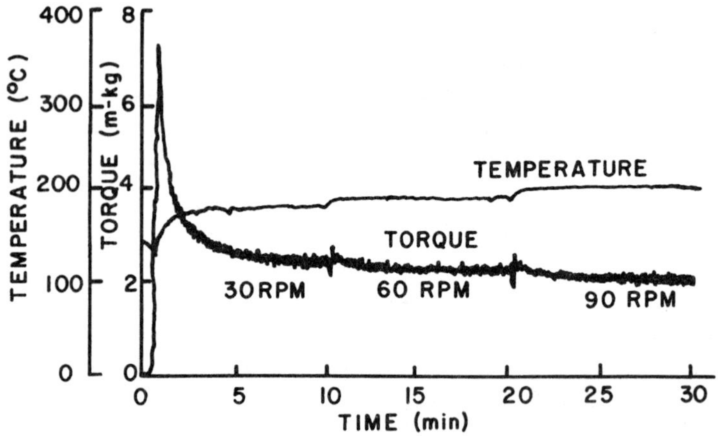

Figure 37 Rheocord can be preprogrammed to provide step or ramp changes in extruder operating conditions such as screw speed and temperature.

NOTATION

a	fractional component of blend
C	constant
C_p	heat capacity
$C(n)$	processability function
E'	storage modulus
E''	loss modulus
F	free energy or force
G^*	shear modulus
G'	in-phase, elastic, or storage modulus
G''	out-of-phase or loss modulus
H	enthalpy
K	power law constant
l	length
ML	Mooney viscosity
n	power law index
P	pressure
S	entropy
s	rotor speed
T	absolute temperature
T_g	glass transition temperature
T_0	reference temperature
t	time

Greek Symbols

α_T	frequency factor
$\dot{\gamma}$	shear rate
δ	phase angle
$\dot{\epsilon}$	strain rate
η	dynamic viscosity
η_ϵ	extensional viscosity
T	torque
τ	shear stress

REFERENCES

Cotten, G. R., and J. L. Thiele, *Rubber Chem. Technol., 51,* 749 (1978).

Gray, A. P., *Thermochim. Acta,* 7(6), 563 (1970).

Macosko, C. W., and W. M. Davis, *Rheol. Acta, 13,* 814 (1974).

Markert, J., *Gummi Asbest. Kunstst., 29,* 568 (1976).

Nakajima, N., and E. A. Collins, *Trans. Soc. Rheol., 20,* 1 (1976).

Nakajima, N., and E. A. Collins, *Polym. Eng. Sci., 14,* 137 (1974).

Perkin-Elmer Corp., Norwalk, Conn., Thermal Analysis Study Report, No. 7 (1973).

Topcik, B., Carbon Black Dispersion and Screen Pack Plugging in Crosslinkable PE, Cities Service Co., *Modern Plastics* (April 1974).

Turetslayetal, S. B., Quantitative Evaluation of Emulsion Styrene-Butane, Rubber and Compounding Variable by Controlled Mixing, Texas–US Chem. Co., Parsippany, NJ, *Rubber Chem., 49,* 401 (1976).

Turner, D. M., M. D. Moore, and R. A. Smith, in *Elastomers: Criteria for Engineering Design,* C. Hepburn and R. J. W. Reynolds, eds., Applied Science Publishers, London, 1979.

White, J. L., *Rubber Chem. Technol., 50,* 163 (1977).

White, J. L., *Rubber Chem. Technol., 42,* 257 (1969).

Willey, S. R., W. M. Davis, C. W. Mecosko, and C. Goldstein, *Trans. Rheol., 18,* 515 (1974).

Wingrave, J. A., Effect of VCM Removal on Commercial Pipe Grade PVC Resins, Chem. Res. Div. Cont. Oil Co., Ponca City, OK, presented SPE Annual Tech. Conf. in Montreal, April 1977, pp. 264–267.

Index